73
Topics in Organometallic Chemistry

Aims and Scope

The series *Topics in Organometallic Chemistry* presents critical overviews of research results in organometallic chemistry. As our understanding of organometallic structure, properties and mechanisms increases, new ways are opened for the design of organometallic compounds and reactions tailored to the needs of such diverse areas as organic synthesis, medical research, biology and materials science. Thus the scope of coverage includes a broad range of topics of pure and applied organometallic chemistry, where new breakthroughs are being achieved that are of significance to a larger scientific audience.

The individual volumes of *Topics in Organometallic Chemistry* are thematic. Review articles are generally invited by the volume editors. All chapters from Topics in Organometallic Chemistry are published Online First with an individual DOI. In references, *Topics in Organometallic Chemistry* is abbreviated as Top Organomet Chem and cited as a journal.

Basker Sundararaju
Editor

Dehydrogenation Reactions with 3d Metals

With contributions by

V. Arora · E. Balaraman · S. Bastin · A. Bisarya · C. Brodie · C. Bruneau · Y. Canac · P. Chakraborty · J. Choudhury · A. Das · K. Das · M. Elghobashy · O. El-Sepelgy · O. Fayafrou · A. Kumar · A. Kumar · B. Maji · T. Mandal · P. G. Nandi · H. Narjinari · M. Pal · R. Pointis · S. Pradhan · K. Sarkar · G. Sivakumar · J.-B. Sortais · B. Sundararaju · A. K. Suresh

Springer

Editor
Basker Sundararaju
Department of Chemistry
Indian Institute of Technology Kanpur
Kanpur, Uttar Pradesh, India

ISSN 1436-6002 ISSN 1616-8534 (electronic)
Topics in Organometallic Chemistry
ISBN 978-3-031-48951-8 ISBN 978-3-031-48952-5 (eBook)
https://doi.org/10.1007/978-3-031-48952-5

This Springer imprint is published by the registered company Springer Nature Switzerland AG
The registered company address is: Gewerbestrasse 11, 6330 Cham, Switzerland

Paper in this product is recyclable.

Preface

Contemporary science is moving towards exploration of resource-economical techniques, the reason being the current rate of depletion of the earth's reservoirs and the dire need to come up with environmentally benign strategies. A promising approach in this direction is the utilization of the transition metal catalyzed dehydrogenation process. This powerful method not only facilitates sustainable chemical production but also plays a pivotal role in generating clean hydrogen gas, with applications spanning various fields. Bringing these strategies into practice contributes significantly towards minimizing the overall carbon footprint on a global scale. Alcohols, being versatile raw materials for industrial processes, can be sourced either from industrial processing or biomass, which serves as a renewable and consistently available alternative in substantial quantities. In contrast to the traditional linear economy, dehydrogenation reactions are integral for achieving a circular economy with closed-loop systems. The limited supply and rapid depletion resulting from high utilization of noble metal-based catalysts, despite their significant catalytic activity, can restrict the practicality of the discovered protocols in the near future. As a result, the utilization of abundant 3d metal-based catalysts in dehydrogenation processes has gained considerable interest over the last decade. Besides, the ability to undergo both one- and two-electron transfer and changes in spin state is a unique characteristic of 3d metals, which might lead to new reactivities. The discovery of earth-abundant metal catalysts and catalytic processes for dehydrogenation reactions is a thriving research area that dwells on the excellent reactivity and selectivity achieved by proper tuning of catalyst along rationally designed ligand systems. To attain carbon neutrality by 2050, it is crucial to dynamically transform the current global systems of energy and material production. Thus, knowledge of the overall developments of 3d-metals in dehydrogenation reactions along with mechanistic understanding of the catalytic cycle is crucial in finding new catalysts or novel reactivity for future developments.

In this book volume, a comprehensive summary of the achievements on 3d (base) metal-catalyzed dehydrogenation reactions are presented and discussed. The book delves into the latest breakthroughs in catalytic applications, explicitly focusing on the dehydrogenative strategy for the production of heterocycles and the execution of tandem-multicomponent reactions involving three or more components. Additionally, it explores the biomass upgrading processes, such as the transformation of ethanol into n-butanol, the synthesis of higher oxidized hydrocarbons including acids, esters, and amides from alcohols, and the α-alkylation of amines, ketones, and amides using alcohols. Incorporating bifunctional and redox-active ligands in 3d metal catalysis has mechanistically opened up new avenues for more sustainable chemical processes. Together, these chapters will give readers a clear perspective on what lies ahead for this field's advancements, and they are appropriately targeted toward graduate students, researchers, and synthetic chemists in both academia and industry.

I extend my sincere gratitude to the authors for their invaluable contributions that have shaped the developments highlighted in this volume. I also would like to extend my thanks to Ms. Priyank Chakraborty for her help throughout this project. My appreciation also goes to the Series Editors, including Pierre H. Dixneuf and Matthias Beller, for their adept organization of this compelling series. Additionally, I express thanks to Alamelu Damodharan and Charlotte Hollingworth (Springer staff) for their constant support and unwavering patience throughout the publication process.

Kanpur, India Basker Sundararaju

Contents

Top Organomet Chem (2024) 73: 1–32
https://doi.org/10.1007/3418_2023_107

Published online: 27 October 2023

Dehydrogenation of Alcohols Using Transition Metal Catalysts: History and Applications

Christian Bruneau

Contents

Abstract Dehydrogenation of primary or secondary alcohols promoted by transition metal catalysts is the key initial step in many transformations involving alcohols. Besides the formation of value-added carbonyl products, the liberation of dihydrogen has been envisioned as a potential source of energy, or as a useful reagent for further metal-catalyzed hydrogenation of unsaturated substrates. The history of alcohol dehydrogenation and its applications in various types of transformations initiated with noble transition metal and continued with 3d base metal catalysts are reported in this chapter.

C. Bruneau (✉)
Univ Rennes, CNRS, ISCR (Institut des Sciences Chimiques de Rennes) – UMR 6226, Rennes, France
e-mail: christian.bruneau@univ-rennes.fr

Keywords Alcohols · Dehydrogenation · Hydrogen borrowing · Transfer hydrogenation

Abbreviations

Ac	Acetyl
Acac	Acetylacetonate
Ar	Aryl
Cp	Cyclopentadienyl
Cp*	Pentamethylcyclopentadienyl
DMSO	Dimethyl sulfoxide
dppf	Diphenylphosphinoferrocene
h	Hour
^{i}Pr	Isopropyl
Ph	Phenyl
py	Pyridine
TBE	*Tert*-butylethylene
^{t}Bu	*Tert*-butyl

1 Introduction

More than 100 years ago, dehydrogenation of alcohols was observed leading to the formation of carbonyl derivatives, namely aldehydes or ketones with liberation of dihydrogen. Initially, the reactions took place at high temperature in the gas phase in the presence of heterogeneous catalysts based on 3d metals such as copper, nickel, and zinc [1–3]. The harsh conditions that were used most often led to higher transformations of the alcohols into olefins, but the equilibrium nature of the reaction and the beneficial effect of a hydrogen acceptor to favorably displace this equilibrium were noted [3]. Interestingly, mechanistic studies based on experimental results were started and the formation of metal hydride species was proposed [4, 5]. Then, efficient transition metal catalysts became available and made possible dehydrogenation of alcohols at lower temperature under homogeneous catalysis conditions. Rapidly, the interest of this reaction became obvious as both the formed carbonyl product and dihydrogen could find useful straightforward applications in organic synthesis and in the field of energy. In Scheme 1 are illustrated the different potential applications of the reaction of dehydrogenation of alcohols ranging from production of hydrogen, synthesis of aldehydes and ketones, condensation of the carbonyl intermediate with various nucleophiles (essentially *O*-, *C*- and in minor extend *S*-nucleophiles) producing unsaturated products to hydrogen borrowing processes leading to saturated products. In this chapter, these different applications will be examined from a chronological point of view highlighting the pioneering role of

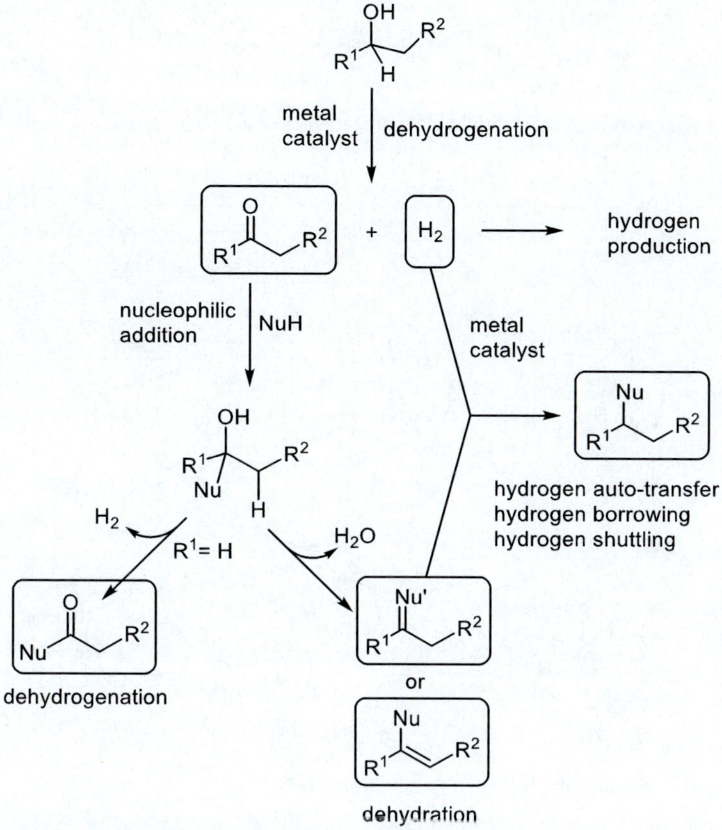

Scheme 1 Simplified possibilities offered by dehydrogenation of alcohols

noble transition metal catalysts in homogeneous catalysis and the promising future of 3d metals. There is a huge number of references dealing with dehydrogenation of alcohols and its applications. This introductory chapter is focused on homogeneous catalysis with transition metals operating under hydride/proton transfer mechanism. In particular, other important emerging processes involving radical processes and organocatalysis are not presented. Some aspects, which are not discussed in the following chapters of this book, have been explored in more detail.

2 Production of Hydrogen from Alcohols

Considering the high potential of hydrogen as energy supplier in the future, in particular for transportation uses, the need for easy to handle and safe liquid organic hydrogen carrier (LOHCs) has triggered intense researches [6–12]. Besides formic acid, saturated hydrocarbons and *N*-heterocycles, primary and secondary alcohols,

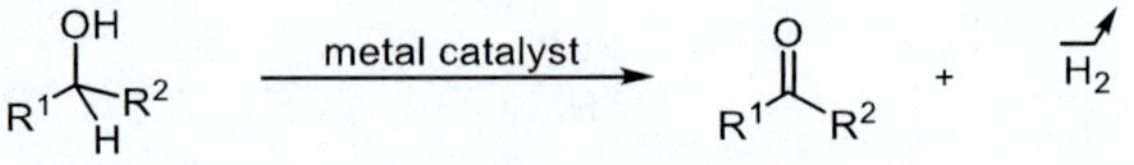

Scheme 2 Generation of hydrogen by dehydrogenation of aliphatic alcohols

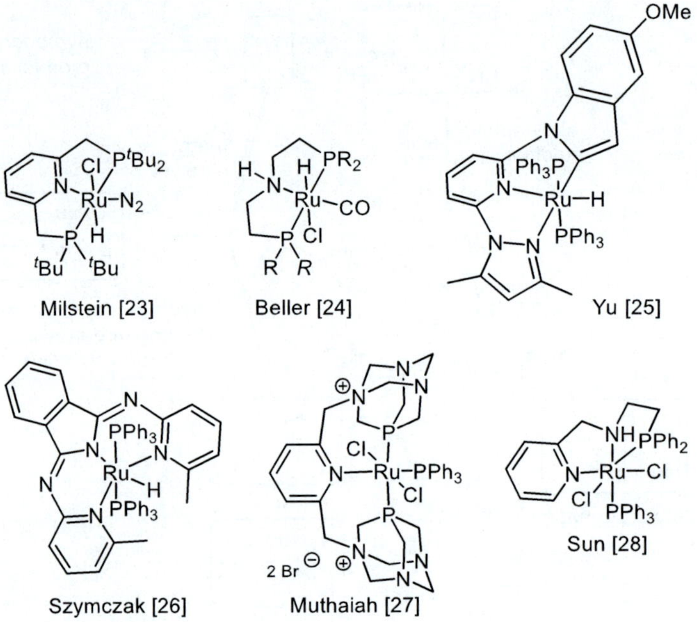

Scheme 3 Short selection of ruthenium catalysts for dehydrogenation of alcohols

especially short easily available alcohols such as methanol and ethanol including bio-derived versions, are key candidates for this purpose considering that their acceptorless dehydrogenation provides a ketone or/and aldehyde and hydrogen (1 mol of each per hydroxyl group) (Scheme 2). In this purpose, hydrogen production with high turnover frequency is essential.

Following seminal results obtained with rhodium chloride where the metal precipitated during the dehydrogenation of isopropanol [13], examples of homogeneous transition metal-catalyzed dehydrogenation of alcohols appeared in the 1980s with ruthenium carboxylate catalysts such as $Ru(OCOCF_3)_2(CO)(PPh_3)_2$ [14, 15], $Ru(OAc)_2Cl(PR_3)_3$ [16], $[Ru(\mu\text{-}OCO\text{-}C_2F_4\text{-}OCO)(CO)(H_2O)(dppf)]_2.H_2O$ [17]. Ruthenium hydride catalysts such as $RuH_2(PPh_3)_4$ [18] and $RuH_2(N_2)(PPh_3)_3$ [19] were also used but in most cases the turnover numbers with short alcohols remained moderate (less than 250 h^{-1} for ethanol). Then, improvement of the productivity of hydrogen was brought with the introduction of ligands equipped with basic sites such as diamines and aminoalcohols to ruthenium centers and operating under basic conditions [20–22]. Further beneficial effects were obtained

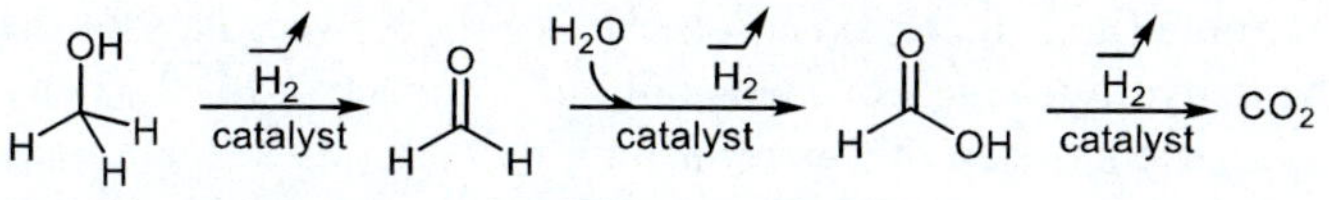

Scheme 4 Production of hydrogen from methanol in water

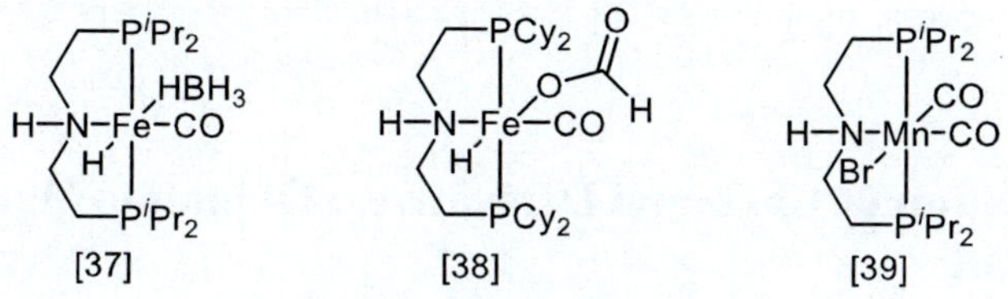

Scheme 5 Methanol reforming catalysts based on 3d metals

by using pincer ligands as exemplified by the ruthenium complexes shown in Scheme 3 [23–28].

However, a more productive dehydrogenation sequence was accessible using dehydrogenation of methanol in water. In that case, the overall transformation corresponding to methanol dehydrogenation, water addition, dehydration into formic acid followed by dehydrogenation of formic acid led to the production of 3 mol of hydrogen per mole of methanol according to Scheme 4.

This strategy was initially developed with noble metal catalysts. In 2013, Beller reported turnover frequencies superior to 2,500 h^{-1} and high productivity (TON >350,000) with a ruthenium complex equipped with a PNP pincer (Scheme 3, R = ^{i}Pr) in the presence of a base trapping the final carbon dioxide [29, 30]. The same year Grützmacher showed that a ruthenium complex featuring a chelating bis (olefin)diazadiene ligand was also able to achieve methanol reforming even without the presence of a base [31]. Other PNN (bipyridine-phosphine) and PNP (phosphine-acridine-phosphine) pincer ruthenium catalysts prepared by Milstein [32, 33] were also efficient with or without extra base. Recently, it was shown that the olefin metathesis catalyst $RuCl_2$(3-bromopyridine)$_2$(=CHPh) (IMes = (*N,N*-bis-(mesityl)-imidazolinylidene) could also perform methanol reforming with a maximum TOF of 158 h^{-1} [34]. Methanol reforming has also been achieved with iridium as noble metal with catalysts introduced by Yamaguchi and Fujita [35], and Beller [36] equipped with a bipyridonate and a PNP ligand, respectively. More recently, 3d metal complexes based on iron and manganese, all of them containing a PNP ligand have been successfully introduced for this reaction (Scheme 5). The iron complex developed by Beller [37] operating in the presence of KOH as a base paved the way for the utilization of non-precious metals in this catalytic transformation. Bernskoetter, Hazari, and Holthausen [38] showed that it was possible to improve the TON up to 51,000 by adding $LiBF_4$ as Lewis acid cocatalyst in the absence of base. Finally, the first example of manganese-catalyzed dehydrogenation of methanol in water was reported by Beller in 2017 [39].

It can be noted that there exist some less productive catalysts that efficiently promote the dehydrogenation of methanol but are inactive for the dehydrogenation of formic acid, the last step of this reforming process [40–42], and that under more drastic conditions formation of carbon monoxide together with hydrogen is highly detrimental for the energy storage application [43].

Mechanistic studies for the dehydrogenation of methanol in water have been proposed in several reviews [29, 33, 34, 41, 44].

3 Formation of Carbonyl Derivatives: Oxidation Without Oxidant

From the equation presented in Scheme 2, at first glance, it is clear that dehydrogenation of alcohols can be exploited either for production of hydrogen or as a synthetic route to aldehydes and ketones. Two strategies have been explored in order to reach high efficiency in carbonyl compound preparations: one with hydrogen gas release, the other with chemical hydrogen trapping by a reactive reagent.

3.1 Acceptorless Dehydrogenation of Alcohols

Many noble metal complexes have been tested for the acceptorless dehydrogenation of primary and secondary alcohols into aldehydes or ketones. Hydride complexes of ruthenium and iridium with no extra basic ligands such as $RuH_2(CO)(PPh_3)_2$ [45], $RuCl_2$(*p*-cymene)(IMes) (IMes = 1,3-*bis*-(mesityl)imidazol-2-ylidene) [46], and $IrH_5(^iPr_3P)_2$ [47] have shown activity in the presence or not of a base, ruthenium catalysts usually requiring high reaction temperature. The catalytic activities of noble metal catalysts have been improved by coordination of non-innocent ligands able to promote beneficial metal-ligand cooperation during outer sphere processes in the catalytic cycles. Some examples of this type of catalysts are shown in Schemes 3 and 6 [22, 23, 26, 48–52]. These ligands mainly contain basic groups able to promote proton transfer or are designed favorably to favor aromatization-dearomatization of a pyridine-containing moiety [53].

Recently, besides the 3d metal catalysts based on iron [37, 38] and manganese [39] that were prepared with the objective of hydrogen production from methanol, other groups developed iron-based catalytic systems such as $FeCl(Cp)(CO)_2$/NaH or Fe(Cp)(Ph)(CO)(Py) operating in refluxing toluene and produced dehydrogenation of 2-pyridylmethanol derivatives that provided a favorable chelated metal intermediate [54], and $Fe(acac)_3$/1,10-phenanthroline in refluxing toluene, which was efficient for secondary benzylic alcohols [55]. Jones introduced a PNP pincer ligand in the hydride complex $FeH(H\text{-}BH_3)(NH(CH_2CH_2P^iPr_2)_2)(CO)$, which was found to be efficient in toluene for a variety of benzylic alcohols and selective toward

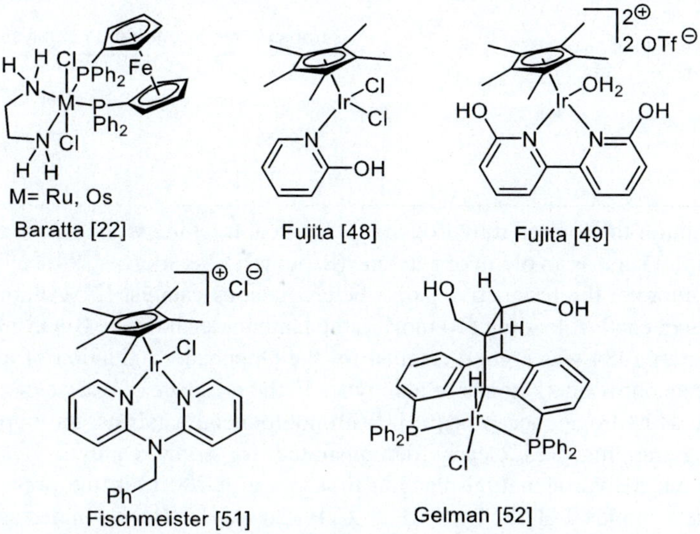

Scheme 6 Selection of noble metal-based catalysts for acceptorless dehydrogenation of alcohols

dehydrogenation of secondary over primary alcohols in 1,2-diols [56]. Hanson used the cobalt complex Co(PNP)(CH_2SiMe_3) (PNP = NH($CH_2CH_2PCy_2$)$_2$) in the presence of $HBrAr^F{}_4$ at 120°C without base for the acceptorless dehydrogenation of benzylic and aliphatic alcohols in good yields [57]. A nickel complex featuring a tris (3,5-dimethylpyrazolyl)borate and a quinolate ligand was shown to catalyze the acceptorless dehydrogenation of benzylic, allylic, and aliphatic alcohols in refluxing toluene in the absence of base [58]. It is noteworthy that this type of nickel catalysts equipped with a cooperative ligand is much more active than other nickel catalysts based on $NiCl_2$(diphosphine) [59]. Very recently, several authors disclosed the acceptorless dehydrogenation of benzylic and aliphatic secondary alcohols at room temperature with the help of a photocatalytic system working under visible light irradiation in the presence of a nickel(II) catalyst [60, 61].

In parallel with the discoveries of new catalytic systems for acceptorless dehydrogenation of alcohols, experimental or computational mechanistic studies have been conducted both with noble metals [53, 62, 63] and earth-abundant metal catalysts [58, 64], the main discussions dealing with inner or outer sphere mechanisms in close relationship with β-H elimination or metal-ligand cooperativity.

3.2 *Alcohol Transfer Dehydrogenation: Oppenauer-Type Oxidation*

It has been frequently observed that the catalytic dehydrogenation of alcohols, which is not thermodynamically favored over its reverse reaction could be facilitated by

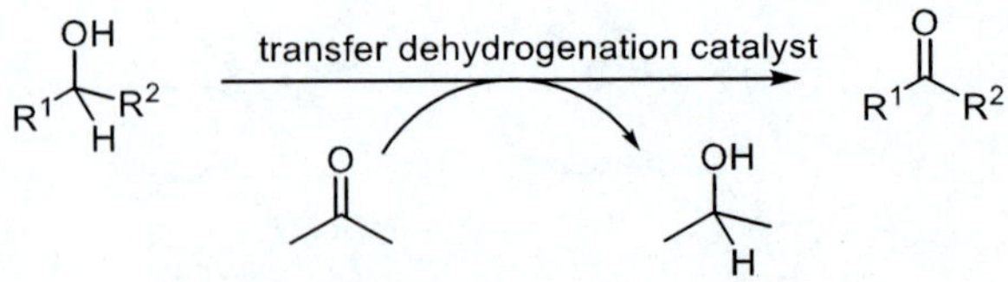

Scheme 7 Transfer dehydrogenation of alcohols with acetone as hydrogen acceptor

elimination of the formed dihydrogen by chemical trapping with a sacrificial hydrogen acceptor such as an olefin or a ketone (Scheme 7). As soon as 1940, ethylene was used to improve the productivity of a heterogeneous catalyst [3]. Although it was used in high catalyst loading (10 mol%), the lanthanide alkoxide ^{t}BuOSmI$_2$ reported by Kagan in 1984 was a good catalyst for the Oppenauer oxidation of a variety of aliphatic alcohols into ketones or aldehydes in the presence of excess of a ketone or aldehyde as hydrogen acceptor [65]. With noble metal catalysts, the importance of hydrogen trapping was clearly demonstrated for some catalytic systems. For instance, no dehydrogenation reaction took place at 200°C in the presence of the dihydride complex IrH$_2$(PCP) (PCP = C$_6$H$_3$-2,6-(CH$_2$P^{t}Bu$_2$)$_2$) under acceptorless conditions whereas the addition of *tert*-butylethylene (TBE) allowed the production of aldehydes and ketones in excellent yields from primary and secondary aliphatic and aromatic alcohols [66]. TBE was also used in the dehydrogenation of isopropanol into acetone at high temperature as a test reaction for evaluation of iridium and rhodium catalysts [67, 68]. Acetone has been used as the simplest ketone as hydrogen acceptor and solvent by Bäckvall in 1996 for the Oppenauer-type oxidation of secondary alcohols to ketones under refluxing acetone with RuCl$_2$(PPh$_3$)$_3$/K$_2$CO$_3$ as catalyst [69]. The same type of catalytic conditions were utilized with various ruthenium and iridium catalysts for the Oppenauer-type oxidation of secondary and primary alcohols [70–73]. It is interesting to note that during the ruthenium-catalyzed dehydrogenation of butan-1-ol the presence of a large excess of acetone improved both the reaction rate and the selectivity toward *n*-butanal disfavoring the formation of butyl butyrate resulting from alcohol dehydrogenation coupling [74].

Oppenauer-type oxidation of secondary and primary alcohols in the presence of acetone has been recently reported with 3d metals, first by Guan with the iron Knölker-type complex [FeH(CO)$_2$(hydroxycyclopentadienyl)] that operated without base at 60°C [75], then by Beller with the iron and manganese complexes ([Fe(H)(BH$_4$)(CO)(HN{CH$_2$CH$_2$P(^{i}Pr)$_2$}$_2$]) [76] (see Scheme 5) and [Mn(NNN)(CO)$_3$]Br (in situ generated from MnBr(CO)$_5$ and *N*-methyl-di-picolylamine) [77, 78] that were active in the presence of a catalytic amount of base at 50–90°C.

4 Transfer Hydrogenation

In catalytic transfer hydrogenation, dihydrogen is substituted by a hydrogen donor, most frequently formic acid or an alcohol for the reduction of carbon–carbon or carbon–heteroatom double bonds. It represents the reverse reaction of the alcohol transfer dehydrogenation in the case of ketone and aldehyde hydrogenation. In the early 1960s, the chemoselective hydrogenation of the C=C bond of conjugated enones was reported with iridium and ruthenium catalysts [79–81]. The hydrogenation of ketones has been more developed, especially the enantioselective transfer hydrogenation with isopropanol as hydrogen donor. The Meerwein-Ponndorf-Verley reduction of saturated ketones with isopropanol was shown with samarium alkoxide [65] and the enantioselective version was achieved at room temperature in high yields and selectivities with a chiral samarium(III) complex [82]. At the same period started a great number of studies involving very efficient asymmetric transfer hydrogenation systems based on noble metals such as ruthenium, iridium, and rhodium, most of them bearing chiral diamine ligands, which have been reported in many reviews and leading articles [83–91]. More recently, about 15 years ago, 3d metal catalysts based on iron [92–100], manganese [93, 101–110], cobalt [92, 93, 111–114], and nickel [92, 93, 115–117] have been extensively investigated in achiral and asymmetric versions. Most of these active 3d metal catalysts were supported by chiral nitrogen- and phosphorus-based polydentate ligands (Scheme 8), some of them leading to very good enantioselectivities.

It can be noted that not only ketones were reduced by hydrogen transfer from alcohols (mainly isopropanol) but also olefins and imine-type substrates as exemplified in Refs. [95, 107, 110, 114].

The catalytic cycles and mechanisms of these transfer hydrogenation reactions have been studied extensively, first experimentally and then with the help of computational methods looking in details into cooperative metal-ligand processes of hydrogen transfer in the presence of bifunctional ligands with noble [118–126] and first row transition metals [95, 97, 127].

5 Acceptorless Dehydrogenative Coupling: Formation of Unsaturated Products

In addition to direct production of carbonyl derivatives and production and uses of hydrogen reported in Sects. 2, 3 and 4, the dehydrogenation of alcohols has found further applications in adding value to the produced carbonyl compounds either with the involvement of the formed dihydrogen or without this contribution. The reactions where the intermediate carbonyl products are reacted with a nucleophile with liberation of water or another molecule of dihydrogen are usually named acceptorless dehydrogenative coupling reactions [128]. They correspond to the reaction sequence described in Scheme 9.

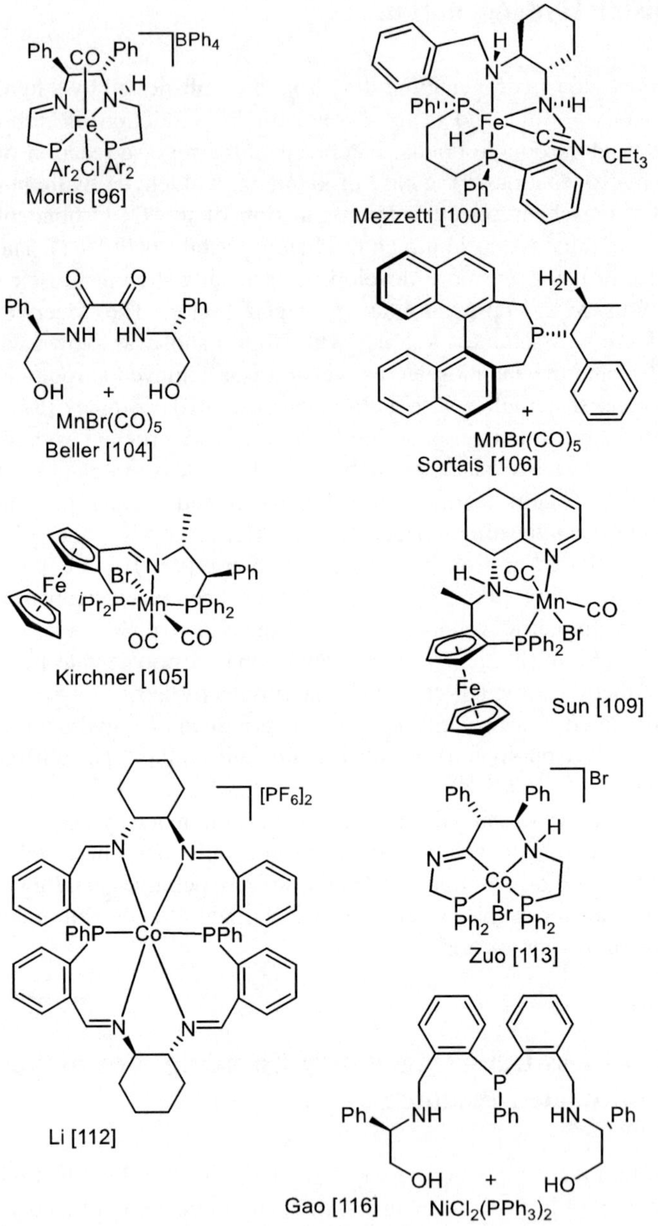

Scheme 8 Selected chiral 3d metal catalysts used in asymmetric transfer hydrogenation of ketones

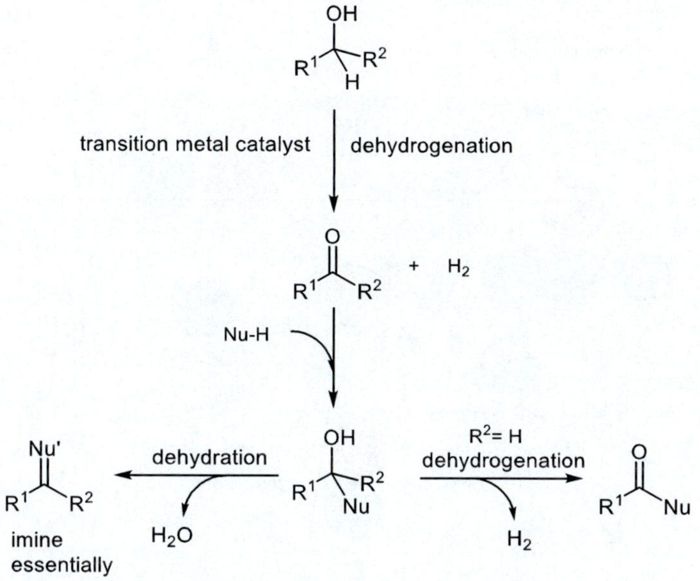

Scheme 9 General scheme for acceptorless dehydrogenative coupling

5.1 Formation of Imines from Primary Amines and Alcohols

Various nucleophiles can be involved in this type of reaction. Here again, the first example was obtained with noble metals, in particular with the ruthenium complex RuH(CO)(PNP) equipped with dearomatized 2,6-bis(di-*tert*-butylphosphinomethyl) pyridine (PNP) pincer ligands developed by Milstein, which allowed the formation of imines from primary alcohols and amines in the absence of extra base (Scheme 10a) [129]. Similar catalysts [130] and other ruthenium complexes featuring various types of ligands such as an *N*-heterocyclic carbene [131], an $NH(CH_2CH_2PPh_2)_2$ pincer ligand [132], a dibenzobarrelene-based cooperating ligand [133], or an assembling naphthyridine -NHC ligand [134] were also competent to achieve the imine preparation. In 2016, Milstein introduced the first manganese complex Mn $(PNP)(CO)_2$ (PNP = bis(di-*tert*-butylphosphinomethyl)pyridine) able to catalyze the formation of aldimines from primary amines and alcohols at 135°C without additional base [135]. Other manganese(I) catalysts with PNP pincer ligands such as those described by Kirchner based on the 2,6-diaminopyridine scaffold [136], or by Kempe based on a triazole core [137], tridentate NNS ligands [138], or bidentate pyrazolylpyridine [139] have been used successfully for imine production. Madsen showed that polydentate porphyrin and salen ligands coordinated to a manganese (III) center could achieve the coupling of alcohols with amines to form imines in the presence of molecular sieves or Ca_3N_2 playing both the role of base and water trap [140, 141]. These manganese-based catalytic systems have been described in more detail in reviews published in 2017 [142] and 2020 [143]. Among 3d metals, a few

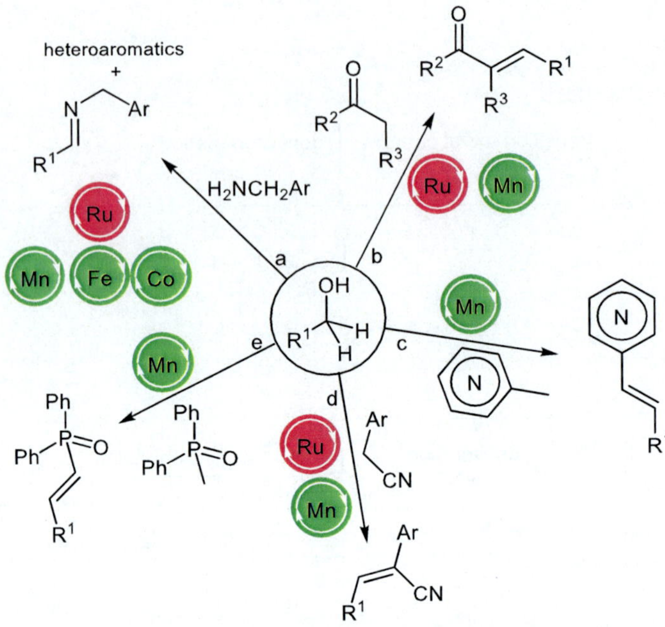

Scheme 10 Examples of acceptorless dehydrogenative coupling of primary alcohols with amines and carbonucleophiles with elimination of hydrogen and water

other examples have been proposed with iron catalysts [144, 145] and with cobalt catalysts featuring PNP pincer ligands [146, 147]. It can be added here that *N*-hetero (poly)aromatic compounds have been recently accessed with cobalt [148, 149], nickel [150–152], and manganese catalysts [153–162].

5.2 *Formation of Alkenes from Alcohols and Carbonucleophiles*

Carbon–carbon double bonds have been generated by dehydrogenation of alcohols followed by condensation with carbonucleophiles generated by deprotonation under basic conditions of carbopronucleophiles (Scheme 10). The first attempts were carried out in the presence of simple ruthenium catalysts such as $RuCl_2(DMSO)_4$ [163, 164] with enolizable ketones and nitriles [165] (Scheme 10b, d). Then, manganese catalysts equipped with pincer ligands, usually prepared from MnBr $(CO)_5$ found efficient applications in the α-olefination of ketones [166], methylated heteroaromatics [167], and nitriles [128, 168–170] (Scheme 10b–d). The alkenylation of methyldiphenylphosphine oxide by a variety of benzylic alcohols was recently achieved with a $MnBr(CO)_2(PNNNP)$ catalyst (Scheme 10e) [171]. A chemoselective one-pot ruthenium-catalyzed synthesis of α,β-unsaturated aldehydes from two alcohol substrates based on dehydrogenation of a benzylic alcohol, condensation with an NH_2 group grafted on silica to form a supported imine followed by a Mannich-type reaction with another aldehyde generated upon dehydrogenation of a primary alcohol has been proposed by Williams, but this sequence required the presence of crotononitrile as hydrogen scavenger [172].

5.3 *Formation of Acetals from Alcohols*

Primary alcohols have also been involved in acceptorless dehydrogenative self-coupling. In this case, the reaction involved 3 molecules of alcohol and the overall reaction led to the formation of dihydrogen and water according to Scheme 11. There are very few examples and all of them result from noble metal catalysis. In 1987, Murahashi reported that at 180°C $RuCl_2(PPh_3)_3$ catalyzed the selective transformations of the aliphatic hexan-1-ol into the corresponding acetal but with modest conversion [173]. Later, Milstein demonstrated that using the ligand-metal cooperativity effect with the acridine-based ruthenium RuHCl(CO)(PNP) pincer complex, the dehydrogenative formation of acetal form hexan-1-ol took place under neutral conditions [174]. Phosphinepyridonate ligands associated with the $[IrCl_2Cp^*]_2$ generated a catalytic system that provides acetals from purely aliphatic and benzylic alcohols in THF at 130–170°C [175].

5.4 *Formation of Carboxylic Esters and Acids*

In this section again, the first results were obtained with noble metal catalysts generating hydride species under relatively harsh temperature conditions. Indeed, in 1981 Shvo reported the formation of esters in the presence of tolane as hydrogen acceptor with $Ru_3(CO)_{12}$ as catalyst (Scheme 12a) [176].

Ruthenium hydrides $RuH_2(PR_3)_4$ were also efficient for the acceptorless formation of esters without a base at high temperature and also for the selective formation of lactones starting from diols [18, 173, 177]. More recently, ruthenium catalysts equipped with electron-rich PNP, PNN, and PCP ligands putting in action the ligand-metal cooperative concept catalyzed the acceptorless dehydrogenative coupling of alcohols to esters and diols to lactones usually in the presence of a base such

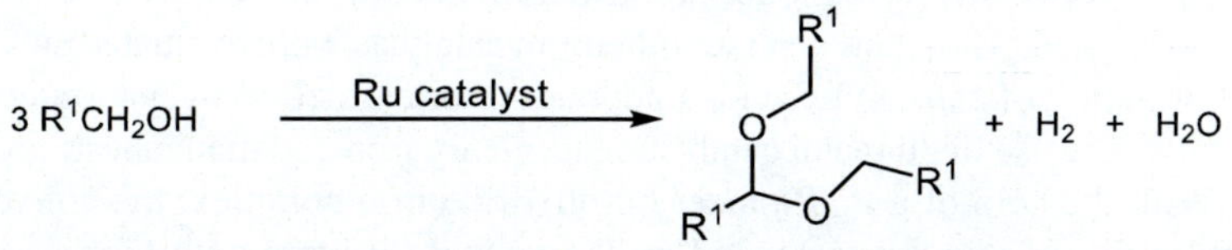

Scheme 11 Acceptorless dehydrogenative formation of acetals from primary alcohols

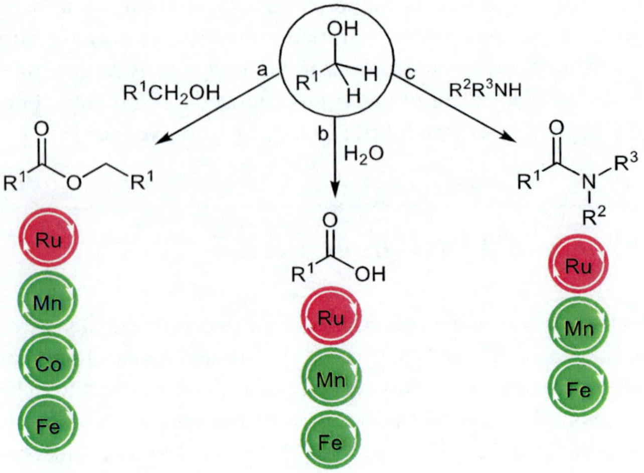

Scheme 12 Examples of acceptorless dehydrogenative coupling reactions

as ^{t}BuOK [120, 133, 178–180]. Very interestingly, the coupling of a primary alcohol with a secondary alcohol to produce mixed esters has been achieved under neutral conditions in refluxing toluene with the RuH(CO)(NNP) catalyst featuring an NNP ligand based on a dearomatized bipyridine-phosphine structure [181]. The same type of NNP ligand coordinated to a manganese(I) dicarbonyl core led to the efficient formation of mixed esters from two different primary alcohols, preferentially a benzylic alcohol with an aliphatic alcohol forming benzoate derivatives in the presence of ^{t}BuOK at 120°C [182]. Similarly, aliphatic PNP pincer-supported manganese(I) dicarbonyl complexes were found as effective catalysts for the acceptorless dehydrogenative coupling of a wide range of alcohols to homo-esters under base-free conditions [183].

Iron catalysis with PNP pincer and cyclopentadienone have also found applications in esters and lactones synthesis [56, 184]. Recently, a catalytic system based on a cobalt complex with an NPPP pincer ligand has shown high efficiency to produce homo-esters from a wide range of benzylic and aliphatic alcohols in the presence of ^{t}BuOK [185].

As far as carboxylic acid formation via dehydrogenation of alcohols is concerned, a similar historical development has been observed. Initially noble metal catalysts were used to perform the condensation with water under basic conditions (Scheme 12b). An example of dehydrogenative condensation in the presence of olefin as hydrogen acceptor involving a domino rhodium/palladium-catalyzed dehydrogenation was reported [186], and diverse ruthenium catalysts were evaluated successfully in the formation of carboxylic acids under acceptorless dehydrogenation conditions [187, 188]. The use of 3d metal catalysts, essentially iron and manganese appeared in 2015. With the help of a PNP pincer borohydride iron complex, the conversion of glycerol to lactic acid was achieved by Haziri and Crabtree with high activity and selectivity [189]. Bio-sourced alcohols including fatty alcohols and terpenols were transformed by Gauvin into the corresponding acids in the presence of iron and manganese catalysts featuring PNP pincer ligands using KOH in the absence of water. The manganese complex were more robust under the catalytic conditions and thus more efficient for the formation of carboxylic acids [190]. The same type of manganese catalyst ($[HN(C_2H_4PPh_2)_2]Mn(CO)_2Br$) was utilized by Maji for the direct formation of α-hydroxy acids from 1,2-diols [191]. The same author proposed a straightforward preparation of α-hydroxy carboxylic acids from ethylene glycol and primary alcohol, based on a sequence of dehydrogenation of the two alcohol substrates followed by aldol condensation under basic conditions, conducted in the presence of a manganese catalyst featuring a PNP ligand constructed from a triazine [192].

5.5 *Formation of Amides via Acceptorless Dehydrogenative Coupling*

This chemistry also started with noble metal catalysts, especially ruthenium (Scheme 12c), and has been reviewed by Verpoort and Wu in 2016 [193] and Milstein in 2023 [194]. After the initial discovery by Milstein that alcohols and amines could be selectively coupled into amides with release of dihydrogen [195] (later including ammonia [196]), a huge number of ruthenium complexes with a variety of ligands were evaluated in this reaction [193]. Only recently, in 2017 the formation of amides and imides from monoalcohols and diols was described with 3d metals, with manganese and iron pincer catalysts [197–199]. Advantage was taken of the special reactivities of methanol and ammonia to form ureas from methanol and two molecules of amine [200], and secondary amides from two molecules of benzylic alcohol and one of ammonia with stoichiometric amount of base [201], with iron and manganese catalysts, respectively.

6 Reactions Proceeding with H-Auto-Transfer (or Borrowing) Processes

All the reactions described in Sect. 5 release at least one molecule of dihydrogen corresponding to the first step, which is the alcohol dehydrogenation. A more atom economical strategy would be to use the generated hydrogen in a catalytic sequence dehydrogenation/condensation/hydrogenation allowing the direct transformation of intermediates into hydrogenated products. From a mechanistic point of view, the last step could be envisioned as a classical homolytic metal-catalyzed hydrogenation or as a sequence involving hydride (originating from transient metal hydride) and proton transfer. This strategy has been developed at the beginning of the 2000s

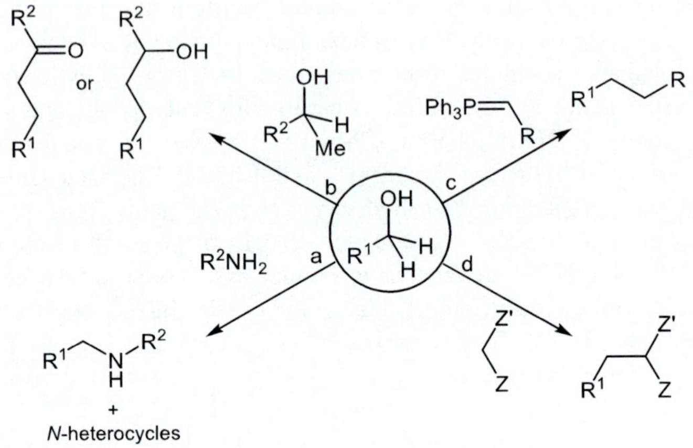

Scheme 13 Selected examples of hydrogen borrowing processes

with noble metals and ruthenium as the main actor, and named hydrogen borrowing [202], hydrogen auto-transfer [203], or hydrogen shuttling [204].

Several types of transformations with creation of C–C and C–N bonds have thus been investigated (Scheme 13). Typical examples are the *N*-alkylation of amines and the formation of *N*-heterocycles (Scheme 13a) and C–C bond formation in aldol condensation (Scheme 13b) as well as with more limited examples such as Wittig-type reactions (Scheme 13c) and condensation with soft carbonucleophiles such as 1,3-keto-esters, keto-nitrile, diesters (Scheme 13d). The applications with noble metal catalysts in these various reactions have been reviewed regularly [202–210]. For alkylation of amines, the use of 3d metal homogeneous catalysts started in the mid-2010s with the preparation of a variety of catalysts featuring cooperative metal-ligand effects and showing very good properties for simple alkylation of amines including methylation from methanol, and for the preparation of *N*-heterocycles from amino alcohols. The main base metal catalysts used for this purpose were based on manganese [211–213], iron [214–216], cobalt [217, 218], and even copper [219]. Some comparisons of 3d metal and noble catalyst activities have been carried out by Kempe [220] and Hultzsch [221].

Concerning the *C*-alkylation reactions based on aldol condensation from ketones, esters, amides, phosphine oxides, and secondary alcohols, they have also known a huge development with 3d metal catalysts equipped with the appropriate cooperative ligands. During the last 10 years they have been developed with manganese [222–228], iron [229–236], cobalt [237–241], nickel [128, 242–244], and copper [245, 246] with synthetic objectives and for the homologation of alcohols to produce fuels [223, 247].

7 Conclusion

Transition metal-catalyzed dehydrogenation of alcohols represents a useful transformation with various applications in energy and chemical synthesis. Initially performed with noble metal catalysts under high temperature conditions and with a hydrogen scavenger, the discovery of cooperative metal-ligand catalysts able to facilitate proton and hydride transfers in the coordination sphere of the metal has led to improved experimental conditions and to the applications in modern synthetic methodologies such as enantioselective transfer hydrogenation, acceptorless dehydrogenative condensation, and hydrogen borrowing processes. During the last two decades, the presence of 3d metal catalysts in this field has increased a lot due to intense researches on manganese, iron, cobalt, nickel, and copper catalysts motivated by the abundance and low cost of these metals and positive impact on green and sustainable catalysis. Despite the increasing number of examples reported in recent reviews on base metal-catalyzed reactions [248–253], some applications of noble metal catalysis could not be achieved with 3d metals at the moment. High catalyst loading and high temperature in some reactions still represent disadvantages of 3d over noble metal catalysts but considering that dehydrogenation of alcohols and its applications with 3d metal catalysis is a very recent topic, there is an avenue for new discoveries in the near future. With this respect, improvements regarding the reaction temperature have been made possible using light-mediated conditions (UV or visible), which have allowed α-alkylation of ketones with an iron catalyst [235], and the formation of carbonyl compounds and amides with manganese [254] at room temperature. Researches have already started in this direction with the development of multi-component transformations including at least one alcohol dehydrogenation step, which facilitate the elaboration of complex molecules [227, 255–259].

Also very promising is a distinct mechanism paradigm that has been recently applied for the dehydrogenation process. Whereas it was established that noble metals operated according to inner or outer sphere proton/hydride mechanisms, it has been demonstrated that 3d metals could involve radical processes in dehydrogenation of alcohols [260–265]. In particular, Adhikari took profit of the redox properties of the non-innocent azo-phenolate ligand coordinated to nickel to achieve various *C*- and *N*-alkylations [262–270]. With the introduction of photocatalysis and dual hybrid catalysis in alcohol dehydrogenation, reactions are now possible at room temperature under radical conditions with 3d metals [271–275], opening new possibilities for dehydrogenation and borrowing reactions.

References

1. Rideal EK (1921) On the catalytic dehydrogenation of alcohols. P Roy Soc Lond A Mat 99: 153–162
2. Constable FH (1925) The mechanism of catalytic decomposition. P Roy Soc Lond A Mat 108: 355–378

3. Reeve W, Adkins H (1940) Catalytic dehydrogenation of alcohols in the liquid phase using ethylene as a hydrogen acceptor. J Am Chem Soc 62:2874–2876
4. Badin EJ, Pacsu E (1964) Mechanism of catalytic hydrogenation and dehydrogenation of aldehydes and alcohols. J Am Chem Soc 66:1963–1968
5. Murahashi SI, Shimamura T, Moritani I (1974) Conversion of alcohols into unsymmetrical secondary or tertiary amines by a palladium catalyst. Synthesis of *N*-substituted pyrroles. JCS Chem Commun:931–932
6. Alsabeth PG, Mellmann D, Junge H, Beller M (2014) Ruthenium-catalyzed hydrogen generation from alcohols and formic acid, including Ru-pincer-type complexes. Top Organomet Chem 48:45–80
7. Sordakis K, Tang C, Vogt LK, Junge H, Dyson PJ, Beller M, Laurenczy G (2018) Homogeneous catalysis for sustainable hydrogen storage in formic acid and alcohols. Chem Rev 2: 372–433
8. Niermann M, Drünert S, Kaltschmitt BK (2019) Liquid organic hydrogen carriers (LOHCs)-technico-economic analysis of LOHCs in defined process chain. Energ Environ Sci 12:290–307
9. Crabtree RH (2008) Hydrogen storage in liquid organic heterocycles. Energ Environ Sci 1: 134–138
10. Crabtree RH (2017) Homogeneous transition metal catalysis of acceptorless dehydrogenative alcohol oxidation: applications in hydrogen storage and to heterocycle synthesis. Chem Rev 117:9228–9246
11. Kumar A, Daw P, Milstein D (2022) Homogeneous catalysis for sustainable energy: hydrogen and methanol economies, fuels from biomass, and related topics. Chem Rev 122:385–441
12. Campos J (2018) Dehydrogenation of alcohols and polyols from a hydrogen production perspective. Phys Sci Rev 3:20170017
13. Charman HB (1967) Hydride transfer reactions catalyzed by metal complexes. J Chem Soc B:629–632
14. Dobscn A, Robinson SD (1975) Catalytic dehydrogenation of primary and secondary alcohols by $Ru(OCOCF_3)_2(CO)(PPh_3)_2$. J Organomet Chem 87:C52–C53
15. Rybak WK, Ziolkowski JJ (1981) Dehydrogenation of alcohols catalyzed by polystyrene-supported ruthenium complexes. J Mol Catal 11:365–370
16. Shinoda S, Itagaki H, Saito Y (1985) Dehydrogenation of methanol in the liquid phase with a homogeneous ruthenium complex catalyst. J Chem Soc Chem Commun:860–861
17. van Buijtenen J, Meuldijk J, Vekemans JAJM, Hulshof LA, Kooijman H, Spek AL (2006) Dinuclear ruthenium complexes bearing dicarboxylate and phosphine ligands. Acceptorless catalytic dehydrogenation of 1-phenylethanol. Organometallics 25:873–881
18. Murahashi SI, Ito KI, Naota T, Maeda Y (1981) Ruthenium catalyzed transformation of alcohols to esters and lactones. Tetrahedron Lett 22:5327–5330
19. Morton D, Cole-Hamilton DJ (1988) Molecular hydrogen complexes in catalysis: highly efficient hydrogen production from alcoholic substrates catalyzed by ruthenium complexes. J Chem Soc Chem Commun:1154–1156
20. Junge H, Loges B, Beller M (2007) Novel improved ruthenium catalysts for the generation of hydrogen from alcohols. Chem Commun:522–524
21. Baratta W, Herdtweck E, Siega K, Toniutti M, Rigo P (2005) 2-(Aminomethyl)pyridine-phosphine ruthenium(II) complexes: novel highly active transfer hydrogenation catalysts. Organometallics 24:1660–1669
22. Baratta W, Bossi G, Putignano E, Rigo P (2011) Pincer and diamine Ru and Os diphosphane complexes as efficient catalysts for the dehydrogenation of alcohols to ketones. Chem A Eur J 17:3474–3481
23. Zhang J, Gandelman M, Shimon LJW, Rozenberg H, Milstein D (2004) Electron-rich, bulky ruthenium PNP-type complexes. Acceptorless catalytic alcohol dehydrogenation. Organometallics 23:4026–4033

24. Nielsen M, Alberico E, Baumann W, Drexler HJ, Junge H, Gladiali S, Beller M (2013) Low-temperature aqueous-phase methanol dehydrogenation to hydrogen and carbon dioxide. Nature 495:85–89
25. Wang Q, Chai H, Yu Z (2018) Acceptorless dehydrogenation of *N*-heterocycles and secondary alcohols by Ru(II)-NNC complexes bearing a pyrazoyl-indolylpyridine ligand. Organometallics 37:584–591
26. Tseng KNT, Kampf JW, Szymczak NK (2013) Base-free, acceptorless, and chemoselective alcohol dehydrogenation catalyzed by an amide-derived *NNN*-ruthenium(II) hydride complex. Organometallics 32:2046–2049
27. Bhatia A, Muthaiah S (2018) Well-defined ruthenium complex for acceptorless alcohol dehydrogenation in aqueous medium. ChemistrySelect 3:3737–3741
28. Wang Z, Pan B, Liu Q, Yue E, Solan GA, Ma Y, Sun WH (2017) Efficient acceptorless dehydrogenation of secondary alcohols to ketones mediated by a PNN-Ru(II) catalyst. Cat Sci Technol 7:1654–1661
29. Nielsen M, Alberico E, Baumann W, Drexler HJ, Junge H, Gladiali S, Beller M (2013) Low-temperature aqueous-phase methanol dehydrogenation to hydrogen and carbon dioxide. Nature 495:85–89
30. Monney A, Barsch E, Sponholz P, Junge H, Ludwig R, Beller M (2014) Base-free hydrogen generation from methanol using a bi-catalytic system. Chem Commun 50:707–709
31. Rodriguez-Lugo RE, Tricado M, Vogt M, Tewes F, Santiso-Quinones G, Grützmacher H (2013) A homogeneous transition metal complex for clean hydrogen production from methanol–water mixtures. Nat Chem 5:342–347
32. Hu P, Diskin-Posner Y, Ben-David Y, Milstein D (2014) Reusable homogeneous catalytic system for hydrogen production from methanol and water. ACS Catal 4:2649–2652
33. Luo J, Kar S, Rauch M, Montag M, Ben-David Y, Milstein D (2021) Efficient base-free aqueous reforming of methanol homogeneously catalyzed by ruthenium exhibiting a remarkable acceleration by added catalytic thiol. J Am Chem Soc 143:17284–17291
34. Wang Q, Lan J, Liang R, Xia Y, Qin L, Chung LW, Zheng Z (2022) New tricks for an old dog: Grubbs catalysts enable efficient hydrogen production from aqueous-phase methanol reforming. ACS Catal 12:2212–2222
35. Fujita KI, Kawahara R, Aikawa T, Yamaguchi R (2015) Hydrogen production from a methanol–water solution catalyzed by an anionic iridium complex bearing a functional bipyridonate ligand under weakly basic conditions. Angew Chem Int Ed 54:9057–9060
36. Prichatz C, Alberico E, Baumann W, Junge H, Beller M (2017) Iridium–PNP pincer complexes for methanol dehydrogenation at low base concentration. ChemCatChem 9:1891–1896
37. Alberico E, Sponholz P, Cordes C, Nielsen M, Drexler HJ, Baumann W, Junge H, Beller M (2013) Selective hydrogen production from methanol with a defined iron pincer catalyst under mild conditions. Angew Chem Int Ed 52:14162–14166
38. Bielinski EA, Förster M, Zhang Y, Bernskoetter WH, Hazari N, Holthausen MC (2015) Base-free methanol dehydrogenation using a pincer-supported iron compound and Lewis acid co-catalyst. ACS Catal 5:2404–2415
39. Andérez-Fernandez M, Vogt LK, Fischer S, Zhou W, Jiao H, Garbe M, Elangovan S, Junge K, Junge H, Ludwig R, Beller M (2016) A stable manganese pincer catalyst for the selective dehydrogenation of methanol. Angew Chem Int Ed 56:559–562
40. Awasthi MK, Rai RK, Behrens S, Singh SK (2021) Low-temperature hydrogen production from methanol over a ruthenium catalyst in water. Cat Sci Technol 11:136–142
41. Arora V, Yasmin E, Tanwar N, Hathwar VR, Wagh T, Dhole S, Kumar A (2023) Pincer-ruthenium-catalyzed reforming of methanol – selective high-yield production of formic acid and hydrogen. ACS Catal 13:3605–3617
42. Campos J, Sharninghausen LS, Manas MG, Crabtree RH (2015) Methanol dehydrogenation by iridium *N*-heterocyclic carbene complexes. Inorg Chem 54:5079–5084

43. Kaithal A, Chatterjee B, Werlé C, Leitner W (2016) Acceptorless dehydrogenation of methanol to carbon monoxide and hydrogen using molecular catalysts. Angew Chem Int Ed 60: 26500–26505
44. Trincado M, Bösken J, Grützmacher H (2021) Homogeneously catalyzed acceptorless dehydrogenation of alcohols: a progress report. Coord Chem Rev 443:213967
45. Muthaiah S, Hong SH (2012) Acceptorless and base-free dehydrogenation of alcohols and amines using ruthenium-hydride complexes. Adv Synth Catal 354:3045–3053
46. Shahane S, Fischmeister C, Bruneau C (2012) Acceptorless ruthenium catalyzed dehydrogenation of alcohols to ketones and esters. Cat Sci Technol 2:1425–1428
47. Lin Y, Ma D, Lu X (1987) Iridium pentahydride complex catalyzed dehydrogenation of alcohols in the absence of a hydrogen acceptor. Tetrahedron Lett 28:3115–3118
48. Fujita KI, Tanino N, Yamaguchi R (2007) Ligand-promoted dehydrogenation of alcohols catalyzed by Cp*Ir complexes. A new catalytic system for oxidant-free oxidation of alcohols. Org Lett 9:109–111
49. Fujita KI, Yoshida T, Imori Y, Yamaguchi R (2011) Dehydrogenative oxidation of primary and secondary alcohols catalyzed by a Cp*Ir complex having a functional C,N-chelate ligand. Org Lett 13:2278–2281
50. Kawahara R, Fujita KI, Yamaguchi R (2012) Dehydrogenative oxidation of alcohols in aqueous media using water-soluble and reusable Cp*Ir catalysts bearing a functional bipyridine ligand. J Am Chem Soc 134:3643–3646
51. Jayaprakash H, Guo L, Wang S, Bruneau C, Fischmeister C (2020) Acceptorless and base-free dehydrogenation of alcohols mediated by a dipyridylamine-iridium(III) catalyst. Eur J Org Chem:4326–4330
52. Musa S, Shaposhnikov I, Cohen S, Gelman D (2011) Ligand-metal cooperation in PCP pincer complexes: rational design and catalytic activity in acceptorless dehydrogenation of alcohols. Angew Chem Int Ed 50:3533–3537
53. Gunanathan C, Milstein D (2011) Metal-ligand cooperation by aromatization-dearomatization: a new paradigm in bond activation and "Green" catalysis. Acc Chem Res 44:588–602
54. Kamitani M, Ito M, Itazaki M, Nakazawa H (2014) Effective dehydrogenation of 2-pyridylmethanol derivatives catalyzed by an iron complex. Chem Commun 50:7941–7944
55. Song H, Kang B, Hong SH (2014) Fe-catalyzed acceptorless dehydrogenation of secondary benzylic alcohols. ACS Catal 4:2889–2895
56. Chakraborty S, Lagaditis PO, Förster M, Bielinski EA, Hazari N, Holthausen MC, Jones WD, Schneider S (2014) Well-defined iron catalysts for the acceptorless reversible dehydrogenation-hydrogenation of alcohols and ketones. ACS Catal 4:3994–4003
57. Zhang G, Hanson SK (2013) Cobalt-catalyzed acceptorless alcohol dehydrogenation: synthesis of imines from alcohols and amines. Org Lett 15:650–653
58. Chakraborty S, Piszel PE, Brennessel WW, Jones WD (2015) A single nickel catalyst for the acceptorless dehydrogenation of alcohols and hydrogenation of carbonyl compounds. Organometallics 34:5203–5206
59. Geetha R, Kumar M, Kulkarni NV, Jones WD (2021) Dehydrogenation of 1-phenylethanol catalyzed by nickel(II)diphosphine complexes. Acta Chim Slov 68:955–960
60. Yang XJ, Zheng YW, Zheng LQ, Wu LZ, Tung CH, Chen B (2019) Visible light-catalytic dehydrogenation of benzylic alcohols to carbonyl compounds by using an eosin Y and nickel-thiolate complex dual catalyst system. Green Chem 21:1401–140560
61. Fuse H, Mitsunuma H, Kanai M (2020) Catalytic acceptorless dehydrogenation of aliphatic alcohols. J Am Chem Soc 142:4493–4499
62. Tseng KNT, Kampf JW, Szymczak NK (2015) Mechanism of N,N,N-amide ruthenium (II) hydride mediated acceptorless alcohol dehydrogenation: inner-sphere β-H elimination versus outer-sphere bifunctional metal-ligand cooperativity. ACS Catal 5:5468–5485

63. Li H, Lu G, Jiang J, Huang F, Wang ZX (2011) Computational mechanistic study on Cp*Ir complex-mediated acceptorless alcohol dehydrogenation: bifunctional hydrogen transfer *vs* β-H elimination. Organometallics 30:2349–2363
64. Zhang G, Vasudevan KV, Scott BL, Hanson SK (2013) Understanding the mechanisms of cobalt-catalyzed hydrogenation and dehydrogenation reactions. J Am Chem Soc 135:8668–8681
65. Namy JL, Souppe J, Collin J, Kagan B (1984) New preparations of lanthanide alkoxides and their catalytical activity in Meerwein-Ponndorf-Verley-Oppenauer reactions. J Org Chem 49: 2045–2049
66. Morales-Morales D, Redon R, Wang Z, Lee DW, Yung C, Magnuson K, Jensen CM (2001) Selective dehydrogenation of alcohols and diols catalyzed by a dihydrido iridium PCP pincer complex. Can J Chem 79:823–829
67. Shi Y, Suguri T, Kojima S, Yamamoto Y (2015) 7-6-7 ring-based transition-metal catalysts for the transfer dehydrogenation of isopropanol. J Organomet Chem 799-800:7–12
68. Polukeev AV, Petrovskii PV, Peregudov AS, Ezernitskaya MG, Koridze AA (2013) Dehydrogenation of alcohols by bis(phosphinite) benzene based and bis(phosphine) ruthenocene based iridium pincer complexes. Organometallics 32:1000–1015
69. Almeida MLS, Beller M, Wang GZ, Bäckvall JE (1996) Ruthenium(II)-catalyzed Oppenauer-type oxidation of secondary alcohols. Chem A Eur J 2:1533–1536
70. Gauthier S, Scopelliti R, Severin K (2004) A heterobimetallic rhodium(I)-ruthenium(II) catalyst for the Oppenauer-type oxidation of primary and secondary alcohols under mild conditions. Organometallics 23:3769–3771
71. Hanasaka F, Fujita KI, Yamaguchi R (2005) Synthesis of new cationic Cp*Ir N-hetrocyclic carbene complexes and their high catalytic activities in the Oppenauer-type oxidation of primary and secondary alcohols. Organometallics 24:3422–3433
72. Labes R, Battilocchio C, Mateos C, Cumming GR, de Frutos O, Rincon JA, Binder K, Ley SV (2017) Chemoselective continuous Ru-catalyzed hydrogen-transfer Oppenauer-type oxidation of secondary alcohols. Org Process Res Dev 21:1419–1422
73. Hill CK, Hartwig JF (2017) Site-selective oxidation, amination and epimerization reactions of complex polyols enabled by transfer hydrogenation. Nat Chem 9:1213–1221
74. Nguyen DH, Morin Y, Zhang L, Trivelli X, Capet F, Paul S, Desset S, Dumeignil F, Gauvin RM (2017) Oxidative transformations of biosourced alcohols catalyzed by earth-abundant transition metals. ChemCatChem 9:2652–2660
75. Coleman MG, Brown AN, Bolton BA, Guan H (2010) Iron-catalyzed Oppenauer-type oxidation of alcohols. Adv Synth Catal 352:967–970
76. Budweg S, Wei Z, Jiao H, Junge K, Beller M (2019) Iron-PNP-pincer-catalyzed transfer dehydrogenation of secondary alcohols. ChemSusChem 12:2988–2993
77. Budweg S, Junge K, Beller M (2019) Transfer-dehydrogenation of secondary alcohols catalyzed by manganese NNN-pincer complexes. Chem Commun 55:14143–14146
78. Budweg S, Junge K, Beller M (2020) Catalytic oxidations by dehydrogenation of alkanes, alcohols and amines with defined (non)-noble metal pincer complexes Catal. Sci Technol 10: 3825–3842
79. Trocha-Grimshaw J, Henbest HB (1967) Catalysis of the transfer of hydrogen from propan-2-ol to α,β-unsaturated ketones by organoiridium compounds. A carbon-iridium compound containing a chelate keto-group. Chem Commun:544
80. Sasson Y, Blum J (1971) Homogeneous catalytic transfer hydrogenation of α,β-unsaturated carbonyl compounds by dichlorotris(triphenylphosphine)ruthenium(II). Tetrahedron Lett 24: 2167–2170
81. Sasson Y, Blum J (1975) Dichlorotris(triphenylphosphine)ruthenium-catalyzed hydrogen transfer from alcohols to saturated and α,β-unsaturated ketones. J Org Chem 40:1887–1896
82. Evans DA, Nelson SG, Gagné MR, Muci AR (1993) A chiral samarium-based catalyst for the asymmetric Meerwein-Ponndorf-Verley reduction. J Am Chem Soc 115:9800–9801

83. Zassinovich G, Mestroni G, Gladiali S (1992) Asymmetric hydrogen transfer reactions promoted by homogeneous transition metal catalysts. Chem Rev 92:1051–1069
84. Hashiguchi S, Fujii A, Takehara J, Ikariya T, Noyori R (1995) Asymmetric transfer hydrogenation of aromatic ketones catalyzed by chiral ruthenium(II) complexes. J Am Chem Soc 117: 7562–7563
85. Noyori R, Hashiguchi S (1997) Asymmetric transfer hydrogenation catalyzed by chiral ruthenium complexes. Acc Chem Res 30:97–102
86. Palmer MJ, Wills M (1999) Asymmetric transfer hydrogenation of C=O and C=N bonds. Tetrahedron Asymmetry 10:2045–2061
87. Murata K, Ikariya T (1999) New chiral rhodium and iridium complexes with chiral diamine ligands for asymmetric transfer hydrogenation of aromatic ketones. J Org Chem 64:2186–2187
88. Gladiali S, Alberico E (2006) Asymmetric transfer hydrogenation: chiral ligands and applications. Chem Soc Rev 35:226–236
89. Ikariya T, Blacker AJ (2007) Asymmetric transfer hydrogenation of ketones with bifunctional transition metal-based molecular catalysts. Acc Chem Res 40:1300–1308
90. Sandoval CA, Li Y, Ding K, Noyori R (2008) The hydrogenation/transfer hydrogenation network in asymmetric reduction of ketones catalyzed by [$RuCl_2$(binap)(pica)] complexes. Chem Asian J 3:1801–1810
91. Saidi O, Williams JMJ (2011) Iridium-catalyzed hydrogen transfer reactions. Top Organomet Chem 34:77–106
92. Li YY, Yu SL, Shen WY, Gao JX (2015) Iron-, cobalt-, and nickel-catalyzed asymmetric transfer hydrogenation and asymmetric hydrogenation of ketones. Acc Chem Res 48:2587–2598
93. Agbossou-Niedercorn F, Michon C (2020) Bifunctional homogeneous catalysts based on first row transition metals in asymmetric hydrogenation. Coord Chem Rev 425:213523
94. Morris RH (2009) Asymmetric hydrogenation, transfer hydrogenation and hydrosilylation of ketones catalyzed by iron complexes. Chem Soc Rev 38:2282–2291
95. Zuo W, Lough AJ, Li YF, Morris RH (2013) Amine(imine)diphosphine iron catalysts for asymmetric transfer hydrogenation of ketones and imines. Science 342:1080–1083
96. Sues PE, Demmans KZ, Morris RH (2014) Rational design of iron catalysts for asymmetric transfer hydrogenation. Dalton Trans 43:7650–7667
97. Morris RH (2015) Exploiting metal-ligand bifunctional reactions in the design of iron asymmetric hydrogenation catalysts. Acc Chem Res 48:1494–1502
98. Bigler R, Huber R, Mezzetti A (2016) Iron chemistry made easy: chiral N_2P_2 ligands for asymmetric catalysis. Synlett:831–847
99. Bigler R, Huber R, Mezzetti A (2015) Highly enantioselective transfer hydrogenation of ketones with chiral $(NH)_2P_2$ macrocyclic iron(II) complexes. Angew Chem Int Ed 54:5171–5174
100. de Luca L, Mezzetti A (2017) Base-free asymmetric transfer hydrogenation of 1,2-di- and monoketones catalyzed by a $(NH)_2P_2$ macrocyclic iron(II) hydride. Angew Chem Int Ed 56: 11949–11953
101. Perez M, Elangovan S, Spannenberg A, Junge K, Beller M (2017) Molecularly defined manganese pincer complexes for selective transfer hydrogenation of ketones. ChemSusChem 10:83–86
102. Bruneau-Voisine A, Wang D, Dorcet V, Roisnel T, Darcel C, Sortais JB (2017) Transfer hydrogenation of carbonyl derivatives catalyzed by an inexpensive phosphine-free manganese precatalyst. Org Lett 19:3656–3659
103. Wang D, Bruneau-Voisine A, Sortais JB (2018) Practical (asymmetric) transfer hydrogenation of ketones catalyzed by manganese with (chiral) diamines ligands. Catal Commun 105:31–36
104. Schneekönig J, Junge K, Beller M (2019) Manganese catalyzed asymmetric transfer hydrogenation of ketones using chiral oxamide ligands. Synlett:503–507

105. Zirakzadeh A, de Aguiar SRMM, Stöger B, Widhalm M, Kirchner K (2017) Enantioselective transfer hydrogenation of ketones catalyzed by a manganese complex containing an unsymmetrical chiral PNP' tridentate ligand. ChemCatChem 9:1744–1748
106. Azouzi K, Bruneau-Voisine A, Vendier L, Sortais JB, Bastin S (2020) Asymmetric transfer hydrogenation of ketones promoted by manganese(I) pre-catalysts supported by bidentate aminophosphines. Catal Commun 142:106040
107. Wei D, Bruneau-Voisine A, Dubois M, Bastin S, Sortais JB (2019) Manganese-catalyzed transfer hydrogenation of aldimines. ChemCatChem 11:5256–5259
108. Jayaprakash H (2021) Mn(I) phosphine-amino-phosphinites: a highly modular class of pincer complexes for enantioselective transfer hydrogenation of aryl-alkyl ketones. Dalton Trans 50: 14115–14119
109. Yang J, Yao L, Wang Z, Zuo Z, Liu S, Gao P, Han M, Liu Q, Solan GA, Sun WH (2023) Manganese(I)-catalyzed asymmetric (transfer) hydrogenation of ketones: an insight into the effect of chiral PNN and NN ligands. J Catal 418:40–50
110. Wang L, Lin J, Xia C, Sun W (2022) Manganese-catalyzed asymmetric transfer hydrogenation of hydrazones. J Catal 413:487–497
111. Zhang G, Hanson SK (2013) Cobalt-catalyzed transfer hydrogenation of C=O and C=N bonds. Chem Commun 49:10151–10153
112. Ruan SH, Fan ZW, Zhang WJ, Xu H, An DL, Wei ZB, Yuan RM, Gao JX, Li YY (2023) Asymmetric transfer hydrogenation of ketones catalyzed by chiral macrocyclic cobalt (II) complexes. J Catal 418:100–109
113. Huo S, Chen H, Zuo W (2021) A cobalt(II) complex bearing the amine(imine)diphosphine PN (H)NP ligand for asymmetric transfer hydrogenation of ketones. Eur. J Inorg Chem:37–42
114. Zhang G, Yin Z, Tan J (2016) Cobalt(II)-catalysed transfer hydrogenation of olefins. RSC Adv 6:22419–22423
115. Castellanos-Blanco N, Arévalo A, García JJ (2016) Nickel-catalyzed transfer hydrogenation of ketones using ethanol as a solvent and a hydrogen donor. Dalton Trans 45:13604–13614
116. Dong ZR, Li YY, Yu SL, Sun GS, Gao JX (2012) Asymmetric transfer hydrogenation of ketones catalyzed by nickel complex with new PNO-type ligands. Chin Chem Lett 23:533–536
117. Chen H, Wang Z, Li M, Zuo W (2023) Amido-ene(amido) Ni(II)-catalyzed highly enantioselective transfer hydrogenations of ketone: dual functions of the ene(amido) group. ACS Catal 13:4261–4271
118. Noyori R, Yamakawa M, Hashiguchi S (2001) Metal-ligand bifunctional catalysis: a nonclassical mechanism for asymmetric hydrogen transfer between alcohols and carbonyl compounds. J Org Chem 66:7931–7944
119. Clapham SE, Hadzovic MRH (2004) Mechanisms of the H_2-hydrogenation and transfer hydrogenation of polar bonds catalyzed by ruthenium hydride complexes. Coord Chem Rev 248:2201–2237
120. Bertoli M, Choualeb A, Lough AJ, Moore B, Gusev DG (2011) Osmium and ruthenium catalysts for dehydrogenation of alcohols. Organometallics 30:3479–3482
121. Dmitry G, Gusev DG (2020) Revised mechanisms of the catalytic alcohol dehydrogenation and ester reduction with the Milstein PNN complex of ruthenium. Organometallics 39:258–270
122. Dub PA, Ikariya T (2013) Quantum chemical calculations with the inclusion of nonspecific and specific solvation: asymmetric transfer hydrogenation with bifunctional ruthenium catalysts. J Am Chem Soc 135:2604–2619
123. Dub PA, Henson NJ, Martin RL, Gordon JC (2014) Unravelling the mechanism of the asymmetric hydrogenation of acetophenone by [RuX_2(diphosphine)(1,2-diamine)] catalysts. J Am Chem Soc 136:3505–3521
124. Dub PA, Gordon JC (2016) The mechanism of enantioselective ketone reduction with Noyori and Noyori–Ikariya bifunctional catalysts. Dalton Trans 45:6756–6781

125. Dub PA, Gordon JC (2017) Metal-ligand bifunctional catalysis: the "Accepted" mechanism, the issue of concertedness, and the function of the ligand in catalytic cycles involving hydrogen atoms. ACS Catal 7:6635–6655
126. Tkachenko NV, Rublev P, Dub PA (2022) The source of proton in the Noyori–Ikariya catalytic cycle. ACS Catal 12:13149–13157
127. Wang Y, Liu S, Yang H, Li H, Lan Y, Liu Q (2022) Structure, reactivity and catalytic properties of manganese-hydride amidate complexes. Nat Chem 14:1233–1241
128. Subaramanian M, Sivakumar G, Balaraman E (2021) First-row transition-metal catalyzed acceptorless dehydrogenation and related reactions: a personal account. Chem Rec 21:3839–3871
129. Gnanaprakasam B, Zhang J, Milstein D (2010) Direct synthesis of imines from alcohols and amines with liberation of H_2. Angew Chem Int Ed 49:1468–1471
130. Rigoli JW, Moyer SA, Simon D, Pearce SD, Schomaker JM (2012) α,β-Unsaturated imines *via* Ru-catalyzed coupling of allylic alcohols and amines. Org Biomol Chem 10:1746–1749
131. Maggi A, Madsen R (2012) Dehydrogenative synthesis of imines from alcohols and amines catalyzed by a ruthenium *N*-heterocyclic carbene complex. Organometallics 31:451–455
132. Oldenhuis NJ, Dong VM, Guan Z (2014) Catalytic acceptorless dehydrogenations: Ru-Macho catalyzed construction of amides and imines. Tetrahedron 70:4213–4218
133. Musa S, Fronton S, Vaccaro L, Gelman D (2013) Bifunctional ruthenium(II) PCP pincer complexes and their catalytic activity in acceptorless dehydrogenative reactions. Organometallics 32:3069–3073
134. Saha B, Rahaman SMW, Daw P, Sengupta G, Bera JK (2014) Metal-ligand cooperation on a diruthenium platform: selective imine formation through acceptorless dehydrogenative coupling of alcohols with amines. Chem A Eur J 20:6542–6551
135. Mukherjee A, Nerush A, Leitus G, Shimon LJW, Ben David Y, Espinosa Jalapa NA, Milstein D (2016) Manganese-catalyzed environmentally benign dehydrogenative coupling of alcohols and amines to form aldimines and H_2: a catalytic and mechanistic study. J Am Chem Soc 138: 4298–4301
136. Mastalir M, Glatz M, Gorgas N, Stöger B, Pittenauer E, Allmaier G, Veiros LF, Kirchner K (2016) Divergent coupling of alcohols and amines catalyzed by isoelectronic hydride Mn^{I} and Fe^{II} PNP pincer complexes. Chem A Eur J 22:12316–12320
137. Fertig R, Irrgang T, Freitag F, Zander J, Kempe R (2018) Manganese-catalyzed and base-switchable synthesis of amines or imines *via* borrowing hydrogen or dehydrogenative condensation. ACS Catal 8:8525–8530
138. Das K, Mondal A, Pal D, Srivastava HK, Srimani D (2019) Phosphine-free well-defined Mn (I) complex-catalyzed synthesis of amine, imine, and 2,3-dihydro-1H-perimidine *via* hydrogen autotransfer or acceptorless dehydrogenative coupling of amine and alcohol. Organometallics 38:1815–1825
139. Chai H, Yu K, Liu B, Tan W, Zhang G (2020) A highly selective manganese-catalyzed synthesis of imines under phosphine-free conditions. Organometallics 39:217–226
140. Samuelsen SV, Santilli C, Ahlquistb MSG, Madsen R (2019) Development and mechanistic investigation of the manganese(III) salen-catalyzed dehydrogenation of alcohols. Chem Sci 10:1150–1157
141. Azizi K, Akrami S, Madsen R (2019) Manganese(III) porphyrin-catalyzed dehydrogenation of alcohols to form imines, tertiary amines and quinolones. Chem A Eur J 25:6439–6446
142. Maji B, Barman MK (2017) Recent developments of manganese complexes for catalytic hydrogenation and dehydrogenation reactions. Synthesis 49:3377–3393
143. Rohit KR, Radhika S, Saranya S, Anilkumara G (2020) Manganese-catalysed dehydrogenative coupling - an overview. Adv Synth Catal 362(1602):1650
144. Nazarahari M, Azizian J (2021) $FeCl_2$-PPh_3 as an efficient catalytic system for the acceptorless dehydrogenation of amines into imines. Chem Papers 75:5705–5710
145. Bala M, Verma PK, Kumar N, Sharma U, Singh B (2013) Highly efficient iron phthalocyanine catalyzed oxidative synthesis of imines from alcohols and amines. Can J Chem 91:732–737

146. Zhang G, Hanson SK (2013) Cobalt-catalyzed acceptorless alcohol dehydrogenation: synthesis of imines from alcohols and amines. Org Lett 15:650–653
147. Bottaro F, Madsen R (2019) *In situ* generated cobalt catalyst for the dehydrogenative coupling of alcohols and amines into imines. ChemCatChem 11:2707–2712
148. Midya SP, Landge VG, Sahoo MK, Rana J, Balaraman E (2018) Cobalt-catalyzed acceptorless dehydrogenative coupling of aminoalcohols with alcohols: direct access to pyrrole, pyridine and pyrazine derivatives. Chem Commun 54:90–93
149. Shee S, Ganguli K, Jana K, Kundu S (2018) Cobalt complex catalyzed atom-economical synthesis of quinoxaline, quinoline and 2-alkylaminoquinoline derivatives. Chem Commun 54:6883–6886
150. Parua S, Sikari R, Sinha S, Chakraborty G, Mondal R, Paul ND (2018) Accessing polysubstituted quinazolines *via* nickel catalyzed acceptorless dehydrogenative coupling. J Org Chem 83:11154–11166
151. Parua S, Sikari R, Sinha S, Das S, Chakraborty G, Paul ND (2018) A nickel catalyzed acceptorless dehydrogenative approach to quinolones. Org Biomol Chem 16:274–284
152. Das S, Maiti D, De Sarkar S (2018) Synthesis of polysubstituted quinolines from α-2-aminoaryl alcohols *via* nickel-catalyzed dehydrogenative coupling. J Org Chem 83:2309–2316
153. Mastalir M, Glatz M, Pittenauer E, Allmaier G, Kirchner K (2016) Sustainable synthesis of quinolines and pyrimidines catalyzed by manganese PNP pincer complexes. J Am Chem Soc 138:15543–15546
154. Kallmeier F, Dudziec B, Irrgang T, Kempe R (2017) Manganese-catalyzed sustainable synthesis of pyrroles from alcohols and amino alcohols. Angew Chem Int Ed 56:7261–7265
155. Deibl N, Kempe R (2017) Manganese-catalyzed multicomponent synthesis of pyrimidines from alcohols and amidines. Angew Chem Int Ed 56:1663–1666
156. Daw P, Kumar A, Espinosa-Jalapa NA, Diskin-Posner Y, Ben-David Y, Milstein D (2018) Synthesis of pyrazines and quinoxalines *via* acceptorless dehydrogenative coupling routes catalyzed by manganese pincer complexes. ACS Catal 8:7734–7741
157. Das K, Mondal A, Srimani D (2018) Selective synthesis of 2-substituted and 1,2-disubstituted benzimidazoles directly from aromatic diamines and alcohols catalyzed by molecularly defined nonphosphine manganese(I) complex. J Org Chem 83:9553–9560
158. Das K, Mondal A, Srimani D (2018) Phosphine free Mn-complex catalysed dehydrogenative C–C and C-heteroatom bond formation: a sustainable approach to synthesize quinoxaline, pyrazine, benzothiazole and quinoline derivatives. Chem Commun 54:10582–10585
159. Mondal A, Sahoo MK, Subaramanian M, Balaraman E (2020) Manganese(I)-catalyzed sustainable synthesis of quinoxaline and quinazoline derivatives with the liberation of dihydrogen. J Org Chem 85:7181–7191
160. Chai H, Tan W, Lu Y, Zhang G, Ma J (2020) Sustainable synthesis of quinolines (pyridines) catalyzed by a cheap metal Mn(I)-NN complex catalyst. Appl Organomet Chem 34:e5685
161. Sivakumar G, Subaramanian M, Balaraman E (2022) Single-molecular Mn(I)-complex-catalyzed tandem double dehydrogenation cross-coupling of (amino)alcohols under solventless conditions with the liberation of H_2 and H_2O. ACS Sustain Chem Eng 10:7362–7373
162. Patra K, Bhattacherya A, Li C, Bera JK, Soo HS (2022) Understanding the visible-light-initiated manganese-catalyzed synthesis of quinolines and naphthyridines under ambient and aerobic conditions. ACS Catal 12:15168–15180
163. Martinez R, Ramon DJ, Yus M (2006) Easy α-alkylation of ketones with alcohols through a hydrogen autotransfer process catalyzed by $RuCl_2(DMSO)_4$. Tetrahedron 62:8988–9001
164. Sharma R, Samanta A, Sardar B, Roy M, Srimani D (2022) Progressive study on ruthenium catalysis for de(hydrogenative) alkylation and alkenylation using alcohols as a sustainable source. Org Biomol Chem 20:7998–8030
165. Thiyagarajan S, Gunanathan C (2018) Ruthenium-catalyzed α-olefination of nitriles using secondary alcohols. ACS Catal 8:2473–2478

166. Gawali SS, Pandia BK, Gunanathan C (2019) Manganese(I)-catalyzed α-alkenylation of ketones using primary alcohols. Org Lett 21:3842–3847
167. Barman MK, Waiba S, Maji B (2018) Manganese-catalyzed direct olefination of methyl-substituted heteroarenes with primary alcohols. Angew Chem Int Ed 57:9126–9130
168. Chakraborty S, Das UK, Ben-David Y, Milstein D (2017) Manganese catalyzed α-olefination of nitriles by primary alcohols. J Am Chem Soc 139:11710–11713
169. Yadav V, Landge VG, Subaramanian M, Balaraman E (2020) Manganese-catalyzed α-olefination of nitriles with secondary alcohols. ACS Catal 10:947–954
170. Lu Y, Zhao R, Guo J, Liu Z, Dagnaw WM, Wang ZX (2019) A unified mechanism to account for manganese- or ruthenium-catalyzed nitrile α-olefinations by primary or secondary alcohols: a DFT mechanistic study. Chem A Eur J 25:3939–3949
171. Pandia BK, Pattanaik S, Gunanathan C (2021) Manganese(I) catalyzed alkenylation of phosphine oxides using alcohols with liberation of hydrogen and water. J Org Chem 86: 17848–17855
172. Mura MG, De Luca L, Taddei M, Williams JMJ, Porcheddu A (2014) Synthesis of α,-β-unsaturated aldehydes based on a one-pot phase-switch dehydrogenative cross-coupling of primary alcohols. Org Lett 16:2586–2589
173. Murahashi SI, Naota T, Ito K, Maeda Y, Taki H (1987) Ruthenium-catalyzed oxidative transformation of alcohols and aldehydes to esters and lactones. J Org Chem 52:4319–4327
174. Gunanathan C, Shimon LJW, Milstein D (2009) Direct conversion of alcohols to acetals and H_2 catalyzed by an acridine-based ruthenium pincer complex. J Am Chem Soc 131:3146–3147
175. Sahoo AR, Jiang F, Bruneau C, Sharma GVM, Suresh S, Achard M (2016) Acetals from primary alcohols with the use of tridentate proton responsive phosphinepyridonate iridium catalysts. RSC Adv 6:100554–100558
176. Blum Y, Reshef D, Shvo Y (1981) H-transfer catalysis with $Ru_3(CO)_{12}$. Tetrahedron Lett 22: 1541–1544
177. Zhao J, Hartwig JF (2005) Acceptorless, neat, ruthenium-catalyzed dehydrogenative cyclization of diols to lactones. Organometallics 24:2441–2446
178. Zhang J, Leitus G, Ben-David Y, Milstein D (2005) Facile conversion of alcohols into esters and dihydrogen catalyzed by new ruthenium complexes. J Am Chem Soc 127:10840–10841
179. Tao J, Wen L, Lv X, Qi Y, Hailiang Yin H (2016) Ruthenium(II)-PNN pincer complex catalyzed dehydrogenation of benzyl alcohol to ester: a DFT study. J Mol Struct 1110:24e31
180. Nielsen M, Junge H, Kammer A, Beller M (2012) Towards a green process for bulk-scale synthesis of ethyl acetate: efficient acceptorless dehydrogenation of ethanol. Angew Chem Int Ed 51:5711–5713
181. Srimani D, Balaraman E, Gnanaprakasam B, Ben-David Y, Milstein D (2012) Ruthenium pincer-catalyzed cross-dehydrogenative coupling of primary alcohols with secondary alcohols under neutral conditions. Adv Synth Catal 354:2403–2406
182. Das UK, Ben-David Y, Leitus G, Diskin-Posner Y, Milstein D (2019) Dehydrogenative cross-coupling of primary alcohols to form cross-esters catalyzed by a manganese pincer complex. ACS Catal 9:479–484
183. Nguyen DH, Trivelli X, Capet F, Paul JF, Dumeignil F, Gauvin RM (2017) Manganese pincer complexes for the base-free, acceptorless dehydrogenative coupling of alcohols to esters: development, scope, and understanding. ACS Catal 7:2022–2032
184. Tang Y, Meador RIL, Malinchak CT, Harrison EE, McCaskey KA, Hempel MC, Funk TW (2020) (Cyclopentadienone)iron-catalyzed transfer dehydrogenation of symmetrical and unsymmetrical diols to lactones. J Org Chem 85:1823–1834
185. Paudel K, Pandey B, Xu S, Taylor DK, Tyer DL, Lopez Torres C, Gallagher S, Kong L, Ding K (2018) Cobalt-catalyzed acceptorless dehydrogenative coupling of primary alcohols to esters. Org Lett 20:4478–4481

186. Trincado M, Grützmacher H, Vizza F, Bianchini C (2010) Domino rhodium/palladium-catalyzed dehydrogenation reactions of alcohols to acids by hydrogen transfer to inactivated alkenes. Chem A Eur J 16:2751–2757
187. Cherepakhin V, Williams TJ (2018) Iridium catalysts for acceptorless dehydrogenation of alcohols to carboxylic acids: scope and mechanism. ACS Catal 8:3754–3763
188. Sarbajna A, Dutta I, Daw P, Dinda S, Rahaman SMW, Sarkar A, Bera JK (2017) Catalytic conversion of alcohols to carboxylic acid salts and hydrogen with alkaline water. ACS Catal 7: 2786–2790
189. Sharninghausen LS, Mercado BQ, Crabtree RH, Hazari N (2015) Selective conversion of glycerol to lactic acid with iron pincer precatalysts. Chem Commun 51:16201–16204
190. Nguyen DH, Morin Y, Zhang L, Trivelli X, Capet F, Paul S, Desset S, Dumeignil F, Gauvin RM (2017) Oxidative transformations of biosourced alcohols catalyzed by earth-abundant transition metals. ChemCatChem 9:2652–2660
191. Waiba S, Maiti M, Maji B (2022) Manganese-catalyzed reformation of vicinal glycols to α-hydroxy carboxylic acids with the liberation of hydrogen gas. ACS Catal 12:3995–4001
192. Waiba S, Maji K, Maiti M, Maji B (2023) Sustainable synthesis of α-hydroxycarboxylic acids by manganese catalyzed acceptorless dehydrogenative coupling of ethylene glycol and primary alcohols. Angew Chem Int Ed 62:e20221832
193. Chen C, Verpoort F, Wu Q (2016) Atom-economic dehydrogenative amide synthesis via ruthenium catalysis. RSC Adv 6:55599–55607
194. Montag M, Milstein D (2023) Sustainable amidation through acceptorless dehydrogenative coupling by pincer-type catalysts: recent advances. Pure Appl Chem 95:109–124
195. Gunanathan C, Ben-David Y, Milstein D (2007) Direct synthesis of amide from alcohols and amines with liberation of H_2. Science 317:790–792
196. Luo J, Zhou QQ, Montag M, Ben-David Y, Milstein D (2022) Acceptorless dehydrogenative synthesis of primary amides from alcohols and ammonia. Chem Sci 13:3894–3901
197. Kumar A, Espinosa-Jalapa NA, Leitus G, Diskin-Posner Y, Avram L, Milstein D (2017) Direct synthesis of amides by dehydrogenative coupling of amines with either alcohols or esters: manganese pincer complex as catalyst. Angew Chem Int Ed 56:14992–14996
198. Espinosa-Jalapa NA, Kumar A, Leitus G, Diskin-Posner Y, Milstein D (2017) Synthesis of cyclic imides by acceptorless dehydrogenative coupling of diols and amines catalyzed by a manganese pincer complex. J Am Chem Soc 139:11722–11725
199. Lane EM, Uttley KB, Hazari N, Bernskoetter W (2017) Iron-catalyzed amide formation from the dehydrogenative coupling of alcohols and secondary amines. Organometallics 36:2020–2025
200. Lane EM, Hazari N, Bernskoetter WH (2018) Iron-catalyzed urea synthesis: dehydrogenative coupling of methanol and amines. Chem Sci 9:4003–4008
201. Daw P, Kumar A, Espinosa-Jalapa NA, Ben-David Y, Milstein D (2019) Direct synthesis of amides by acceptorless dehydrogenative coupling of benzyl alcohols and ammonia catalyzed by a manganese pincer complex: unexpected crucial role of base. J Am Chem Soc 141:12202–12206
202. Hamid MHSA, Slatford PA, Williams JMJ (2007) Borrowing hydrogen in the activation of alcohols. Adv Synth Catal 349:1555–1575
203. Guillena G, Ramón DJ, Yus M (2007) Alcohols as electrophiles in C-C bond-forming reactions: the hydrogen autotransfer process. Angew Chem Int Ed 46:2358–2364
204. Pingen D, Müller C, Vogt D (2010) Direct amination of secondary alcohols using ammonia. Angew Chem Int Ed 49:8130–8133
205. Nixon TD, Whittlesey MK, Williams JMJ (2009) Transition metal catalysed reactions of alcohols using borrowing hydrogen methodology. Dalton Trans:753–762
206. Dobereiner GE, Crabtree RH (2010) Dehydrogenation as a substrate-activating strategy in homogeneous transition-metal catalysis. Chem Rev 110:681–703
207. Guillena G, Ramón DJ, Yus M (2010) Hydrogen autotransfer in the *N*-alkylation of amines and related compounds using alcohols and amines as electrophiles. Chem Rev 110:1611–1641

208. Yamaguchi R, Fujita KI, Zhu M (2010) Recent progress of new catalytic synthetic methods for nitrogen heterocycles based on hydrogen transfer reactions. Heterocycles 81:1093–1140
209. Yang Q, Wang Q, Yu Z (2015) Substitution of alcohols by *N*-nucleophiles *via* transition metal-catalyzed dehydrogenation. Chem Soc Rev 44:2305–2329
210. Corma A, Navas J, Sabater MJ (2018) Advances in one-pot synthesis through borrowing hydrogen catalysis. Chem Rev 118:1410–1459
211. Elangovan S, Neumann J, Sortais JB, Junge K, Darcel C, Beller M (2016) Efficient and selective *N*-alkylation of amines with alcohols catalysed by manganese pincer complexes. Nat Commun 7:12641
212. Bruneau-Voisine A, Wang D, Dorcet V, Roisnel T, Darcel C, Sortais JB (2017) Mono-*N*-methylation of anilines with methanol catalyzed by a manganese pincer-complex. J Catal 347: 57–62
213. Das K, Mondal A, Pal D, Srivastava HK, Srimani D (2019) Phosphine-free well-defined Mn (I) complex-catalyzed synthesis of amine, imine, and 2,3-dihydro-1H-perimidine *via* hydrogen autotransfer or acceptorless dehydrogenative coupling of amine and alcohol. Organometallics 38:1815–1825
214. Yan T, Feringa BL, Barta K (2014) Iron catalysed direct alkylation of amines with alcohols. Nat Commun 5:5602
215. Yan T, Barta K (2016) Sustainable pathways to pyrroles through iron-catalyzed *N*-heterocyclization from unsaturated diols and primary amines. ChemSusChem 9:2321–2325
216. Lator A, Gaillard S, Poater A, Renaud JL (2018) Well-defined phosphine-free iron-catalyzed *N*-ethylation and *N*-methylation of amines with ethanol and methanol. Org Lett 20:5985–5990
217. Rösler S, Ertl M, Irrgang T, Kempe R (2015) Cobalt-catalyzed alkylation of aromatic amines by alcohols. Angew Chem Int Ed 54:15046–15050
218. Zhang G, Yin Z, Zheng S (2016) Cobalt-catalyzed *N*-alkylation of amines with alcohols. Org Lett 18:300–303
219. Xu Z, Wang DS, Yu X, Yang Y, Wanga D (2017) Tunable triazole-phosphine-copper catalysts for the synthesis of 2-aryl-1H-benzo[d]imidazoles from benzyl alcohols and diamines by acceptorless dehydrogenation and borrowing hydrogen reactions. Adv Synth Catal 359: 3332–3340
220. Irrgang T, Kempe R (2019) 3d-metal catalyzed *N*- and *C*-alkylation reactions *via* borrowing hydrogen or hydrogen autotransfer. Chem Rev 119:2524–2549
221. Hofmann N, Hultzsch KC (2021) Borrowing hydrogen and acceptorless dehydrogenative coupling in the multicomponent synthesis of *N*-heterocycles: a comparison between base and noble metal catalysis. Eur J Org Chem:6206–6223
222. Pena-Lopez M, Piehl P, Elangovan S, Neumann H, Beller M (2016) Manganese-catalyzed hydrogen-autotransfer C-C bond formation: α-alkylation of ketones with primary alcohols. Angew Chem Int Ed 55:14967–14971
223. Fu S, Shao Z, Wang Y, Liu Q (2017) Manganese-catalyzed upgrading of ethanol into 1-butanol. J Am Chem Soc 139:11941–11948
224. Liu T, Wang L, Wu K, Yu Z (2018) Manganese-catalyzed β-alkylation of secondary alcohols with primary alcohols under phosphine-free conditions. ACS Catal 8:7201–7207
225. Chakraborty S, Daw P, Ben David Y, Milstein D (2018) Manganese-catalyzed α-alkylation of ketones, esters, and amides using alcohols. ACS Catal 8:10300–10305
226. Bruneau-Voisine A, Pallova L, Bastin S, César V, Sortais JB (2019) Manganese catalyzed α-methylation of ketones with methanol as a C1 source. Chem Commun 55:314–317
227. Borghs JC, Azofra LM, Biberger T, Linnenberg O, Cavallo L, Rueping M, El-Sepelgy O (2019) Manganese-catalyzed multicomponent synthesis of pyrroles through acceptorless dehydrogenation hydrogen autotransfer catalysis: experiment and computation. ChemSusChem 12:3083–3088
228. El-Sepelgy O, Matador E, Brzozowska A, Rueping M (2019) C-alkylation of secondary alcohols by primary alcohols through manganese-catalyzed double hydrogen autotransfer. ChemSusChem 12:3099–3102

229. Yang J, Liu X, Meng DL, Chen HY, Zong ZH, Feng TT, Sun K (2012) Efficient iron-catalyzed direct β-alkylation of secondary alcohols with primary alcohols. Adv Synth Catal 354:328–334
230. Polidano K, Williams JMJ, Morrill LC (2019) Iron-catalyzed borrowing hydrogen β-C(sp^3)-methylation of alcohols. ACS Catal 9:8575–8580
231. Dambatta MB, Polidano K, Northey AD, Williams JMJ, Morrill LC (2019) Iron-catalyzed borrowing hydrogen C-alkylation of oxindoles with alcohols. ChemSusChem 12:2345–2349
232. Bettoni L, Seck C, Mbaye MD, Gaillard S, Renaud JL (2019) Iron-catalyzed tandem three-component alkylation: access to α-methylated substituted ketones. Org Lett 21:3057–3061
233. Alanthadka A, Bera S, Banerjee D (2019) Iron-catalyzed ligand free α-alkylation of methylene ketones and β-alkylation of secondary alcohols using primary alcohols. J Org Chem 84: 11676–11686
234. Bettoni L, Gaillard S, Renaud JL (2020) Iron-catalyzed α-alkylation of ketones with secondary alcohols: access to β-disubstituted carbonyl compounds. Org Lett 22:2064–2069
235. Abdallah MS, Joly N, Gaillard S, Poater A, Renaud JL (2022) Blue-light-induced iron-catalyzed α-alkylation of ketones. Org Lett 24:5584–5589
236. Wu J, Narayanasamy SN, Darcel C (2022) Late-stage modifications of phosphine oxide ligands by iron-catalyzed hydrogen borrowing reactions. J Organomet Chem 979:122510
237. Deibl N, Kempe R (2016) General and mild cobalt-catalyzed C-alkylation of unactivated amides and esters with alcohols. J Am Chem Soc 138:10786–10789
238. Zhang G, Wu J, Zeng H, Zhang S, Yin Z, Zheng S (2017) Cobalt-catalyzed α-alkylation of ketones with primary alcohols. Org Lett 19:1080–1083
239. Freitag F, Irrgang T, Kempe R (2017) Cobalt-catalyzed alkylation of secondary alcohols with primary alcohols via borrowing hydrogen/hydrogen autotransfer. Chem A Eur J 23:12110–12113
240. Pandey B, Xu S, Ding K (2021) Switchable β-alkylation of secondary alcohols with primary alcohols by a well-defined cobalt catalyst. Organometallics 40:1207–1212
241. Chakraborty P, Gangwar MK, Emayavaramban B, Manoury E, Poli R, Sundararaju B (2019) α-Alkylation of ketones with secondary alcohols catalyzed by well-defined Cp*Co(III)-complexes. ChemSusChem 12:3463–3467
242. Tang G, Cheng CH (2011) Synthesis of α-hydroxy carboxylic acids via a nickel(II)-catalyzed hydrogen transfer process. Adv Synth Catal 353:1918–1922
243. Zhang MJ, Li HX, Young DJ, Lia HY, Lang JP (2019) Reaction condition controlled nickel (II)-catalyzed C–C cross-coupling of alcohols. Org Biomol Chem 17:3567–3574
244. Li WZ, Wang ZX (2021) Nickel-catalyzed coupling of $R_2P(O)Me$ (R= aryl or alkoxy) with (hetero)arylmethyl alcohols. Org Biomol Chem 19:2233–2242
245. Tan DW, Li HX, Zhu DL, Li HY, Young DJ, Yao JL, Lang JP (2018) Ligand-controlled copper(I)-catalyzed cross-coupling of secondary and primary alcohols to α-alkylated ketones, pyridines, and quinolones. Org Lett 20:608–611
246. Nguyen NK, Nam DH, Phuc BV, Nguyen VH, Trịnh QT, Hung TQ, Dang TT (2016) Efficient copper-catalyzed synthesis of C3-alkylated indoles from indoles and alcohols. Mol Catal 505: 111462
247. Aitchison H, Wingad RL, Wass DF (2016) Homogeneous ethanol to butanol catalysis-Guerbet renewed. ACS Catal 6:7125–7132
248. Alig L, Fritz M, Schneider S (2019) First-row transition metal (de)hydrogenation catalysis based on functional pincer ligands. Chem Rev 119:2681–2751
249. Filonenko GA, van Putten R, Hensen EMJ, Pidko EA (2018) Catalytic (de)hydrogenation promoted by non-precious metals - Co, Fe and Mn: recent advances in an emerging field. Chem Soc Rev 47:1459–1483
250. Reed-Berendt BG, Polidano K, Morrill LC (2019) Recent advances in homogeneous borrowing hydrogen catalysis using earth-abundant first row transition metals. Org Biomol Chem 17: 1595–1607

251. Nad P, Mukherjee A (2021) Acceptorless dehydrogenative coupling reactions by manganese pincer complexes. Asian J Org Chem 10:1958–1985
252. Borthakur I, Sau A, Kundu S (2022) Cobalt-catalyzed dehydrogenative functionalization of alcohols: Progress and future prospect. Coord Chem Rev 451:214257
253. Das K, Waiba S, Jana A, Maji B (2022) Manganese-catalyzed hydrogenation, dehydrogenation, and hydroelementation reactions. Chem Soc Rev 51:4386–4464
254. Joshi A, Kumari S, Kundu S (2022) Photoredox (*NN*)Mn(I)-catalysed acceptorless dehydrogenation: synthesis of amides, aldehydes and ketones. Adv Synth Catal 364:4371–4383
255. Paul B, Maji M, Chakrabarti K, Kundu S (2020) Tandem transformations and multicomponent reactions utilizing alcohols following dehydrogenation strategy. Org Biomol Chem 18:2193–2214
256. Bettoni L, Joly N, Lohier JF, Gaillard S, Poater A, Renaud JL (2021) Ruthenium-catalyzed three-component alkylation: a tandem approach to the synthesis of nonsymmetric *N,N*-dialkyl acyl hydrazides with alcohols. Adv Synth Catal 363:4009–4017
257. Bettoni L, Seck C, Mbaye MD, Gaillard S, Renaud JL (2019) Iron-catalyzed tandem three-component alkylation: access to α-methylated substituted ketones. Org Lett 21:3057–3061
258. Chakraborty G, Sikari R, Mondal R, Mandal S, Paul ND (2020) Nickel-catalyzed synthesis of pyrimidines *via* dehydrogenative functionalization of alcohols. Asian J Org Chem 9:431–436
259. Dang TT, Seayad AM (2017) Efficient [Cu(NHC)]-catalyzed multicomponent synthesis of pyrroles. Chem Asian J 12:2383–2387
260. Lyaskovskyy V, de Bruin B (2012) Redox non-innocent ligands: versatile new tools to control catalytic reactions. ACS Catal 2:270–279
261. Sengupta D, Bhattacharjee R, Pramanick R, Rath SP, Chowdhury NS, Datta A, Goswami S (2016) Exclusively ligand-mediated catalytic dehydrogenation of alcohols. Inorg Chem 55: 9602–9610
262. Bains AK, Adhikari D (2020) Mechanistic insight into the azo radical-promoted dehydrogenation of heteroarene towards *N*-heterocycles. Cat Sci Technol 10:6309–6318
263. Singh K, Kundu A, Adhikari D (2022) Ligand-based redox: catalytic applications and mechanistic aspects. ACS Catal 12:13075–13107
264. Bains AK, Kundu A, Adhikari D (2023) Mechanistic elucidation of a radical-promoted hydrogenation relevant to borrowing hydrogen catalysis. ChemCatChem:e202300586
265. Singh R, Bains AK, Kundu A, Jain H, Yadav S, Dey D, Adhikari D (2023) Mechanistic elucidation of an alcohol oxidation reaction promoted by a nickel azophenolate complex. Organometallics 42:1759–1765
266. Bains AK, Biswas A, Kundu A, Adhikari D (2022) Nickel-catalysis enabling α-alkylation of ketones by secondary alcohols. Adv Synth Catal 364:2815–2821
267. Bains AK, Kundu A, Yadav S, Adhikari S (2019) Borrowing hydrogen-mediated *N*-alkylation reactions by a well-defined homogeneous nickel catalyst. ACS Catal 9:9051–9059
268. Bains AK, Kundu A, Maiti D, Adhikari D (2021) Ligand-redox assisted nickel catalysis toward stereoselective synthesis of (n+1)-membered cycloalkanes from 1,n-diols with methyl ketones. Chem Sci 12:14217–14223
269. Bains AK, Biswas A, Adhikari D (2020) Nickel-catalysed chemoselective C-3 alkylation of indoles with alcohols through a borrowing hydrogen method. Chem Commun 56:15442–15445
270. Biswas A, Bains AK, Adhikari D (2022) Ligand-assisted nickel catalysis enabling sp^3C-H alkylation of 9H-fluorene with alcohols. Cat Sci Technol 12:4211–4216
271. Verma PK (2022) Advancement in photocatalytic acceptorless dehydrogenation reactions: opportunity and challenges for sustainable catalysis. Coord Chem Rev 472:214805
272. West JG, Huang D, Sorensen EJ (2015) Acceptorless dehydrogenation of small molecules through cooperative base metal catalysis. Nat Commun 6:10093

273. Yang XJ, Zheng YW, Zheng LQ, Wu LZ, Tung CH, Chen B (2019) Visible light-catalytic dehydrogenation of benzylic alcohols to carbonyl compounds by using an eosin Y and nickel–thiolate complex dual catalyst system. Green Chem 21:1401–1405
274. Fuse H, Mitsunuma H, Kanai M (2020) Catalytic acceptorless dehydrogenation of aliphatic alcohols. J Am Chem Soc 142:4493–4499
275. Li G, Liu Y, Wang W, Zhuo Z, Huang Y (2023) A cobalt redox switch driving alcohol dehydrogenation by redox coupled molecular swing. Chin Chem Lett 34:107630

Top Organomet Chem (2024) 73: 33–62
https://doi.org/10.1007/3418_2023_110

Published online: 17 November 2023

3d-Metal Catalyzed C–C Bond Formation Through α-Alkylation of Ketones

Romane Pointis, Oussama Fayafrou, Yves Canac, Stéphanie Bastin, and Jean-Baptiste Sortais

Contents

R. Pointis, O. Fayafrou, Y. Canac, S. Bastin, and J.-B. Sortais (✉)
LCC-CNRS, Université de Toulouse, CNRS, UPS, Toulouse, France
e-mail: stephanie.bastin@lcc-toulouse.fr; jean-baptiste.sortais@lcc-toulouse.fr

Abstract The formation of C–C bonds is at the very heart of synthetic chemistry for building the carbon skeleton of molecules. This chapter gives an overview of the catalytic systems developed in recent years based on 3d metals for the α-alkylation of ketones with primary and secondary alcohols by hydrogen auto-transfer.

Keywords 3d metals · Alcohol · Alpha-alkylation · C–C bond formation · Ketone

1 Introduction

The formation of C–C bonds is at the very heart of synthetic chemistry for building the carbon skeleton of molecules. Among the range of methods developed, alkylation of enolates with electrophiles, mainly alkyl (pseudo) halides, is a method of choice [1]. However, this route relies on potentially toxic, carcinogenic electrophiles such as iodomethane, and produces a stoichiometric amount of waste. The generation of enolate from ketones also requires a stoichiometric quantity of base. These aspects are not in line with the development of a more sustainable chemistry that is at the heart of current concerns. On the other hand, the aldolization reaction, using aldehydes as electrophiles, is also a convenient method for forming C–C bonds. Aldolization reaction combined with a crotonization step leading to the corresponding enone, followed by a reduction step of the conjugated C=C double bond, leads to the same type of product as those obtained by alkylation with halogenated electrophiles. The major disadvantage lies in the fact that the aldehydes are not quite stable and the reduction step requires either a reducing agent in a stoichiometric quantity or hydrogen pressure.

Since the pioneer contributions of Grigg et al. [2, 3], hydrogen borrowing reactions, based on dehydrogenation/hydrogenation shuttle processes, have emerged as a powerful method for coupling alcohols with many nucleophiles through oxidation of alcohols to carbonyl compounds [4]. It is a gentle method that meets the criteria of green chemistry. Alcohols are stable compounds, readily available and generally of low toxicity. The only side product is one molecule of water, and no excess hydrogen pressure is generated.

Alkylation of ketones with alcohols, by hydrogen auto-transfer, is an ideal solution to all the limitations mentioned below. The key steps of the α-alkylation of ketones with alcohols are (i) the dehydrogenation of the alcohol to the corresponding aldehydes, (ii) the aldol condensation followed by the crotonization and finally, (iii) the chemo-selective reduction of the α-β-unsaturated enones to the final alkylated ketones (Scheme 1).

The overall reaction seems quite straightforward, but several issues need to be overcome in order to develop an efficient process. First, as the catalysts used need to be efficient in both dehydrogenation of alcohols and hydrogenation of enones, often, undesired transfer hydrogenation of the substrate does occur leading to a mixture of alcohols. The selective reduction of the enone can also be an issue leading to allylic

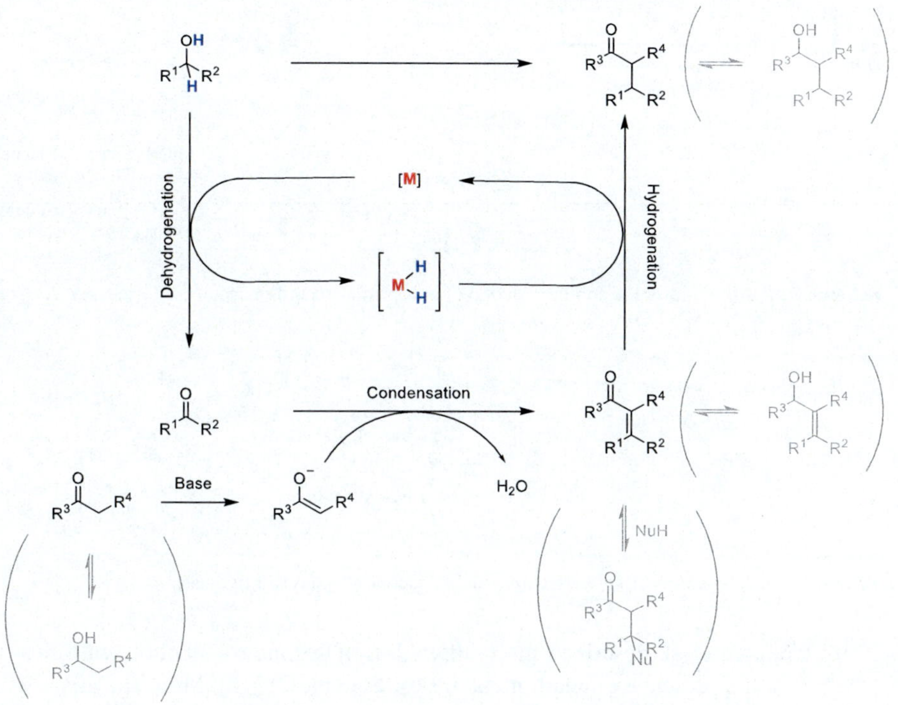

Scheme 1 General mechanism of α-alkylation of ketones with alcohols via hydrogen auto-transfer reaction

alcohol. Then the transient α,β, unsaturated enone is also a Michael acceptor, with which the enolate, or other nucleophiles, can react, affording in the former case 1,5-diketones. Therefore, the role of the catalyst, as well as the experimental conditions, is crucial in directing the successive reactions towards the selective formation of the desired product.

The α-alkylation of ketones was initially developed with precious transition metals, mainly ruthenium and iridium [5, 6]. One of the challenges at the beginning of the twenty-first century is to replace those rare and expensive transition metals by more abundant 3d metals [7]. Great progress has been accomplished in the last decade [8–14]. In this chapter, the development of α-alkylation of ketones with iron, manganese, cobalt, nickel, and copper is discussed in details.

2 α-C-Alkylation of Ketones Catalyzed by Bases

Before going any further in the chapter, we thought important to mention known reactions catalyzed only by bases in the absence of metal complexes, as a non-negligible amount of base is often used in ketone alkylation reactions [15]. In

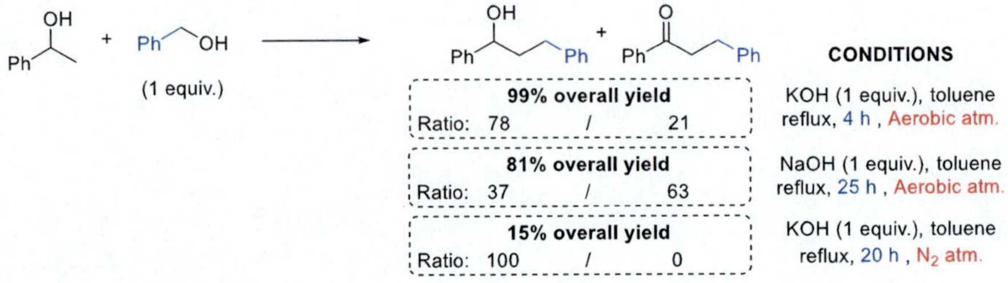

Scheme 2 β-Alkylation of secondary alcohols with primary alcohols catalyzed by bases

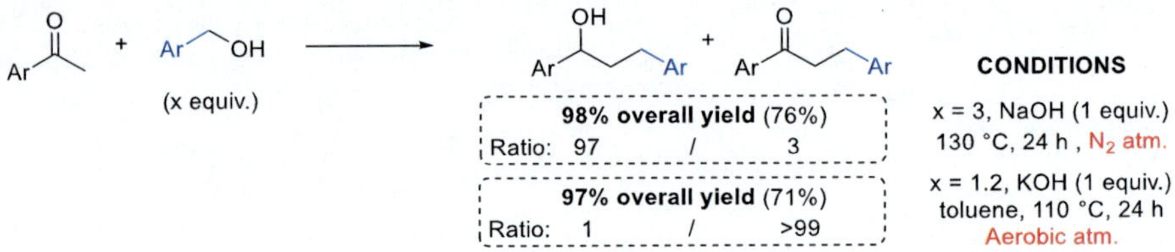

Scheme 3 α-Alkylation of ketones with primary alcohols catalyzed by bases

2010, Crabtree et al. described the β-alkylation of secondary alcohols with benzyl alcohols in the presence of alkali metal bases (Scheme 2) [16]. More specifically, in the presence of a stoichiometric amount of potash (KOH), sodium hydroxide (NaOH) or potassium tertbutylate (^{t}BuOK), in toluene, at reflux under an air atmosphere, one equivalent of 1-phenylethanol reacts with one equivalent of benzyl alcohol to form the corresponding β-alkylated product in yields of over 70%, with a marked preference for the alcohol at short times (from 78:21 to 99:1), then for ketones at longer times, probably due to aerobic oxidation of the product. Under a nitrogen atmosphere, conversion and yield dropped significantly (18% and 15%, respectively) with the formation of alcohol exclusively. The authors propose a mechanism based on an Oppenauer oxidation and a Merwein-Pondorff-Verley type reduction promoted by alkali metal bases [17, 18]. It is important to note that weak bases such as Cs_2CO_3, K_2CO_3, and K_3PO_4 did not promote such reaction. The same reaction was recently reinvestigated by Johnson et al. who found that performing the alkylation of secondary alcohols with benzylic alcohols in a sealed pressure tube under air allowed to decreased the amount of KOH to 25 mol% (toluene, 120°C, 18 h) while keeping high conversions and yields [19].

In the same vein, Xu et al. described the α-alkylation of ketones with primary alcohols in the presence of bases (Scheme 3) [20]. Under an inert atmosphere at 130° C over 24 h, in the presence of one equivalent of sodium hydroxide (NaOH) and an excess of primary alcohol (3 equivalents), reduced alkylation products were obtained with good selectivity. The absence of O_2 prevents the oxidation of alcohols and the excess of reactants favored the reduction of the alkylated product via MPV reaction. The alkylated ketones, on the other side, were obtained in the presence of one equivalent of KOH, at 110°C in toluene under aerobic conditions over 24 h. It is

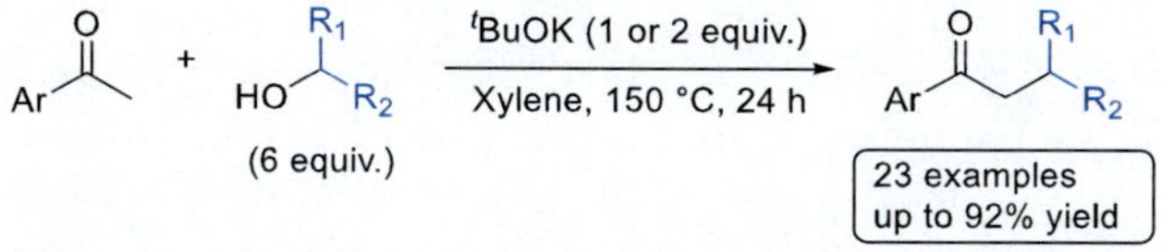

Scheme 4 α-Alkylation of ketones with secondary alcohols catalyzed by bases

interesting to note that this "metal-free" reaction, as the one of Crabtree, is limited to benzyl alcohols, as with hexan-1-ol, only 40% yields were obtained after 3 days at 160°C in the presence of one equivalent of KOH.

The alkylation of ketones with secondary alcohols was also described under metal-free conditions by Morrill et al. (Scheme 4) [21]. 1-(2,3,4,5,6-pentamethylphenyl) ethenone, as model substrate, was alkylated in good yield by a variety of secondary alcohols (6 equiv.) at 150°C in xylene in the presence of ^{t}BuOK (1 or 2 equiv.). Under these conditions, ^{t}BuONa and ^{t}AmONa bases also promoted the reaction, while KOH and K_2CO_3 were found to be inactive.

3 α-C-Alkylation of Ketones Catalyzed by Iron

3.1 Activation of Primary Alcohols

The first example of α-C-alkylation of ketones with 3d transition metals catalysts was accomplished with iron-based catalysts, namely Knölker-type complexes [22]. Knölker-hydride complex was first described in 1999 by Knölker et al. as an intermediate in the demetalation of tricarbonyl(cyclopentadienone) iron for the synthesis of five membered ring by [2 + 2 + 1] cycloaddition of two alkynes and one CO [23]. In 2007, a breakthrough was accomplished, when Casey et al. demonstrated that Knölker-hydride complex, the iron structural analogue of the ruthenium Shvo complex, was an efficient catalyst for hydrogenation and transfer hydrogenation of carbonyl derivatives [24, 25]. Soon after, cyclopentadienone iron tricarbonyl analogues were used as air stable precatalysts and Oppenauer oxidation were developed, opening the way to its use in hydrogen borrowing [26, 27]. Quintard et al. developed a cooperative iron-catalyzed borrowing-hydrogen/iminium-activation strategy for the transformation of allylic alcohols into β-chiral-alcohols [28]. Barta et al. pioneered the first direct N-alkylation of amines with alcohols [29]. A year later, Sortais et al. applied Knölker-type iron complexes for α-C-alkylation of ketones [30].

Using tricarbonyl cyclopentadienone iron complex **Fe1** (2 mol%) in combination with PPh_3 (2 mol%), in the presence of Cs_2CO_3 (10 mol%) in toluene, at 140°C, the alkylation of a variety of ketones was achieved in moderate to good yield with both benzylic and aliphatic alcohols (Scheme 5). A slight excess of alcohols (1.3 equiv.) was used to improve the overall yield, but in all the cases, a small amount of alcohols

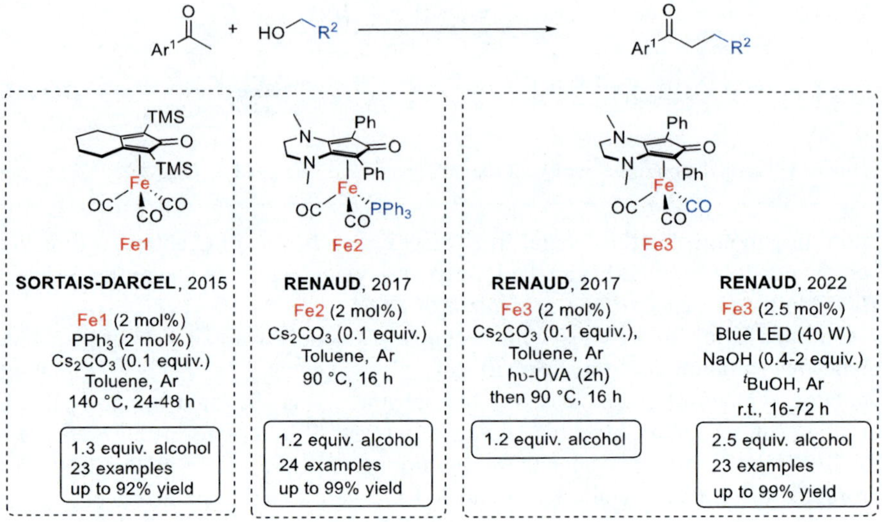

Scheme 5 Knölker type iron complexes for α-alkylation of ketones

arising from the reduction of the keto-group of both starting materials or products was detected. The same catalytic system was also proven to be efficient in Friedländer annulation to yield quinolines.

The overall efficiency of the α-alkylation of ketones was remarkably improved by Renaud et al. with modified Knölker complexes (Scheme 5). Considering the ambiphilic character of the active 16 electrons complex, namely dicarbonyl cyclopentadienone iron complex, that can be regarded as a "transition metal frustrated Lewis pair," Renaud et al. prepared an electron enriched N,N′-dimethyl 3,4-ethylenediamino-substituted cyclopentadienone as ligand [31–33]. Combining a thorough theoretical and experimental study, Renaud et al. showed that both iron complexes **Fe2** and **Fe3** bearing electron rich ligand were highly active in the α-alkylation of ketones [34]. The typical conditions were 90°C for 16 h in toluene in the presence of Cs_2CO_3 (10 mol%), **Fe2** (2 mol%), one equivalent of ketones and 1.3 equivalent of alcohol. With complex **Fe2** bearing a thermally labile PPh_3, thermal activation was sufficient whereas for **Fe3** with three carbonyl ligands, a preactivation step, i.e., irradiation by UV-A light, in order to decoordinate one CO was necessary. The main advantage over previous system, in addition to the lower temperature, was that no overreduction was detected, even in the presence of excess of alcohols.

Lately, the activity of diaminocyclopentadienone iron tricarbonyl complex was reinvestigated by the same authors under blue-light irradiation in order to develop a photoinduced process based on a single catalyst able to harvest the visible light and promote the chemical reaction [35]. Complex **Fe3** was able to play both role by promoting the α-alkylation of a variety of aromatic and aliphatic ketones with

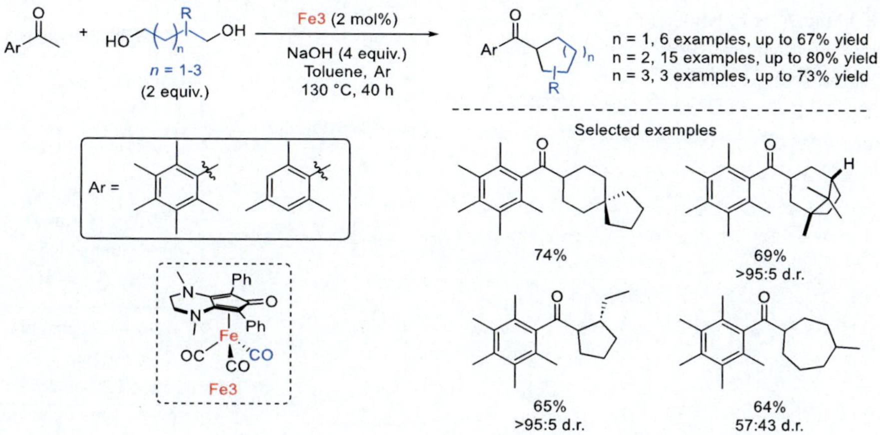

Scheme 6 Synthesis of cycloalkanes via hydrogen borrowing reaction promoted by iron

aliphatic or benzylic alcohols in ^{t}BuOH (or toluene) at room temperature in 16–72 h. The key of this success was the irradiation with 40 W Kessil blue LED lamp, lower power led to lower conversion. It should also be noticed that, unlike the thermal conditions described above, at least 40 mol% of NaOH were requested and that 2.5 equivalents of primary alcohols were used. Nonetheless, the process is quite efficient as even 1-butanol, a model aliphatic alcohol, can be used as electrophile, although it requires a stoichiometric amount of base (1 equivalent of NaOH, 72 h, 45% yield).

Preliminary insights were collected on the mechanism. It was shown that the visible light irradiation induced both the chemoselective reduction of enone to ketones and the dehydrogenation step at r.t, but did not promote the decoordination of one CO. The catalytic activity was rationalized by an initial Hieber-type activation in the presence of NaOH, followed by the release of H_2 by reaction with ^{t}BuOH to generate the active di-carbonyl 16 electrons iron species.

Double α, α-alkylation of ketones with 1,n-diols is more challenging than mono-α-alkylation of substituted methylketones as the 1,n-diols under hydrogen borrowing conditions can undergo various type of self-condensation, oligomerization, or polymerization. Inspired by the seminal contribution of Donohoe et al. with iridium catalysts [36], Renaud et al. succeeded in synthesizing cycloalkanes from terminal diols and 1-mesitylethan-1-one via hydrogen borrowing strategy [37]. The key parameters were: diaminocyclopentadienone iron tricarbonyl complex **Fe3** (2 mol%), a temperature of 130°C, NaOH (4 equivalents) as the base, an excess of diol (2 equivalents) and a concentrated reaction mixture (5 M) in toluene for 40 h (Scheme 6). Under such conditions, 5, 6, and, 7 membered cycloalkanes were constructed, including substituted diols.

Apart from catalytic systems based on Knölker-type complexes, few examples have been described in the literature. Banerjee et al. reported the alkylation of substituted methylketones with primary alcohols (1.25 equiv. excess) in toluene, at 140°C under N_2 atmosphere in the presence of $Fe_2(CO)_9$ (2.5 mol%) as catalyst and

Scheme 7 Non-Knölker type iron complex for α-alkylation of ketones

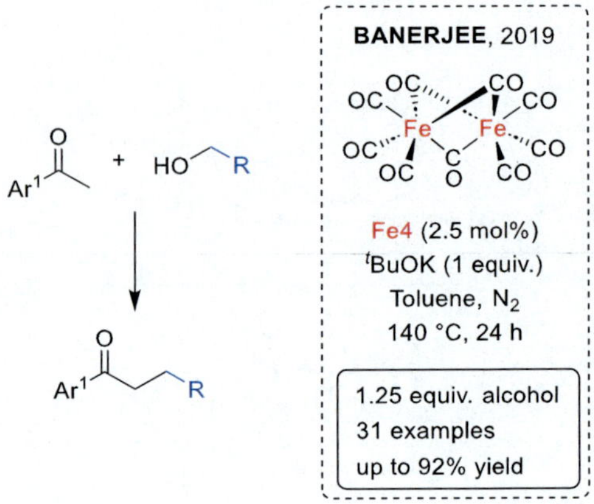

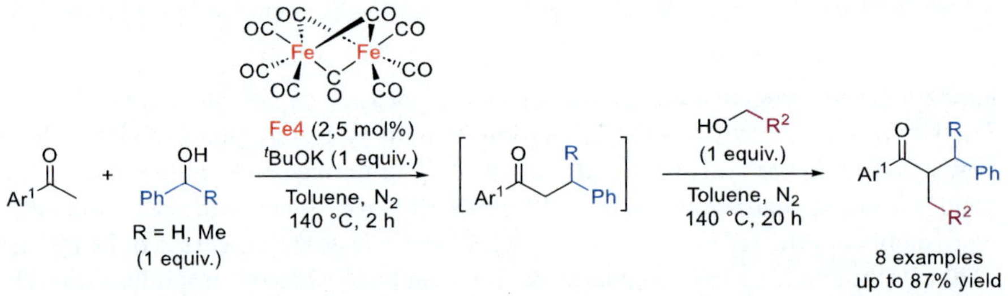

Scheme 8 Synthesis of dissymmetric α,α-disubstituted ketones from methylketones promoted by iron catalyst

^{t}BuOK (1 equiv.) (Scheme 7) [38]. A selectivity of 18:1 towards the α-substituted ketone vs the corresponding alcohol was recorded for the α-alkylation of propiophenone with benzyl alcohol. Blank experiments carried out in the absence of iron catalyst revealed the formation the product in low yield (24% in ketone, but 40% hydrogenated product). This transformation is not limited to benzylic alcohol as cyclopropyl methanol, cyclohexyl methanol or 1-butanol were amenable for this transformation.

The scope of the transformation was extended first to the cross-coupling reaction of secondary alcohols with primary alcohols and then to a one-pot sequential double alkylation of acetophenone derivatives with two different alcohols affording dissymmetric α,α-disubstituted ketones from methylketones (Scheme 8).

Yang et al. used $FeCl_2$ (1 mol%) in toluene at 150°C in the presence of ^{t}BuOK (10–50 mol%) under an argon atmosphere to achieve the alkylation of acetophenones derivatives with aliphatic and benzylic alcohols in moderate to good yield (60–90% in most cases) [39].

It also worth mentioning that an example of a heterogeneous system was reported by Namitharan et al. based on iron oxide nano-Fe_2O_3 (30 mol% catalyst, 30 mol% ^{t}BuOK, 135°C, toluene, argon, 24 h) [40].

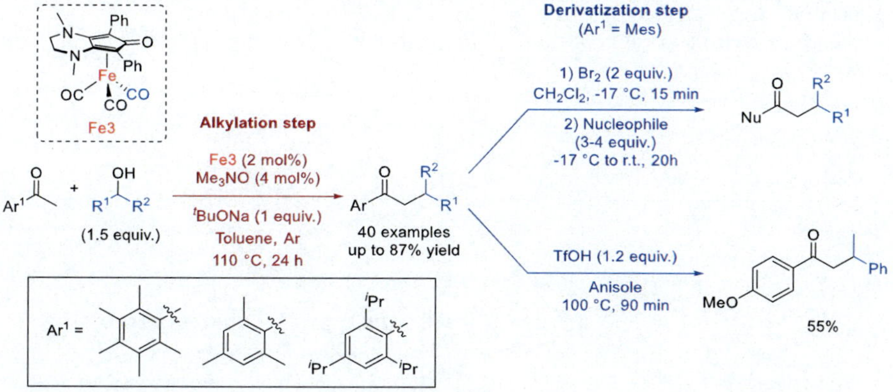

Scheme 9 Iron catalyzed α-alkylation of ketones with secondary alcohols

3.2 *Activation of Secondary Alcohols*

The alkylation of ketones with secondary alcohols is more complex than with primary alcohols. In addition to the difficulty of dehydrogenating secondary alcohols and reducing enones with triple-substituted C=C double bonds, the starting ketones and those generated by oxidation can self-condense, giving rise to a multitude of secondary products. Inspired by the pioneering work of Donohoe et al. [41], using aryl ketones bearing a 2,3,4,5,6-pentamethylphenyl (Ph*) group, Renaud et al., with the aid of the **Fe3** (diamino) complex, extended the alkylation of ketones to secondary alcohols (Scheme 9) [42]. The use of di-ortho-substituted ketones is a key element in obtaining products from cross-alkylations. In addition to 1-mesitylethan-1-one, 2,3,4,5,6-pentamethylphenyl ethan-1-one and 2,4,6-triisopropylphenyl ethan-1-one were also successfully submitted to the alkylation reaction. In toluene at reflux, in the presence of Me_3NO as decarbonylating agent (4 mol%), **Fe3** (2 mol%) and one equivalent of NaO^{t}Bu as base, the hindered di-orthosubstituted aryl ketones were alkylated with a series of benzyl secondary alcohols as well as cyclic or acyclic aliphatic alcohols affording β-disubstituted carbonyl compounds (yields ranging from 60 to 80% in most cases). The synthetic utility of the Ph* group is that it can be substituted by various nucleophiles, via a retro-Friedel-Crafts acylation reaction with dibromine followed by a nucleophilic addition on the acyl bromide intermediate [43]. Renaud et al. demonstrated that the same strategy could be applied to trimethylarylketones leading to the formation of esters, amides and even a ketone, by performing a retro-Friedel Craft/Friedel Craft sequence promoted by triflic acid in anisole.

Scheme 10 Catalysts used to promote α-methylation of ketones

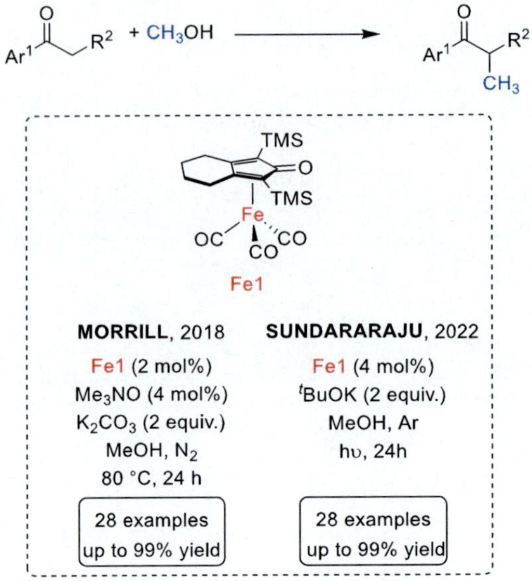

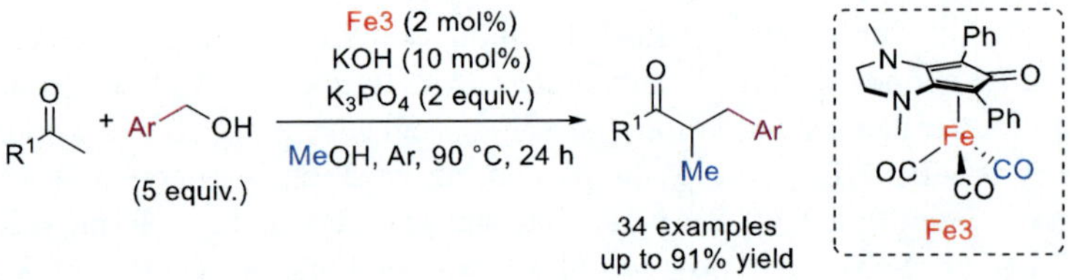

Scheme 11 Tandem three-component alkylation reaction yielding α-methyl-α-alkyl ketones

3.3 *Activation of Methanol*

Methanol is the simplest aliphatic alcohol from a structural point of view, but also the most difficult to activate by hydrogen auto-transfer, as the energy barrier of the dehydrogenation step is the highest among all the alcohols. Although methylation reactions are the subject of a specific chapter in this book, this chapter would not be complete without at least mentioning the systems that have been able to activate methanol.

Morrill et al. used Knölker type catalyst to promote the α-methylation of a large variety of ketones, under mild conditions (80°C in MeOH, 2 equiv. of K_2CO_3, **Fe1** 2 mol% and Me_3NO (4 mol%)) (Scheme 10) [44]. Sundararaju et al. succeeded in decreasing the temperature to 40°C, performing the reaction under continuous irradiation with LED bulbs (4 x 7 W) in the presence of catalyst **Fe1** (4 mol%) and ^{t}BuOK as a base (2 equiv.) (Scheme 10) [45].

Finally, Renaud et al. has developed a tandem three-component alkylation reaction combining the α-alkylation of methylketones followed by the α-methylation of the resulting ketones yielding α-methyl-α-alkyl ketones (Scheme 11) [46]. (Hetero)-aromatic and alkyl ketones proved amenable to this reaction.

4 α-C-Alkylation of Ketones Catalyzed by Manganese

4.1 Activation of Primary Alcohols

After iron and titanium, manganese is the third most abundant transition metal in the earth's crust, making it an excellent candidate for developing sustainable chemistry. It is an affordable metal and it exists in many oxidation states in complexes, offering numerous possibilities in catalysis. Its use in hydrogenation reactions and hydrogen transfer for the reduction of polar unsaturated derivatives is relatively recent, as are the first examples in hydrogen borrowing [47].

In 2016, just after highlighting the potential of manganese (I) carbonyl complexes in the hydrogenation of ketones, nitriles and aldehydes [48] and N-alkylation of amines [49], Beller et al. described the first example of the alkylation of ketones with alcohols using [(PN(H)P)Mn(CO)$_3$Br] complex [50]. In the presence of only 2 mol% of complex **Mn1** stabilized by the ^{i}Pr$_2$P(CH$_2$)$_2$N(H)(CH$_2$)$_2$P^{i}Pr$_2$ ligand (also known as the MACHO-^{IPr} ligand), 5 mol% of Cs_2CO_3 and *tert*-amyl alcohol at 140°C, a variety of acetophenone derivatives were alkylated with benzyl and aliphatic alcohols (Scheme 12). It should be noted that no by-products were detected under these conditions. The robustness of the system was demonstrated by applying it to the late-stage functionalization of hormones such as estrone and testosterone.

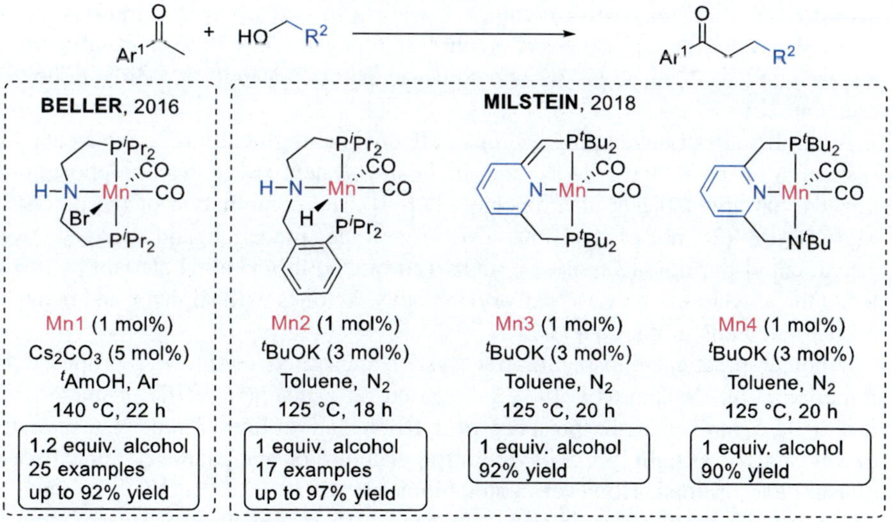

Scheme 12 Well-defined manganese catalysts for the α-alkylation of ketones

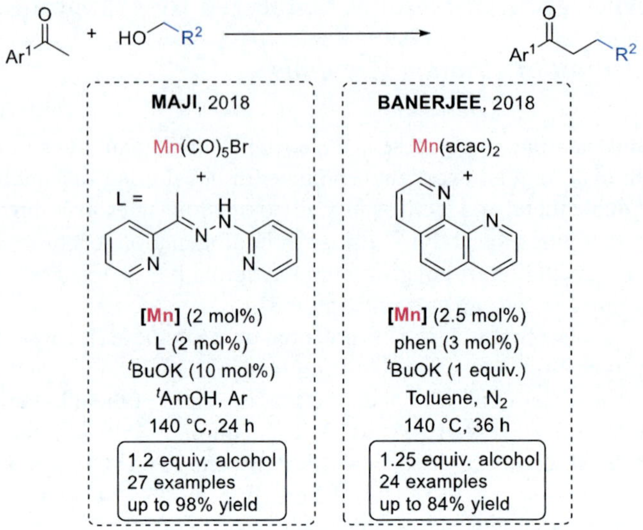

Scheme 13 In situ phosphorus-free manganese based catalytic system for α-alkylation of ketones

For their part, Milstein et al., after carrying out the first manganocatalyzed dehydrogenative coupling of amines and alcohols [51], used an asymmetric PN (H)P complex **Mn2** to couple ketones, as well as non-activated amides and esters, with alcohols [52]. It should be noted that the dearomatized Mn-PNP^{t}Bu **Mn3** and deprotonated Mn-PNN **Mn4** complexes were also active for this transformation (acetophenone (1 equiv.), benzyl alcohol (1 equiv.), 97%, 92%, and 90% yield, respectively, in 20 h at 125°C and toluene, ^{t}BuOK (3 mol%), [Mn] (1 mol%)) (Scheme 12).

With the aim of developing a simple, effective, and globally affordable catalytic system, Maji et al. have developed an in-situ system based on non-phosphorus ligands that are stable in air (Scheme 13) [53]. The combination of the precursor $Mn(CO)_5Br$ (2 mol%) and the hydrazone-type pincer ligand derived from 2-hydrazinyl pyridine (2 mol%), ^{t}BuOK (10 mol%) in *tert*-amyl alcohol at 140°C led to the alkylation of a variety of aryl and alkyl ketones with aliphatic and benzylic primary alcohols in good yields.

Another practical phosphorus-free system generated in-situ was proposed by Banerjee et al. (Scheme 13) [54]. It is based on a manganese (II) precursor, Mn $(acac)_2$ (2.5 mol%) in the presence of 1,10-phenanthroline (3 mol%) (as for the nickel catalytic system [55, 56], vide infra). The use of manganese (II) precursor is unusual and original. However, a stoichiometric quantity of ^{t}BuOK is required to obtain the branched ketones in good yields at 140°C in toluene. This system allowed to perform a sequential one pot double alkylation using the same catalyst, by adding a second alcohol to the reaction mixture once the first alkylation was finished. More recently, the same group developed a heterogeneous catalyst, by immobilization of homogeneous $[Mn(phen)Cl_2]$ into MOF pores [57].

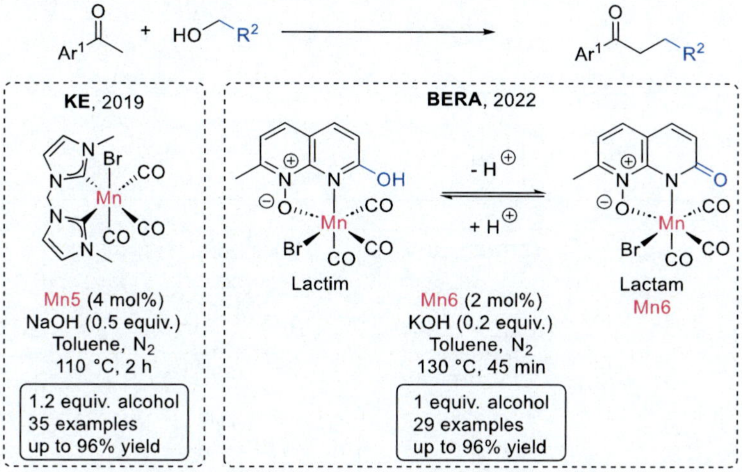

Scheme 14 Manganese catalysts with bidendate ligands

An original system based not only on a bidentate ligand, instead of the tridentate ligands that are largely dominant with manganese, but also on a non-bifunctional ligand, was proposed by the group of Liu and Ke (Scheme 14) [58]. Several manganese complexes incorporating bidentate ligands (pyridine-NHC, bipyridine, NHC-NHC) were tested and the complex bearing a di-carbon bis-NHC^{Me} ligand **Mn5** [59, 60] was found to be the most active. The reaction was carried out rapidly (2 h) at 110°C in toluene with 0.5 equivalent of sodium hydroxide. The theoretical mechanistic study concluded that the catalytic system proceeds according to an outer-sphere mechanism without decoordination of any of the three CO ligands.

Following a different strategy, Bera et al. prepared a manganese (I) complex displaying a 1,8-naphthyridine-N-oxide backbone ligand, bearing a proton responsive hydroxy unit capable of promoting the alkylation of ketones in 45 min (**Mn6** (2 mol%), KOH (20 mol%), toluene, 130°C) (Scheme 14) [61]. The equilibrium between the protonated lactim and deprotonated lactam forms, was established by NMR and UV spectroscopy and favors proton/hydride transfers during the catalytic cycle. This catalytic system was applied to the derivatization of pregnenolone and progesterone.

The synthesis of substituted cycloalkanes (C_5, C_6, and C_7) through the coupling of ketones, or secondary alcohols, with α-ω diols, was achieved by Leitner et al. under the catalysis of Mn-MACHO-^{iPr}. In the presence of 4 equivalents of diols and ^{t}BuOK, **Mn1** (2 mol%) at 150°C for 32 h in toluene, 1-(methylpolysubstituted-phenyl)-ethanones were alkylated without over reduction. The seven membered ring derivatives were produced in higher yields compared to six and five ones (Scheme 15) [62].

Maji et al. carried out the same reaction with an air-stable catalyst **Mn7** based on a tridendate phosphine-free NNS ligand, an aminomethylpyridine with a thiophene arm (Scheme 15) [63]. This system can be used not only to alkylate ketones with a

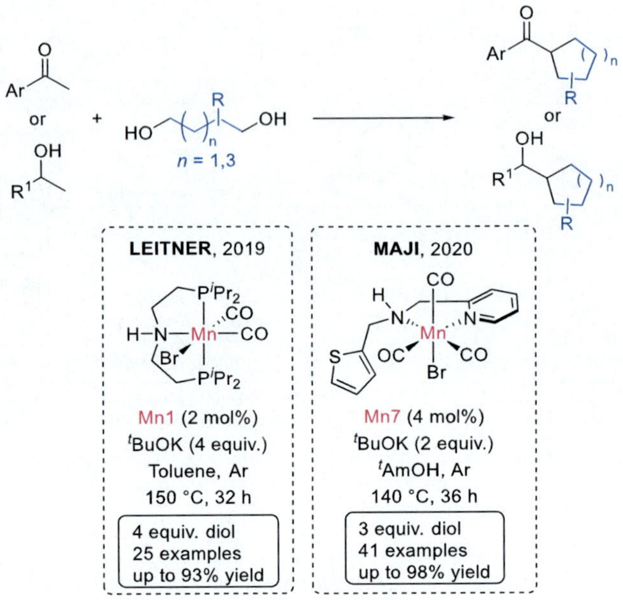

Scheme 15 Synthesis of cycloalkanes promoted by manganese catalysts

high degree of steric hindrance on the phenyl group, but also slightly hindered acetophenone derivatives. In the latter case, the corresponding alcohols were obtained. In addition to the bifunctional character of the picolylamine moiety, the role of the sulphur sidearm is essential in obtaining high yields, compared with its furan analogue. The hemilabile nature of the thiophenyl fragment has been demonstrated by theoretical studies, particularly in the key step of dehydrogenation by β-H elimination of the manganese-coordinated alkoxides to form the aldehydes.

Finally, three heterogeneous catalyst-based systems were also described for this transformation, based on Mn-N-graphene [64], Mn-MgO/Al_2O_3 [65], and δ-MnO_2 nanoparticles [66], respectively.

4.2 Activation of Secondary Alcohols

Following on from his study on the synthesis of substituted cycloalkanes, Maji et al. developed the only known manganese system for the synthesis of β-branched carbonyl compounds via the α-alkylation of ketones with secondary alcohols [67]. Using the same catalyst **Mn7** (Scheme 15) with the thiophene arm and the picolylamine moiety (2 mol%), hindered ketones, such as 2,3,4,5,6-pentamethyl acetophenone for example, were alkylated with a wide variety of secondary alcohols (various cyclic, acyclic, symmetrical, and unsymmetrical alcohols) at 140°C over 24 h in the presence of one equivalent of ^{t}BuOK.

4.3 Activation of Methanol

The α-alkylation of ketones was reported in 2019 with two different PNP-manganese catalysts. El-Sepelgy et al. found that cationic diphenylphosphine-based complex featuring a pyridine core **Mn8** promoted efficiently the methylation of propiophenone, while aliphatic MACHO analogue led to moderate yield (Scheme 16) [68]. In the presence of 2 equivalent of Cs_2CO_3, at 85°C, with a catalytic charge of 2.5 mol%, propiophenone derivatives as well as heteroaryl, cyclic and alkyl ketones were mono-methylated and acetophenone derivatives were di–αmethylated in good yields. Particular attention was paid to the trideuteromethylation of ketones with deuterated methanol for the biological properties conferred by the CD_3 fragment [69]. A non-classical carbonylic carbon-centered mechanism was proposed by Schaefer for this transformation [70].

A similar cationic tricarbonyl PN^3P Mn catalyst **Mn9** (Scheme 16) [71], also active for the selective mono-N-methylation of anilines [72], was used by Sortais et al. for the methylation of a series of aromatic ketones (Mn catalyst 3 mol%; NaO^{t}Bu 50 mol%, MeOH/toluene, 0.08 M, 120°C) [73]. This system was particularly suitable for dihydrochalcone derivatives. It should be noted that CD_3OD was also amenable to this catalytic system. Interestingly, when the reaction was carried out in a more concentrated solution (0.1 M) with one equivalent of NaO^{t}Bu, the selectivity towards 1,5-diketones was reversed. In addition, the range of applications was extended to the α-methylation of carboxylic acid esters.

Recently, in 2023, a manganese hydride complex based on an original $PC_{NHC}P$ platform **Mn10**, with a central carbene as the coordination site, was described by de Ruiter et al. (Scheme 16) [74]. Unlike previous systems, this catalyst has no acidic NH or aromatic heterocyclic ring that could participate in cooperative methanol

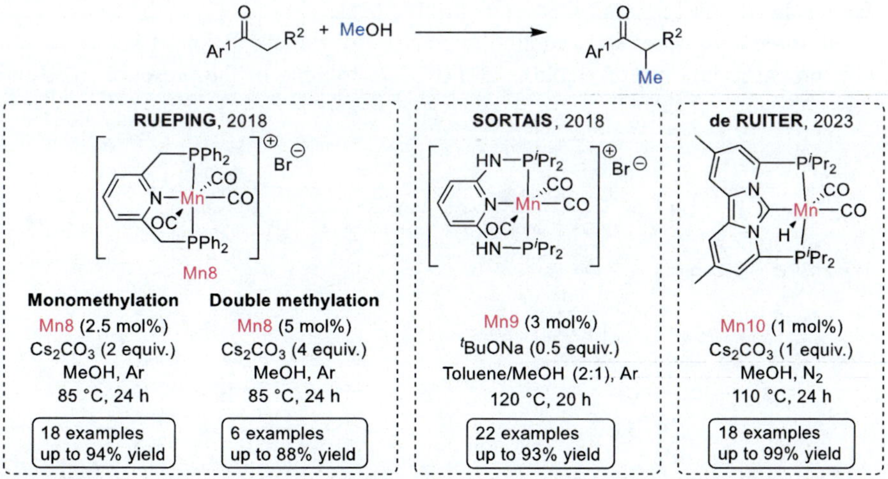

Scheme 16 Manganese catalysts for the α-methylation of ketones

activation, but efficiently promotes methylation of ketones in the presence of Cs_2CO_3 (1 equiv.) at 110°C in methanol.

5 α-C-Alkylation of Ketones Catalyzed by Cobalt

5.1 Activation of Primary Alcohols

The formation of C–C bond with cobalt catalysts was initiated by Kempe et al. with the alkylation of unactivated amides and esters with PN_5P-pincer type ligands (see the corresponding chapter for more details) [75]. Regarding reactions with ketones, Zhang et al. reported the first example of α-alkylation with primary alcohols (Scheme 17) [76]. The catalytic system which relies on an ionic cobalt(II)-PNP complex $[(PNHPCy)Co(CH_2SiMe_3)][BAr^F_4]$ **Co1**, was a highly active catalyst initially developed by Hanson for the hydrogenation of alkenes and carbonyl derivatives [77] and reversible (de)hydrogenation of N-heterocycles [78]. Following his work on the N-alkylation of amines with alcohols [79] and amines [80], Zhang et al. studied the α-alkylation of ketones with the same catalyst. With a catalyst loading of 2 mol%; ^{t}BuOK (5 mol%), in toluene at 120°C for 24 h, acetophenone derivatives, including ketones with pyridine ring, were alkylated with both benzylic and aliphatic alcohols in moderate to good yields (26 examples). The system was tolerant toward halogens, even 4-iodobenzylic alcohol was successfully coupled.

Pincer ligands with phosphorus-based chelating groups, such as the Macho-type ligands in the previous example, are very commonly found in hydrogen auto-transfer systems. By contrast, non-phosphorus tridentate ligands are much rarer. Thus, Ghosh et al. prepared a series of three NNN-type ligands, with a central pyridine and two lateral imines as the coordination site, and the corresponding series of cobalt dichloride complexes (**Co2-Co4,** Scheme 18) [81].

All three complexes showed similar activity for the alkylation of ketones with a stoichiometric amount of alcohols at 110°C in toluene in the presence of ^{t}BuOK

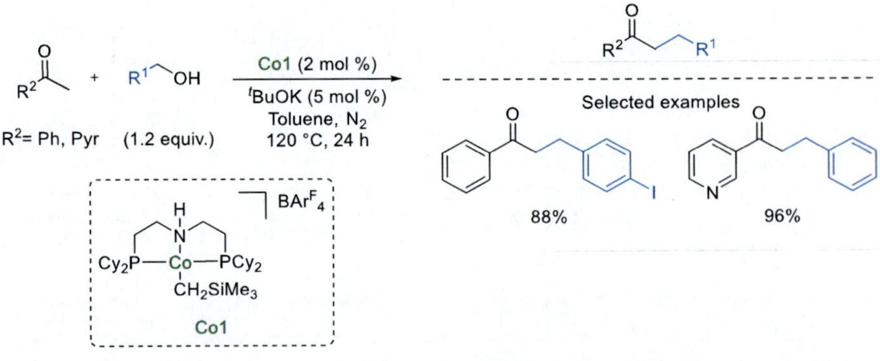

Scheme 17 α-Alkylation of ketones with primary alcohols promoted by PNP-Cobalt catalyst

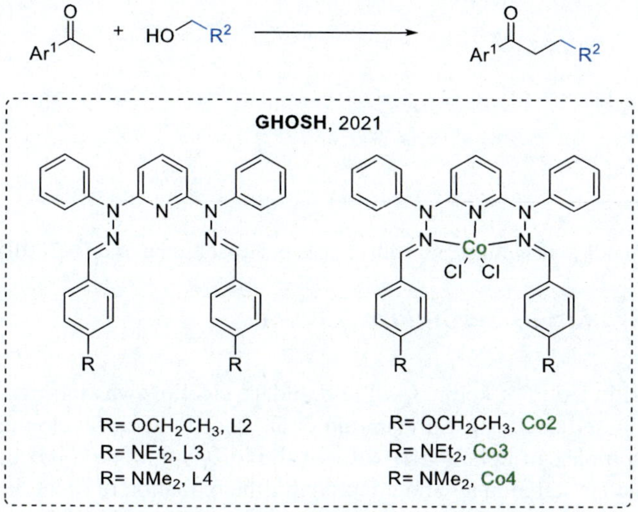

Scheme 18 Phosphine-free tridentate ligands suitable for the α-alkylation of ketones with cobalt

(50 mol%) and 2 mol% of catalyst. With benzyl alcohols, the yields are good (over 80% in most cases, 44 examples). Under these conditions, the mono-methylation and monoethylation of acetophenone could be carried out in good yields (69–78%).

An original approach was highlighted by Wang et al. to develop reusable catalysts [82]. Cobalt coordination polymer material (Co-CIA) and porous oval polymer material (Co-NCIA) have been prepared with an indole-based diacid moiety ligand. The alkylation of ketones with benzylic alcohols was achieved in water. Unlike previous systems, no strong alcoholic base was required, but instead one equivalent of both $AgNTf_2$ and KF, as well as TBAB as phase transfer catalyst (20 mol%) were found necessary to reach high conversion and yield. Both materials Co-CIA and Co-Co-NCIA are active, the former exhibiting slightly better performance than the latter. Interestingly, the catalyst can be recycled and reused, as the yield dropped only from 93% to 87% after five runs.

As an alternative to molecular defined organometallic complexes, Hara et al. developed a heterogeneous MgO-co-deposited Co on TiO_2 (Co-MgO/TiO_2) catalyst able to promote the C–C bond formation in toluene, under argon at 110°C without any base or any additive [83]. A series of control experiments demonstrated that both metal oxide support and co-deposited MgO, and direct contact between each component, are necessary for the reaction sequence to proceed well.

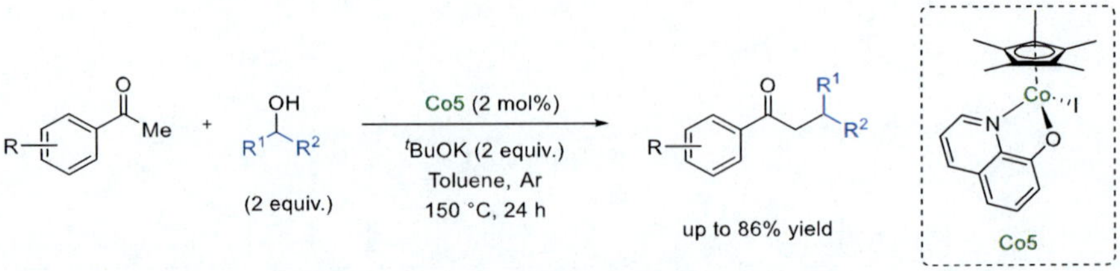

Scheme 19 α-Alkylation of ketones with secondary alcohols promoted by Co(III) catalyst

5.2 *Activation of Secondary Alcohols*

The first α-alkylation of ketones with secondary alcohols was indeed achieved with cobalt (III) based catalysts. Sundararaju et al., shortly after developing a new half-sandwich complex of cobalt (III) with an 8-hydroxyquinone (NO) ligand [Cp*Co (NO)I], which is stable in air and effective for the oxidation of secondary alcohols in acetone [84], investigated the usefulness of this complex for self-hydrogen transfer reactions, focusing on the coupling of ketones with secondary alcohols (Scheme 19) [85]. For unsubstituted aryl-methyl ketones, yields were low due to self-condensation processes. Using Donohoe's trick [41], i.e., by hindering the aromatic ring of the substrate, the desired β-branched C-alkylated products were obtained in moderate yields by working at 150°C, in toluene, with two equivalents of ^{t}BuOK and **Co5** (2 mol%).

In collaboration with the group of Poli and Manoury, the mechanism of the transformation was elucidated via a combined computational and experimental investigations, revealing that the dehydrogenation of the bound alkoxide in [Cp*Co(Oquin)(OR')] proceeded via a transfer of proton to the hydroxyquinoline ligand instead of a classical β-H elimination [86]. In the final rehydrogenation step, the proton from the coordinated ligand HOquin is transferred back to the product leading to the formation of the enolate product, evolving to the final ketone.

5.3 *Activation of Methanol*

To conclude this paragraph on cobalt catalyzed reactions, three contributions involving methanol as C1 source should be mentioned. The first one is the α,-α-dimethylation of methylketones reported by Liu et al., involving a very convenient catalytic system composed of commercially available tetradentate P $(CH_2CH_2PPh_2)_3$ ligand (1 mol%), $Co(BF_4)_2.6H_2O$ (1 mol%), K_2CO_3 (1 equiv.) in methanol as solvent at 100°C for 24 h (Scheme 20) [87].

The last two contributions are both based on the reactivity of the enone intermediate resulting from the addition of the ketone to the methanal formed in-situ and not reduced by the catalyst (Scheme 21). Xiao et al. developed α-methoxymethylation and α-aminomethylation reactions using $CoCl_2.6H_2O$ and tert-butyl hydroperoxide

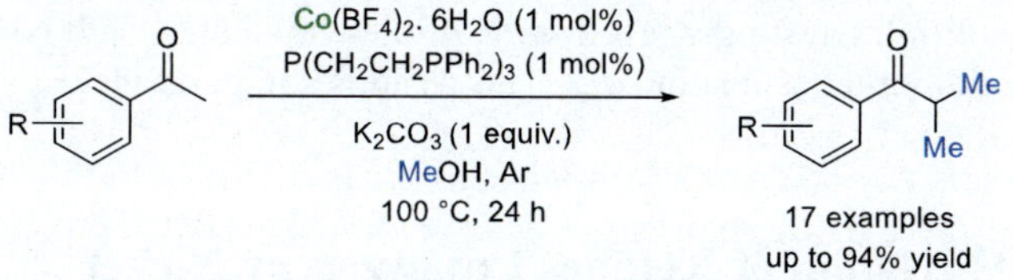

Scheme 20 α,α-Dimethylation of methylketones promoted by cobalt catalyst

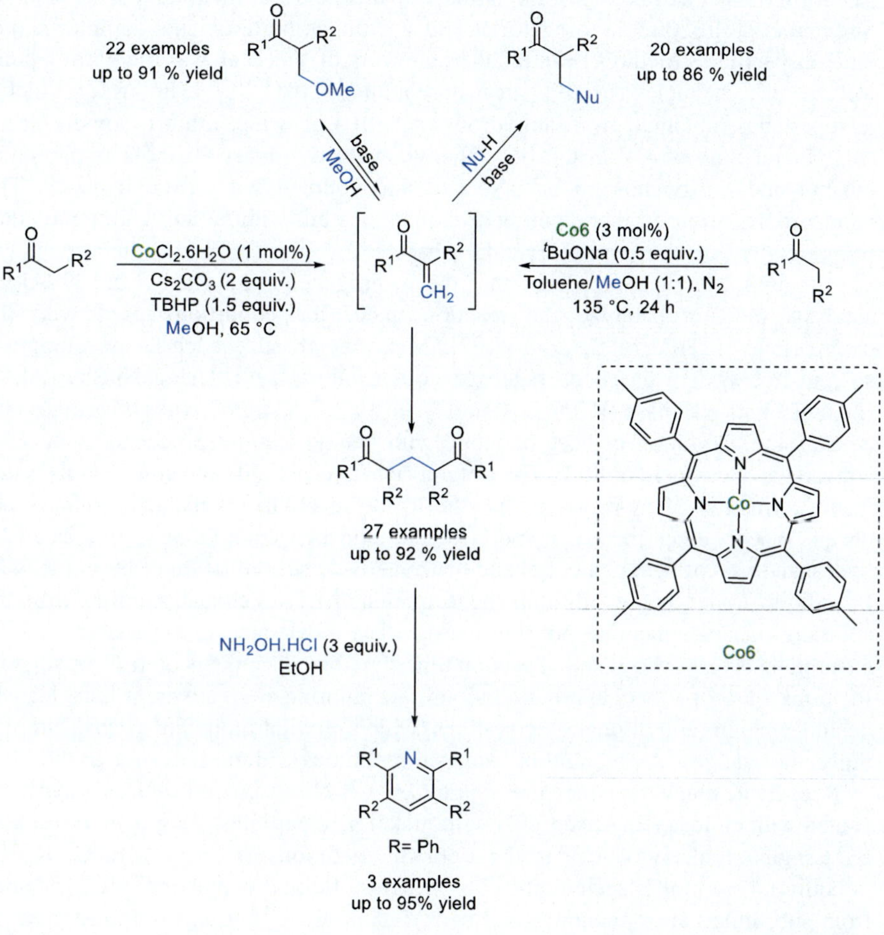

Scheme 21 α-Methoxymethylation, α-aminomethylation reactions and synthesis of 1,5-diketones

(TBHP) as the oxidant, resulting from the addition of methanol or amine to the enone intermediate (Scheme 21, top) [88]. Chandrasekhar and Venkatasubbaiah selectively obtained 1,5-diketones by addition of the enolate to the same intermediate (Scheme

21; bottom) [89]. Interestingly, cyclization of these diketones with $NH_2OH.HCl$ led to the formation of tetrasubstituted 2,3,5,6-pyridines in good yields.

6 α-C-Alkylation of Ketones Catalyzed by Nickel

6.1 Activation of Primary Alcohols

The formation of ketones by α-alkylation with alcohols was mentioned in a patent by Sugitani et al. in 2002, in a reactor at 180°C with supported nickel complexes, but with unspecified structures [90]. In 2007, the work of Yus et al. was the starting point for the use of nickel in this area of chemistry [91, 92]. The use of nickel nanoparticles, obtained from anhydrous nickel(II) chloride, lithium powder and a catalytic quantity of 4,4′-di-tert-butylphenyl, enabled primary alcohols, in particular ethanol and n-propanol, to be activated and coupled with acetophenones. The nanoparticles are used in stoichiometric quantities but without any additional additives, i.e., hydrogen acceptor, ligand, or base.

Although Yus et al. reported that the heterogeneous Ni Raney and Ni/Al_2O_3 catalysts did not catalyze the reaction under the conditions used with the nanoparticles (THF, 76°C, 24 h) [92], Métay et al. succeeded in developing a solvent-free system based on heterogeneous catalysts [93]. Using Ni/SiO_2-Al_2O_3 (20 mol%), a weak base (K_3PO_4, 10 mol%), at 175°C, a slight excess of ketones (1.2 equiv.), the alkylation of acetophenone with benzyl alcohol proceeded with total conversion and a yield of 86%. The difference between yields and conversions is due to the formation of by-products specific to the reactivity of nickel, generally not observed with other metals, namely toluene and benzene. Toluene is formed by hydrogenolysis of benzyl alcohol and benzene by decarbonylation of benzaldehyde. In addition, under their conditions, the formation of 1,5-diketones, resulting from the addition of acetophenone to the intermediate chalcone, is reversible, as the by-product is observed at short reaction times but then disappears on full conversion. In terms of scope, the main limitations are chlorine derivatives, which lead to dehalogenation, and aliphatic derivatives, which provide moderate yields. Interestingly, the catalytic cycle could be recycled five times without loss of activity.

Recyclable magnetic nanocatalysts Fe_3O_4@CS-Ni, based on Fe_3O_4 nanoparticles coated with chitosan and decorated with nickel nanoparticles, have been developed by Eshghi et al. [94]. Under the optimal conditions (toluene, 110°C, K_3PO_4 (1 equiv.), Ni 4 mol%, sealed tube, 7 h), a series of dihydrochalcones were obtained from substituted acetophenones and benzyl alcohols. Using an external magnet to remove the nanoparticles, the catalyst was reused up to six runs.

With in-situ generated molecular catalysts, Banerjee initially focused on the synthesis of gem-bis(alkyl) branched ketones (α,α-disubstituted ketones) [55]. Using a relatively simple catalyst system, $NiBr_2$ (5 mol%), 1,10-phenanthroline (6 mol%), a slight excess of alcohol (1.5 equivalents) and one equivalent of ^{t}BuOK at 140°C under N_2 in toluene, a series of propiophenones were alkylated with benzyl

Scheme 22 Nickel catalyzed synthesis of α,α-disubstituted ketones

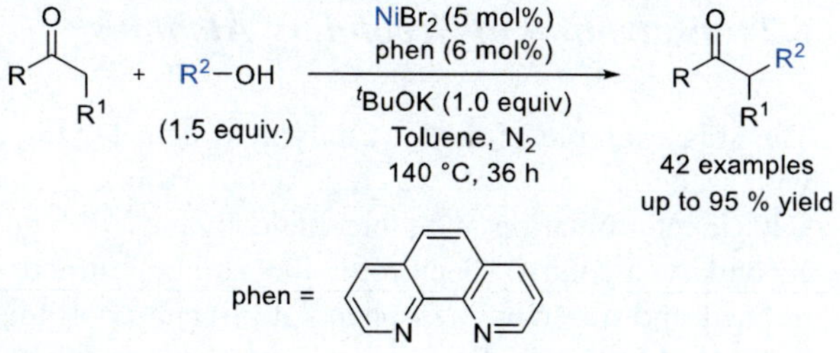

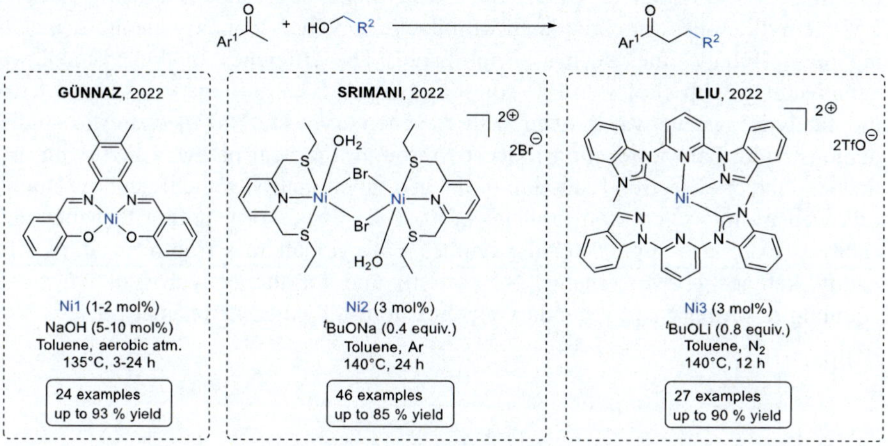

Scheme 23 Well-defined nickel catalysts for α-alkylation of ketones

alcohols in moderate to good yields (Scheme 22). Switching to aliphatic ketones or aliphatic alcohols, including methanol, led to more modest yields. Nevertheless, this system is tolerant towards double and triple substituted double bonds such as those present in oleic alcohol and citronellol.

The same group reoptimized the catalytic system to obtain the mono-α-alkylation products of methyl-ketones selectively [56]. The use of a weaker base Cs_2CO_3, in catalytic quantity (10 mol%) in 1,4-dioxane with more catalyst (10% $NiBr_2$) or less strong base (^{t}BuOK, 20 mol%) was the key to not observing over-alkylation of acetophenones.

For their part, the groups of Günnaz [95], Srimani [96], and Liu [97] have used well-defined nickel complexes to promote this transformation (Scheme 23). Günnaz's system is based on tetradentate salen ligands, ONNO, Srimani's on tridentate SNS ligands with a pyridine core and two thioether arms and Wu's on tridentate CNN, benzimidazole/pyridine/pyrrole ligands. All three systems were operated in toluene at 135–140°C in the presence of a catalytic amount of base (NaOH, 5 mol%, NaO^{t}Bu 40 mol% and LiO^{t}Bu, 80 mol%, respectively).

6.2 Activation of Secondary Alcohols

The sole example of nickel catalytic system for the difficult alkylation of ketones with secondary alcohols was reported by Adhikari et al. in 2022 [98]. Simple nickel salts in combination with bidentate ligands did not promote the coupling with secondary alcohols. Therefore, the authors turned their attention towards well-defined and air-stable azo-phenolate ligand-coordinated nickel catalyst, originally developed for N-alkylation of amines (Scheme 24) [99–101]. With a catalyst charge of 5 mol%; 1 equivalent of ^{t}BuOK and 2 equivalents of alcohols, sterically hindered aryl methyl ketones were alkylated with cyclic, acyclic secondary alcohols, including pure aliphatic ones such as isopropanol. The efficiency of the protocol was demonstrated with cholesterol as coupling partner. The homogeneous character of the catalytic system was proven with the mercury test. The mechanistic studies demonstrated that the reaction proceeds through a radical pathway involving as a first step the one electron reduction of the azo functionality of the ligand by ^{t}BuOK, followed by hydrogen atom transfer [100, 102]. It is worth noting that the same catalysts were also applied for the synthesis of cycloalkanes from (1,n)-diols with methyl-ketones [103] (Scheme 24, bottom) and for the dehydrogenative cross-coupling of primary and secondary alcohols to α-alkylated ketones [102].

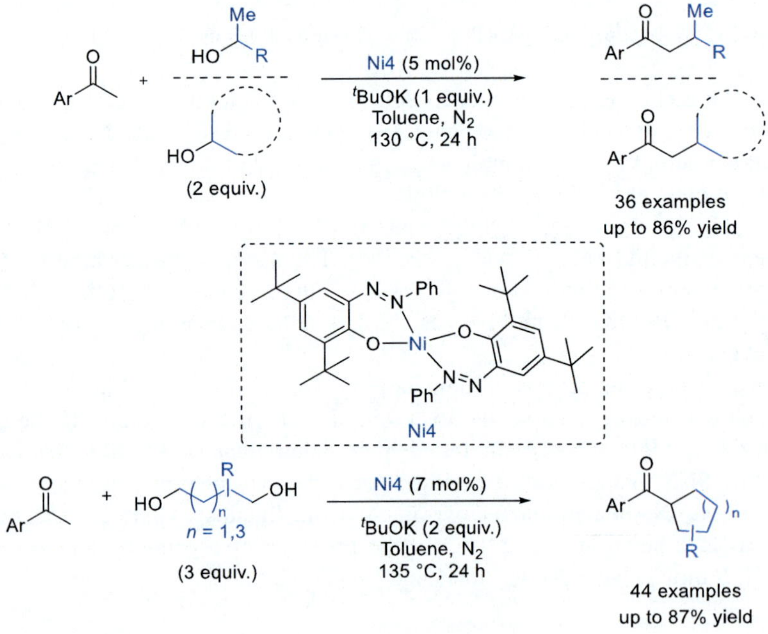

Scheme 24 Nickel catalyzed α-alkylation of ketones with secondary alcohols

6.3 Activation of Methanol

Examples of nickel catalyzed α-methylation of ketones are rare. During the reduction of dibenzylidene acetone with [(dippe)Ni(μ-H)]$_2$ by transfer hydrogenation with methanol as hydrogen donor, Garcia et al. observed the formation of the β-monomethylated ketone, resulting from the alkylation of the α,β-unsatured ketones (Scheme 25) [104]. The catalytic system is likely composed of both homogeneous Ni(0) complexes and nickel nanoparticles.

For so far, only Métay and Duguet's heterogeneous Ni/SiO$_2$-Al is capable of satisfactorily activating methanol [105]. The conditions had to be made more severe than for aliphatic alcohols (185°C, neat, 21 equivalents of methanol) but good yields of di-methylation of acetophenones and mono-methylation of tetralones have been obtained. Methylation of aliphatic ketones was more tedious, leading to mixtures of mono, di and tri-methylation. The efficiency of the system was also demonstrated by carrying out a three-component cross-benzylation-methylation reaction of acetophenones (Scheme 26).

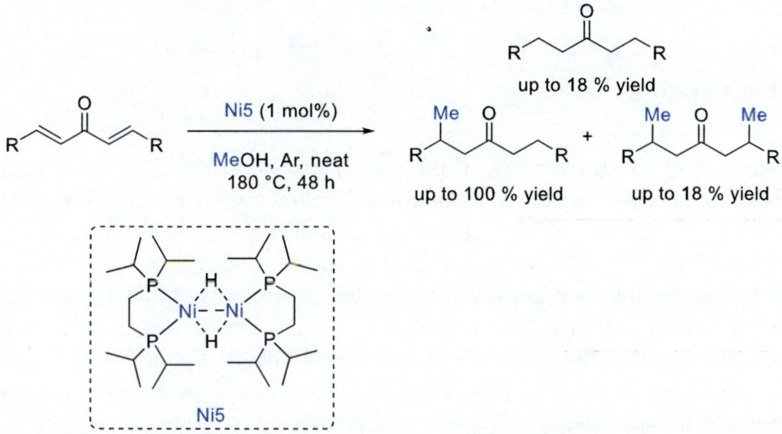

Scheme 25 Nickel promoted formation of β-methylated ketones from dibenzylidene acetone

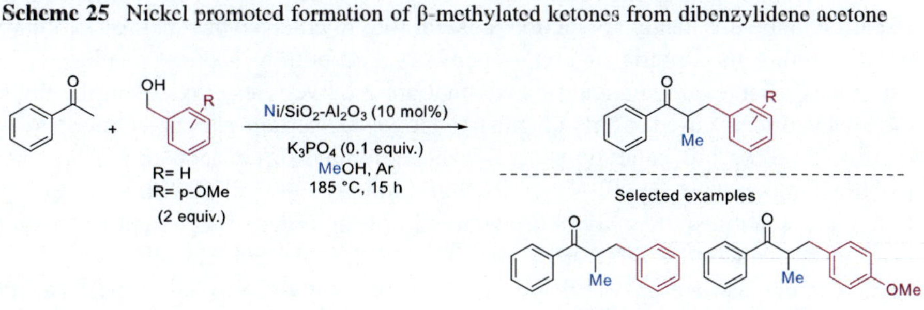

Scheme 26 Three-component cross-benzylation-methylation reaction of acetophenones promoted by nickel

7 α-C-Alkylation of Ketones Catalyzed by Copper

7.1 *Activation of Primary Alcohols*

Copper has very rarely been used in ketone alkylation reactions. In 2012, Xu et al. developed a very simple system based on $Cu(OAc)_2.H_2O$ (1 mol%), NaOH (90 mol %), in air, at 120°C in 12 h to couple acetophenones and alcohols (3 equiv.) to the corresponding β-alkylated alcohols, without any solvent [106]. The first stage of the reaction is based on copper-promoted aerobic oxidation of the alcohol to aldehyde, followed by aldolization/crotonization and finally a hydrogen transfer step. In 2013, Mishra et al. used a copper catalyst supported over Mg-Al hydrotalcite to promote the formation of C–C bonds at 180°C in 15 h [107]. Lately, Bala et al. used Cu (I) complexes based on triazolium ligands in the presence of 2 equivalents of KOH in refluxing THF for the coupling of benzyl alcohols and acetophenones [108]. Finally, Sang et al. immobilized $[(Binap)CuI]_2$ complexes on hydrotalcite, forming a heterogeneous system that can be recycled several times, for the formation of mono-alkylated ketones with benzyl alcohols [109].

8 Conclusion

This chapter gives an overview of the catalytic systems developed in recent years based on 3d metals for the α-alkylation of ketones with primary and secondary alcohols by hydrogen auto-transfer. Well-defined complexes of iron, manganese, cobalt, and nickel have demonstrated their ability to effectively promote this reaction under homogeneous conditions, whereas in the case of copper mainly heterogeneous systems have been applied. The structural diversity of the ligands efficiently applied in the field, whether bidentate, tridentate, phosphine-containing or not, as well as displaying or not cooperative properties, leave a vast field for the optimization of the future catalytic systems.

The excellent selectivities and reactivities of the catalytic systems reported in the literature make this catalytic reaction a useful tool to enhance the chemist's toolbox while meeting the criteria of green chemistry and current societal challenges by using inexpensive and abundant metals that are relatively less toxic than the noble metals historically used in this chemistry. However, certain parameters need to be improved before this catalytic process can be applied on an industrial scale. The reaction temperatures are still generally high (around 130–150°C with a few exceptions), the quantities of catalysts are generally of the order of a percent and would need to be reduced by a factor of 10–100, and the quantities of additives and in particular of base are also still too high. In our opinion, this last aspect can be explained by the generation of water in stoichiometric quantities, which neutralizes part of the base. It should also be noted that to date no catalytic system based on titanium, vanadium, or chromium has been reported for the α-alkylation of ketones.

We can therefore still expect significant advances in this area of chemistry in the coming years, leading to interesting spin-offs in the field of synthesis and potential applications.

References

1. Clayden J, Greeves N, Warren S (2012) Organic chemistry. Oxford University Press
2. Grigg R, Mitchell TRB, Sutthivaiyakit S, Tongpenyai N (1981) Transition metal-catalysed N-alkylation of amines by alcohols. J Chem Soc Chem Commun:611–612
3. Grigg R, Mitchell TRB, Sutthivaiyakit S, Tongpenyai N (1981) Oxidation of alcohols by transition metal complexes part V. Selective catalytic monoalkylation of arylacetonitriles by alcohols. Tetrahedron Lett 22:4107–4110
4. Corma A, Navas J, Sabater MJ (2018) Advances in one-pot synthesis through borrowing hydrogen catalysis. Chem Rev 118:1410–1459
5. Nixon TD, Whittlesey MK, Williams JMJ (2009) Transition metal catalysed reactions of alcohols using borrowing hydrogen methodology. Dalton Trans 753–762:753
6. Obora Y (2014) Recent advances in α-alkylation reactions using alcohols with hydrogen borrowing methodologies. ACS Catal 4:3972–3981
7. Irrgang T, Kempe R (2019) 3d-metal catalyzed N- and C-alkylation reactions via borrowing hydrogen or hydrogen autotransfer. Chem Rev 119:2524–2549
8. Yang D-Y, Wang H, Chang C-R (2022) Recent advances for alkylation of ketones and secondary alcohols using alcohols in homogeneous catalysis. Adv Synth Catal 364:3100–3121
9. Borthakur I, Sau A, Kundu S (2022) Cobalt-catalyzed dehydrogenative functionalization of alcohols: progress and future prospect. Coord Chem Rev 451:214257
10. Zhang M-J, Ge X-L, Young DJ, Li H-X (2021) Recent advances in Co-catalyzed C–C and C–N bond formation via ADC and ATH reactions. Tetrahedron 93:132309
11. Reed-Berendt BG, Latham DE, Dambatta MB, Morrill LC (2021) Borrowing hydrogen for organic synthesis. ACS Central Sci 7:570–585
12. Hofmann N, Hultzsch KC (2021) Borrowing hydrogen and acceptorless dehydrogenative coupling in the multicomponent synthesis of N-heterocycles: a comparison between base and Noble metal catalysis. Eur J Org Chem 2021:6206–6223
13. Waiba S, Maji B (2020) Manganese catalyzed acceptorless dehydrogenative coupling reactions. ChemCatChem 12:1891–1902
14. Reed-Berendt BG, Polidano K, Morrill LC (2019) Recent advances in homogeneous borrowing hydrogen catalysis using earth-abundant first row transition metals. Org Biomol Chem 17: 1595–1607
15. Porcheddu A, Chelucci G (2019) Base-mediated transition-metal-free dehydrative C−C and C−N bond-forming reactions from alcohols. Chem Rec 19:2398–2435
16. Allen LJ, Crabtree RH (2010) Green alcohol couplings without transition metal catalysts: base-mediated [small beta]-alkylation of alcohols in aerobic conditions. Green Chem 12: 1362–1364
17. Ouali A, Majoral J-P, Caminade A-M, Taillefer M (2009) NaOH-promoted hydrogen transfer: does NaOH or traces of transition metals catalyze the reaction? ChemCatChem 1:504–509
18. Polshettiwar V, Varma RS (2009) Revisiting the Meerwein–Ponndorf–Verley reduction: a sustainable protocol for transfer hydrogenation of aldehydes and ketones. Green Chem 11: 1313–1316
19. Garg NK, Tan M, Johnson MT, Wendt OF (2023) Highly efficient base catalyzed N-alkylation of amines with alcohols and β-alkylation of secondary alcohols with primary alcohols. ChemCatChem:e202300741

20. Xu Q, Chen J, Tian H, Yuan X, Li S, Zhou C, Liu J (2014) Catalyst-free dehydrative α-alkylation of ketones with alcohols: green and selective autocatalyzed synthesis of alcohols and ketones. Angew Chem Int Ed 53:225–229
21. Dambatta MB, Santos J, Bolt RRA, Morrill LC (2020) Transition metal free α-C-alkylation of ketones using secondary alcohols. Tetrahedron 76:131571
22. Adrien Q, Jean R (2014) Iron cyclopentadienone complexes: discovery, properties, and catalytic reactivity. Angew Chem Int Ed 53:4044–4055
23. Knölker H-J, Baum E, Goesmann H, Klauss R (1999) Demetalation of tricarbonyl (cyclopentadienone)iron complexes initiated by a ligand exchange reaction with NaOH – X-ray analysis of a complex with nearly square-planar coordinated sodium. Angew Chem Int Ed 38:2064–2066
24. Casey CP, Guan H (2007) An efficient and chemoselective iron catalyst for the hydrogenation of ketones. J Am Chem Soc 129:5816–5817
25. Casey CP, Guan H (2009) Cyclopentadienone iron alcohol complexes: synthesis, reactivity, and implications for the mechanism of iron-catalyzed hydrogenation of aldehydes. J Am Chem Soc 131:2499–2507
26. Moyer SA, Funk TW (2010) Air-stable iron catalyst for the Oppenauer-type oxidation of alcohols. Tetrahedron Lett 51:5430–5433
27. Plank TN, Drake JL, Kim DK, Funk TW (2012) Air-stable, nitrile-ligated (cyclopentadienone)iron dicarbonyl compounds as transfer reduction and oxidation catalysts. Adv Synth Catal 354:597–601
28. Quintard A, Constantieux T, Rodriguez J (2013) An iron/amine-catalyzed Cascade process for the enantioselective functionalization of allylic alcohols. Angew Chem Int Ed 52:12883–12887
29. Yan T, Feringa BL, Barta K (2014) Iron catalysed direct alkylation of amines with alcohols. Nat Commun 5:5602
30. Elangovan S, Sortais J-B, Beller M, Darcel C (2015) Iron-catalyzed α-alkylation of ketones with alcohols. Angew Chem Int Ed 54:14483–14486
31. Pagnoux-Ozherelyeva A, Pannetier N, Mbaye MD, Gaillard S, Renaud J-L (2012) Knölker's iron complex: an efficient in situ generated catalyst for reductive amination of alkyl aldehydes and amines. Angew Chem Int Ed 51:4976–4980
32. Moulin S, Dentel H, Pagnoux-Ozherelyeva A, Gaillard S, Poater A, Cavallo L, Lohier J-F, Renaud J-L (2013) Bifunctional (cyclopentadienone)iron–tricarbonyl complexes: synthesis, computational studies and application in reductive amination. Chem Eur J 19:17881–17890
33. Thai T-T, Mérel DS, Poater A, Gaillard S, Renaud J-L (2015) Highly active phosphine-free bifunctional iron complex for hydrogenation of bicarbonate and reductive amination. Chem Eur J 21:7066–7070
34. Seck C, Mbaye MD, Coufourier S, Lator A, Lohier JF, Poater A, Ward TR, Gaillard S, Renaud JL (2017) Alkylation of ketones catalyzed by bifunctional iron complexes: from mechanistic understanding to application. ChemCatChem 9:4410–4416
35. Abdallah M-S, Joly N, Gaillard S, Poater A, Renaud J-L (2022) Blue-light-induced iron-catalyzed α-alkylation of ketones. Org Lett 24:5584–5589
36. Akhtar WM, Armstrong RJ, Frost JR, Stevenson NG, Donohoe TJ (2018) Stereoselective synthesis of cyclohexanes via an iridium catalyzed (5 + 1) annulation strategy. J Am Chem Soc 140:11916–11920
37. Bettoni L, Gaillard S, Renaud J-L (2020) A phosphine-free iron complex-catalyzed synthesis of cycloalkanes via the borrowing hydrogen strategy. Chem Commun 56:12909–12912
38. Alanthadka A, Bera S, Banerjee D (2019) Iron-catalyzed ligand free α-alkylation of methylene ketones and β-alkylation of secondary alcohols using primary alcohols. J Org Chem 84: 11676–11686
39. Ibrahim JJ, Reddy CB, Zhang S, Yang Y (2019) Ligand-free FeCl2-catalyzed α-alkylation of ketones with alcohols. Asian J Org Chem 8:1858–1861

40. Nallagangula M, Sujatha C, Bhat VT, Namitharan K (2019) A nanoscale iron catalyst for heterogeneous direct N- and C-alkylations of anilines and ketones using alcohols under hydrogen autotransfer conditions. Chem Commun 55:8490–8493
41. Akhtar WM, Cheong CB, Frost JR, Christensen KE, Stevenson NG, Donohoe TJ (2017) Hydrogen borrowing catalysis with secondary alcohols: a new route for the generation of β-branched carbonyl compounds. J Am Chem Soc 139:2577–2580
42. Bettoni L, Gaillard S, Renaud J-L (2020) Iron-catalyzed α-alkylation of ketones with secondary alcohols: access to β-Disubstituted carbonyl compounds. Org Lett 22:2064–2069
43. Frost JR, Cheong CB, Akhtar WM, Caputo DFJ, Stevenson NG, Donohoe TJ (2015) Strategic application and transformation of ortho-disubstituted phenyl and cyclopropyl ketones to expand the scope of hydrogen borrowing catalysis. J Am Chem Soc 137:15664–15667
44. Polidano K, Allen BDW, Williams JMJ, Morrill LC (2018) Iron-catalyzed methylation using the borrowing hydrogen approach. ACS Catal 8:6440–6445
45. Emayavaramban B, Chakraborty P, Dahiya P, Sundararaju B (2022) Iron-catalyzed α-methylation of ketones using methanol as the C1 source under Photoirradiation. Org Lett 24:6219–6223
46. Bettoni L, Seck C, Mbaye MD, Gaillard S, Renaud J-L (2019) Iron-catalyzed tandem three-component alkylation: access to α-methylated substituted ketones. Org Lett 21:3057–3061
47. Sortais J-B (2021) Manganese catalysis in organic synthesis. Weinheim, WILEY-VCH GmbH
48. Elangovan S, Topf C, Fischer S, Jiao H, Spannenberg A, Baumann W, Ludwig R, Junge K, Beller M (2016) Selective catalytic hydrogenations of nitriles, ketones and aldehydes by well-defined manganese pincer complexes. J Am Chem Soc 138:8809–8814
49. Elangovan S, Neumann J, Sortais J-B, Junge K, Darcel C, Beller M (2016) Efficient and selective *N*-alkylation of amines with alcohols catalysed by manganese pincer complexes. Nat Commun 7:12641
50. Peña-López M, Piehl P, Elangovan S, Neumann H, Beller M (2016) Manganese-catalyzed hydrogen-autotransfer C−C bond formation: α-alkylation of ketones with primary alcohols. Angew Chem Int Ed 55:14967–14971
51. Mukherjee A, Nerush A, Leitus G, Shimon LJW, Ben David Y, Espinosa Jalapa NA, Milstein D (2016) Manganese-catalyzed environmentally benign dehydrogenative coupling of alcohols and amines to form aldimines and H2: a catalytic and mechanistic study. J Am Chem Soc 138: 4298–4301
52. Chakraborty S, Daw P, Ben David Y, Milstein D (2018) Manganese-catalyzed α-alkylation of ketones, esters, and amides using alcohols. ACS Catal:10300–10305
53. Barman MK, Jana A, Maji B (2018) Phosphine-free NNN-manganese complex catalyzed α-alkylation of ketones with primary alcohols and Friedländer quinoline synthesis. Adv Synth Catal 360:3233–3238
54. Kabadwal LM, Das J, Banerjee D (2018) Mn(ii)-catalysed alkylation of methylene ketones with alcohols: direct access to functionalised branched products. Chem Commun 54:14069–14072
55. Das J, Singh K, Vellakkaran M, Banerjee D (2018) Nickel-catalyzed hydrogen-borrowing strategy for α-alkylation of ketones with alcohols: a new route to branched gem-Bis(alkyl) ketones. Org Lett 20:5587–5591
56. Das J, Vellakkaran M, Banerjee D (2019) Nickel-catalyzed alkylation of ketone Enolates: synthesis of monoselective linear ketones. J Org Chem 84:769–779
57. Dey G, Saifi S, Sk M, Sinha ASK, Banerjee D, Aijaz A (2022) Immobilizing a homogeneous manganese catalyst into MOF pores for α-alkylation of methylene ketones with alcohols. Dalton Trans 51:17973–17977
58. Lan X-B, Ye Z, Huang M, Liu J, Liu Y, Ke Z (2019) Nonbifunctional outer-sphere strategy achieved highly active α-alkylation of ketones with alcohols by N-heterocyclic Carbene manganese (NHC-Mn). Org Lett 21:8065–8070

59. Franco F, Pinto MF, Royo B, Lloret-Fillol J (2018) A highly active N-heterocyclic carbene manganese(I) complex for selective electrocatalytic CO2 reduction to CO. Angew Chem Int Ed 57:4603–4606
60. Pinto M, Friães S, Franco F, Lloret-Fillol J, Royo B (2018) Manganese N-heterocyclic carbene complexes for catalytic reduction of ketones with silanes. ChemCatChem 10:2734–2740
61. Patra K, Laskar RA, Nath A, Bera JK (2022) A protic Mn(I) complex based on a naphthyridine-N-oxide scaffold: protonation/deprotonation studies and catalytic applications for alkylation of ketones. Organometallics 41:1836–1846
62. Kaithal A, Gracia L-L, Camp C, Quadrelli EA, Leitner W (2019) Direct synthesis of cycloalkanes from diols and secondary alcohols or ketones using a homogeneous manganese catalyst. J Am Chem Soc 141:17487–17492
63. Jana A, Das K, Kundu A, Thorve PR, Adhikari D, Maji B (2020) A phosphine-free manganese catalyst enables stereoselective synthesis of (1 + n)-membered cycloalkanes from methyl ketones and 1,n-diols. ACS Catal:2615–2626
64. Ahmad MS, Inomata Y, Kida T (2022) Heterogenized manganese catalyst for C-, and N-alkylation of ketones and amines with alcohols by pyrolysis of molecularly defined complexes. Mol Catal 526:112390
65. Kita Y, Kuwabara M, Kamata K, Hara M (2022) Heterogeneous low-valent Mn catalysts for α-alkylation of ketones with alcohols through borrowing hydrogen methodology. ACS Catal 12:11767–11775
66. Ghosh A, Hegde RV, Limaye AS, Thrilokraj R, Patil SA, Dateer RB (2023) Biogenic synthesis of δ-MnO2 nanoparticles: a sustainable approach for C-alkylation and quinoline synthesis via acceptorless dehydrogenation and borrowing hydrogen reactions. Appl Organomet Chem 37:e7119
67. Waiba S, Jana SK, Jati A, Jana A, Maji B (2020) Manganese complex-catalysed α-alkylation of ketones with secondary alcohols enables the synthesis of β-branched carbonyl compounds. Chem Commun 56:8376–8379
68. Sklyaruk J, Borghs JC, El-Sepelgy O, Rueping M (2019) Catalytic C1 alkylation with methanol and isotope-labeled methanol. Angew Chem Int Ed 58:775–779
69. Sun Q, Soulé J-F (2021) Broadening of horizons in the synthesis of CD3-labeled molecules. Chem Soc Rev 50:10806–10835
70. Wu Z, Li L, Li W, Lu X, Xie Y, Schaefer HF (2021) Carbonylic-carbon-centered mechanism for catalytic α-methylation. Organometallics 40:2420–2429
71. Bruneau-Voisine A, Wang D, Roisnel T, Darcel C, Sortais J-B (2017) Hydrogenation of ketones with a manganese PN^3P pincer pre-catalyst. Catal Commun 92:1–4
72. Bruneau-Voisine A, Wang D, Dorcet V, Roisnel T, Darcel C, Sortais J-B (2017) Mono-*N*-methylation of anilines with methanol catalyzed by a manganese pincer-complex. J Catal 347: 57–62
73. Bruneau-Voisine A, Pallova L, Bastin S, César V, Sortais J-B (2019) Manganese catalyzed α-methylation of ketones with methanol as a C1 source. Chem Commun 55:314–317
74. Thenarukandiyil R, Kamte R, Garhwal S, Effnert P, Fridman N, de Ruiter G (2023) α-Methylation of ketones and indoles catalyzed by a manganese(I) PCNHCP pincer complex with methanol as a C1 source. Organometallics 42:62–71
75. Deibl N, Kempe R (2016) General and mild cobalt-catalyzed C-alkylation of unactivated amides and esters with alcohols. J Am Chem Soc 138:10786–10789
76. Zhang G, Wu J, Zeng H, Zhang S, Yin Z, Zheng S (2017) Cobalt-catalyzed α-alkylation of ketones with primary alcohols. Org Lett 19:1080–1083
77. Zhang G, Scott BL, Hanson SK (2012) Mild and homogeneous cobalt-catalyzed hydrogenation of C=C, C=O, and C=N bonds. Angew Chem Int Ed 51:12102–12106
78. Xu R, Chakraborty S, Yuan H, Jones WD (2015) Acceptorless, reversible dehydrogenation and hydrogenation of N-heterocycles with a cobalt pincer catalyst. ACS Catal 5:6350–6354
79. Zhang G, Yin Z, Zheng S (2016) Cobalt-catalyzed N-alkylation of amines with alcohols. Org Lett 18:300–303

80. Yin Z, Zeng H, Wu J, Zheng S, Zhang G (2016) Cobalt-catalyzed synthesis of aromatic, aliphatic, and cyclic secondary amines via a "Hydrogen-Borrowing" strategy. ACS Catal 6: 6546–6550
81. Singh A, Maji A, Joshi M, Choudhury AR, Ghosh K (2021) Designed pincer ligand supported Co(ii)-based catalysts for dehydrogenative activation of alcohols: studies on N-alkylation of amines, α-alkylation of ketones and synthesis of quinolines. Dalton Trans 50:8567–8587
82. Tao R, Yang Y, Zhu H, Hu X, Wang D (2020) Ligand-tuned cobalt-containing coordination polymers and applications in water. Green Chem 22:8452–8461
83. Suarsih E, Kita Y, Kamata K, Hara M (2022) A heterogeneous cobalt catalyst for C–C bond formation by a borrowing hydrogen strategy. Cat Sci Technol 12:4113–4117
84. Gangwar MK, Dahiya P, Emayavaramban B, Sundararaju B (2018) Cp*CoIII-catalyzed efficient dehydrogenation of secondary alcohols. Chem Asian J 13:2445–2448
85. Chakraborty P, Gangwar MK, Emayavaramban B, Manoury E, Poli R, Sundararaju B (2019) α-Alkylation of ketones with secondary alcohols catalyzed by well-defined Cp*CoIII-complexes. ChemSusChem 12:3463–3467
86. Chakraborty P, Sundararaju B, Manoury E, Poli R (2021) New borrowing hydrogen mechanism for redox-active metals. ACS Catal 11:11906–11920
87. Liu Z, Yang Z, Yu X, Zhang H, Yu B, Zhao Y, Liu Z (2017) Methylation of C(sp3)–H/C (sp2)–H bonds with methanol catalyzed by cobalt system. Org Lett 19:5228–5231
88. Yang J, Chen S, Zhou H, Wu C, Ma B, Xiao J (2018) Cobalt-catalyzed α-Methoxymethylation and Aminomethylation of ketones with methanol as a C1 source. Org Lett 20:6774–6779
89. Biswal P, Samser S, Nayak P, Chandrasekhar V, Venkatasubbaiah K (2021) Cobalt(II)-porphyrin-mediated selective synthesis of 1,5-Diketones via an interrupted-borrowing hydrogen strategy using methanol as a C1 source. J Org Chem 86:6744–6754
90. Hisamura K, Yamanaka N Sugitani T (2002) The ketone production process. 173460
91. Alonso F, Riente P, Yus M (2007) The α-alkylation of methyl ketones with primary alcohols promoted by nickel nanoparticles under mild and ligandless conditions. Synlett 2007:1877–1880
92. Alonso F, Riente P, Yus M (2008) Alcohols for the α-alkylation of methyl ketones and indirect Aza-Wittig reaction promoted by nickel nanoparticles. Eur J Org Chem 2008:4908–4914
93. Charvieux A, Giorgi JB, Duguet N, Métay E (2018) Solvent-free direct α-alkylation of ketones by alcohols catalyzed by nickel supported on silica–alumina. Green Chem 20:4210–4216
94. Taheri-Torbati M, Eshghi H (2022) Fe3O4@CS-Ni: an efficient and recyclable magnetic nanocatalyst for α-alkylation of ketones with benzyl alcohols by borrowing hydrogen methodology. Appl Organomet Chem 36:e6838
95. Genç S, Arslan B, Gülcemal D, Gülcemal S, Günnaz S (2022) Nickel-catalyzed alkylation of ketones and nitriles with primary alcohols. Org Biomol Chem 20:9753–9762
96. Sharma R, Mondal A, Samanta A, Biswas N, Das B, Srimani D (2022) Well-defined Ni−SNS complex catalysed borrowing hydrogenative α-alkylation of ketones and dehydrogenative synthesis of quinolines. Adv Synth Catal 364:2429–2437
97. Wu D, Wang Y, Li M, Shi L, Liu J, Liu N (2022) Nickel-catalyzed α-alkylation of ketones with benzyl alcohols. Appl Organomet Chem 36:e6493
98. Bains AK, Biswas A, Kundu A, Adhikari D (2022) Nickel-catalysis enabling α-alkylation of ketones by secondary alcohols. Adv Synth Catal 364:2815–2821
99. Bains AK, Kundu A, Yadav S, Adhikari D (2019) Borrowing hydrogen-mediated N-alkylation reactions by a well-defined homogeneous nickel catalyst. ACS Catal 9:9051–9059
100. Bains AK, Adhikari D (2020) Mechanistic insight into the azo radical-promoted dehydrogenation of heteroarene towards N-heterocycles. Cat Sci Technol 10:6309–6318
101. Bains AK, Biswas A, Adhikari D (2020) Nickel-catalysed chemoselective C-3 alkylation of indoles with alcohols through a borrowing hydrogen method. Chem Commun 56:15442–15445

102. Bains AK, Biswas A, Adhikari D (2022) Nickel-catalyzed selective synthesis of α-alkylated ketones via Dehydrogenative cross-coupling of primary and secondary alcohols. Adv Synth Catal 364:47–52
103. Bains AK, Kundu A, Maiti D, Adhikari D (2021) Ligand-redox assisted nickel catalysis toward stereoselective synthesis of (n+1)-membered cycloalkanes from 1,n-diols with methyl ketones. Chem Sci 12:14217–14223
104. Castellanos-Blanco N, Flores-Alamo M, García JJ (2012) Nickel-catalyzed alkylation and transfer hydrogenation of α,β-unsaturated enones with methanol. Organometallics 31:680–686
105. Charvieux A, Duguet N, Métay E (2019) α-Methylation of ketones with methanol catalyzed by Ni/SiO2-Al2O3. Eur J Org Chem 2019:3694–3698
106. Liao S, Yu K, Li Q, Tian H, Zhang Z, Yu X, Xu Q (2012) Copper-catalyzed C-alkylation of secondary alcohols and methyl ketones with alcohols employing the aerobic relay race methodology. Org Biomol Chem 10:2973–2978
107. Dixit M, Mishra M, Joshi PA, Shah DO (2013) Clean borrowing hydrogen methodology using hydrotalcite supported copper catalyst. Catal Commun 33:80–83
108. Lawal NS, Ibrahim H, Bala MD (2021) Cu(I) mediated hydrogen borrowing strategy for the α-alkylation of aryl ketones with aryl alcohols. Chem Monthly 152:275–285
109. Xu Z, Yu X, Sang X, Wang D (2018) BINAP-copper supported by hydrotalcite as an efficient catalyst for the borrowing hydrogen reaction and dehydrogenation cyclization under water or solvent-free conditions. Green Chem 20:2571–2577

Top Organomet Chem (2024) 73: 63–92
https://doi.org/10.1007/3418_2023_97

Published online: 28 September 2023

3d-Metal Catalyzed C–C Bond Formation Through α-Alkylation of Ester, Amide, and Nitriles with Alcohol via Dehydrogenative Coupling

Koushik Sarkar, Animesh Das, and Biplab Maji

Contents

Abstract The borrowing hydrogenation (BH) and acceptorless dehydrogenative coupling (ADC) methods emerged as sustainable tools for forging carbon–carbon bonds at remarkably benign conditions. These protocols utilize alcohol as an alkylating reagent, producing water (and hydrogen gas) as by-products. This makes them advantageous over the traditional C-alkylation procedure, which

K. Sarkar, A. Das, and B. Maji (✉)
Department of Chemical Sciences, Indian Institute of Science Education and Research Kolkata, Mohanpur, India
e-mail: bm@iiserkol.ac.in

requires an organohalide reagent and harsh conditions. Considering the importance of alkylated carboxylate derivatives in pharmaceutical research, their cost-effective preparation is highly demanding. Moreover, due to the less acidity of the α-hydrogen atoms of unactivated esters, amides, and nitriles, their C-alkylation remained challenging. Toward this aim, hydrogen transfer-mediated α-alkylation of these carboxylate derivatives is an attractive choice. Significant progress has been made using catalysts based on precious transition metals, including rhodium, iridium, osmium, and ruthenium. Due to the toxicity, high cost, and limited availability of these noble metals, replacing them with 3d metals would make these processes more sustainable and economically attractive. In this chapter, we discuss the progress made in the hydrogen transfer-mediated C–C bond formation reactions through α-alkylation/olefination of esters, amides, and nitriles using alcohols as carbon sources with the aid of well-defined 3d metal complexes as the catalyst.

Keywords Acceptorless dehydrogenation coupling · Alcohol · Borrowing hydrogenation reaction · C-C bond formation reaction · 3d metal catalysis

1 Introduction

New C–C bond forging methodologies for making molecules of diverse applications, including pharmaceuticals, agrochemicals, polymers, and materials, are always in demand [1–3]. Despite the creative exertions of the researcher, selective synthesis of a target molecule has often faced challenges of lower yield, poor selectivity (chemo-, and stereo-), and copious waste production. Among the tremendous efforts forsaken for making C–C bonds, the C-alkylation of the carbonyl compounds using an organic electrophile, such as halides, is highly notable [4]. However, the use of carcinogenic alkyl halides as electrophiles, the stoichiometric amount of sacrificial base, cryogenic reaction conditions, and the production of halide waste deviate significantly from the "principle of green chemistry." [5] In this regard, alcohols, often derived from renewable sources, emerged as sustainable carbon source in C–C bond formation reactions [5, 6]. However, the direct use of alcohol is rugged due to the poor leaving ability of the hydroxyl group. By converting the hydroxyl group into a better-leaving group like halide or pseudohalide, the challenge is overcome. However, this additional step is detrimental not only to the overall process economy but also to environmental constraints on waste management. A sustainable solution to this problem is the use of a catalyst for activating the alcohol in situ without producing any waste. Hydrogen transfer-mediated coupling is one such process (Scheme 1) [3]. Based on the overall redox economy, they are categorized as net redox-neutral borrowing hydrogenation (BH) or hydrogen auto-transfer reaction and net reductive acceptorless dehydrogenative coupling (ADC).

The first step of these strategies involves the activation of alcohol via a dehydrogenation reaction with the aid of a metal catalyst. The produced carbonyl compound

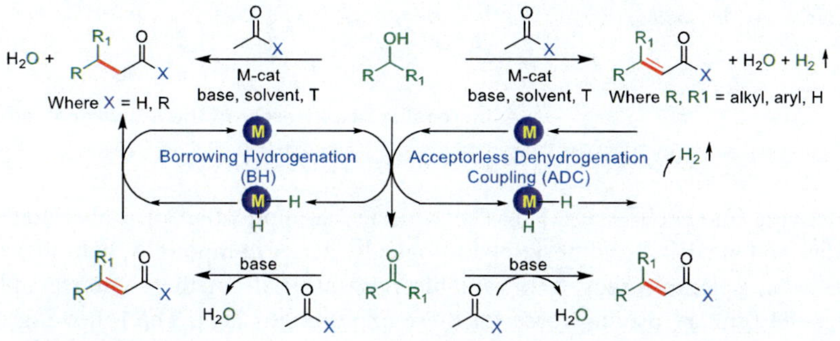

Scheme 1 General mechanism for the C–C bond formation reactions via the hydrogen transfer-mediated coupling reactions

is then coupled with carbon/nitrogen-based nucleophile (condensation with the help of base) in the second step. In the final stage, a bifurcation could happen. The hydrogenation of the in situ formed unsaturated C=C/C=N bonds by the borrowed hydrogen would produce the saturated product as in the BH strategy (Scheme 1, left). In comparison, acceptorless liberation of hydrogen yields the net unsaturated product in ADC reactions (Scheme 1, right). The by-products of these processes are only water for the former and water and hydrogen gas for the latter reactions, which makes them particularly attractive.

The development of these methods is largely dependent on the catalyst used for the initial alcohol dehydrogenation and final hydrogenation (or acceptorless hydrogen liberation) stage. Early exploration has been made with the 4d and 5d transition metals as catalysts [6]. 3d metals, such as manganese, iron, cobalt, and nickel, catalyzed BH and ADC reactions have been emphasized in the last decades [7]. Multifunctional ligand design often plays a crucial role in mitigating the inherent challenges of 3d metal-mediated two-electron redox processes [8]. The higher earth abundance of these metals makes them economically attractive and makes the process more sustainable [9]. Besides, their less toxicity added higher bio-compatibility. Thus, the combination of using alcohols as a carbon source and the earth's abundant early transition metals as a catalyst brings an excellent solution in diversifying these compounds via the C-alkylation reaction.

Another critical step in these reactions is the aldol-type condensation reaction. The acidity of the α-protons (Scheme 2) (https://www.chem.ucalgary.ca/courses/351/Carey5th/Ch21/ch21-2.html) plays a vital role in capturing the in situ-produced carbonyl compounds and preventing by-products from Cannizzaro and oligomerization reactions. Due to higher acidity, the C-alkylation of ketones is highly explored [10]. However, similar reactions with less-acidic esters, amides, and nitriles are

Scheme 2 Acidity order

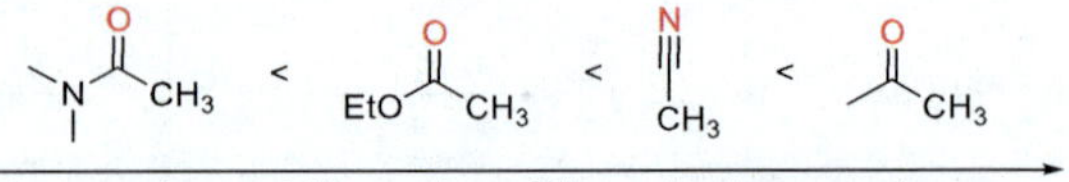

challenging and progressing [3]. The widespread application of α-alkylated esters, amides, and nitriles in synthesizing biologically active compounds, natural products, pharmaceutical molecules, and valuable raw materials in the chemical industries brings the primary driving force for these explorations [11]. The following chapter describes the recent development of the hydrogen transfer-mediated C–C bond formation reactions of esters, amides, and nitriles using alcohols as carbon sources [12].

2 α-Alkylation of Amides and Thioamides

Amides are valuable intermediates and products in organic synthesis [3]. Catalytic α-alkylation of readily available smaller amides with alcohols could significantly advance their synthesis. However, forming the C–C bond at the α-position of the amide is challenging due to comparatively low C-H-acidity as a result of $n_N \rightarrow \pi_{CO}^*$ stabilization. α-Alkylation of unactivated [13] and activated [14] amides with alcohols has been reported using iridium and ruthenium metal-based catalysts [13, 15]. The first-row metal-catalyzed reactions have recently been developed [12, 16]. This section describes the C-alkylation of unactivated amide (cyclic or acyclic) with alcohols as the alkylating agents using diverse 3d metal catalysts (Scheme 3). The sections are subdivided according to the metal used as the catalyst.

2.1 Mn-Catalyzed C-Alkylation of Amides

Over the last decades, homogeneous manganese catalysts have been explored in hydrogenation, dehydrogenation, and hydroelementation reactions [7]. In 2016, the Beller group first demonstrated a few examples of the C-alkylation of oxindole while investigating the C-alkylation of ketones using a well-defined ^{Ph}MACHO-Mn complex **Mn1** as the catalyst (Scheme 3a) [17]. Besides less Brønsted acidity, a competitive amination of free amine in the cyclic amide is another hurdle in this transformation [15, 18, 19]. Oxindoles are among the most targeted amide due to their large availability in a diverse array of naturally occurring compounds [20–22]. Moreover, C3-mono and disubstituted oxindoles are commonly employed in drug discovery [19, 23]. The complex **Mn1** (2 mol%) in the presence of a weak base Cs_2CO_3 (0.05 equiv.) drives the C3-alkylation of 2-oxindoles at 140°C in ^{t}AmOH (Scheme 3a). Four examples are reported that gave excellent 81–92% yields.

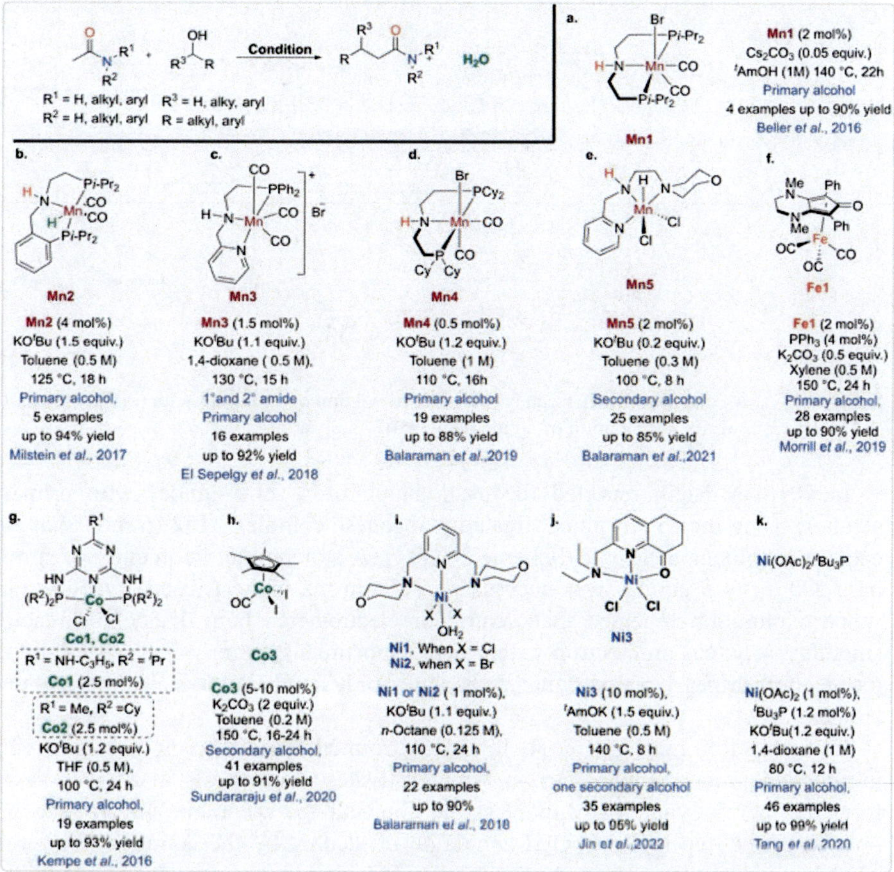

Scheme 3 α-Alkylation of amides with alcohol through 3d-metal catalyzed hydrogen borrowing method

However, despite the limited scope, it opened up new opportunities for further improvement in this field. In the proposed mechanism, the base (Cs_2CO_3) mediated N-H abstraction from **Mn1** provides the active imido complex (**Mn1-a**), which has a distorted trigonal bipyramidal geometry (TBP) (Scheme 4a). Next, the coordination of alcohol gives rise to a coordinatively saturated intermediate **Mn1-b** via the proton transfer process. This is the key step where the dehydrogenation of alcohol takes place to afford the aldehyde and generates the Mn-H intermediate **Mn1-e** via **Mn1-d**. An outer-sphere pathway via N-H bond bifunctional mechanism was invoked. The free aldehyde undergoes a nucleophilic attack by the oxindole carbanion generated via deprotonation at the α-position followed by water elimination. Subsequently, the hydrogenation of the produced α,β-unsaturated 2-oxindole by **Mn1-e** delivers the desired α-alkylated product and the catalyst **Mn1-a** is regenerated.

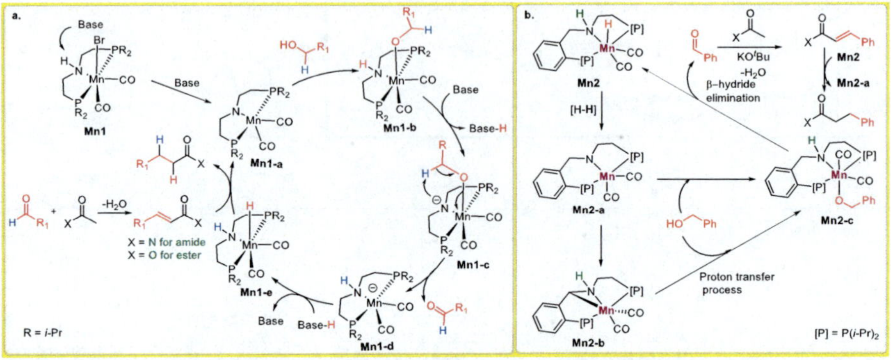

Scheme 4 (**a**) Mechanism of **Mn1**-catalyzed α(C3)-alkylation of oxindoles with primary alcohols. (**b**) Mechanism of the α-alkylation of amide using **Mn2** catalyst

In 2018, Milstein reported the α-alkylation of acyclic amides with primary alcohols using their PNP-pincer ligated manganese complex **Mn2** (5 mol%) as the catalyst in toluene at 125°C (Scheme 3b) [24]. A stoichiometric quantity of strong base KO^{t}Bu (1.5 equiv.) was necessary to obtain the highest reactivity. Substrate scope exploration revealed that neutral or electron-rich benzyl alcohols reacted smoothly, whereas an electron-withdrawing substituent decreased the reactivity of the alcohol partner. However, the protocol was only applied to five alcohols and two amides.

In the reaction mechanism, H_2 liberation from **Mn2** leads to the coordinatively unsaturated amido complex **Mn2-a**, which activates an intramolecular C-H bond to form the thermodynamically more stable C-metallated complex **Mn2-b** (Scheme 4b). The next step includes O-H activation of alcohol by the complex **Mn2-a** or **Mn2-b** via a similar proton transfer process which results in the formation of the alkoxy complex **Mn2-c**. The β-hydride elimination step releases aldehyde along with the regeneration of the active Mn-hydride complex **Mn2**. The base-catalyzed aldol condensation then produced α,β-unsaturated amide that underwent hydrogenation with **Mn2** to yield the product. However, the mechanism for the later step has not been described.

Later in the consecutive years, the El-Sepelgy and Balaraman groups independently reported the C-alkylation of the amide using Mn catalysts (Scheme 3c, d). El-Sepelgy utilized a bench-stable (PNN)Mn(CO)$_3$Br complex **Mn3** as the catalyst (Scheme 3c) [25]. The catalyst activation and mechanism are similar as described above (Scheme 4a). The reaction was performed in the 1,4-dioxane solvent at 130°C in the presence of KO^{t}Bu (1.1 equiv.). The reaction scope includes secondary and tertiary amides and various benzylic, heteroaryl, and aliphatic alcohols. Sixteen examples were shown. The reaction was found to be sensitive to bases. Except for KO^{t}Bu, other bases such as KOH, NaO^{t}Bu, LiO^{t}Bu, Na_2CO_3, and Cs_2CO_3 gave poor yields or no product formation. Balaraman utilized the ^{Cy}MACHO-Mn(I)-complex **Mn4** for the same reaction in toluene at 130°C in the presence of KO^{t}Bu (1.2 equiv.) (Scheme 3d) [26]. Notably, a low 0.5 mol% of the catalyst was used. The

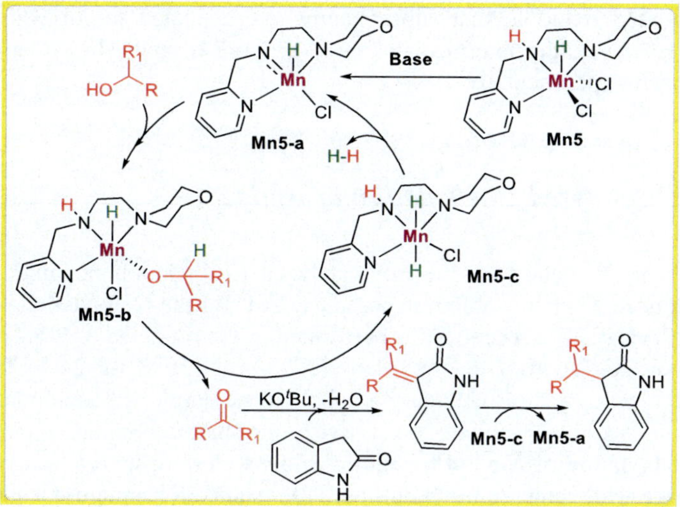

Scheme 5 Mechanism of the **Mn5**-catalyzed C3-alkylation of oxindole with secondary alcohol

scope demonstrates 19 examples with a similar range of functional group compatibility.

C-Alkylation of amides using secondary alcohols is considerably challenging. This is due to the difficulties in all three elementary steps of the BH pathway shown in Scheme 1. The secondary alcohols dehydrogenated slowly [27]. The aldol condensation is problematic with the resultant ketone. The hydrogenation of the produced β,β-disubstituted vinyl amide is also challenging. In 2021, the Balaraman group developed a phosphine-free Mn(II)-complex **Mn5** and utilized it for the α-alkylation of oxindoles using secondary alcohols (Scheme 3e) [28]. **Mn5** was readily prepared from the corresponding NNN-pincer ligand and $MnCl_2{\cdot}4H_2O$ in acetonitrile at room temperature. Unlike previous reaction conditions, a catalytic quantity of KO^{t}Bu (30 mol%) could complete the reaction in 8–10 h. Weak bases such as Na_2CO_3 and K_2CO_3 could not promote this reaction. A diverse range of secondary alcohols containing cyclic, acyclic, long-chain aliphatic, aromatic, and heteroaromatic groups could be successfully applied to this protocol. At this stage, an obvious question arises. Is the **Mn5** catalyzed α-alkylation of oxindoles proceeds via the metal-ligand cooperativity (MLC) pathway? In search for the answer, EPR and X-ray photoelectron spectroscopy (XPS) analysis of the **Mn5** were done before and after the reaction. These measurements indicated that the metal oxidation state remained unchanged throughout the reaction, thus suggesting the participation of MLC during catalysis. Scheme 5 depicts the mechanism of the reaction, which is similar to other Mn(I)-complex catalyzed BH pathways shown in Scheme 4. At first, the active catalyst (**Mn5-a**) is generated from **Mn5** through proton abstraction by KO^{t}Bu. The active imido complex **Mn5-a** activates the alcohol partner via the MLC through the probable intermediate **Mn5-b**. Base-mediated aldol condensation reaction between the in situ generated ketone and oxindole produced the α,β-unsaturated

C3-alkenylated oxindole that subsequently hydrogenated to furnish the desired product. However, the intermediates **Mn5-a**, **Mn5-b**, and **Mn5-c** have not been characterized experimentally.

2.2 *Fe-Catalyzed C-Alkylation of Amide*

There is a single report on the iron-catalyzed C-alkylation of amides. In 2019, Morrill utilized a bench-stable iron complex, **Fe1**, for the C-alkylation of oxindoles (Scheme 3f) [19]. The reaction was performed in the presence of PPh_3 (4 mol%) as an activator, K_2CO_3 (0.5 eq.) in xylene (0.2 M) at 150°C for 24 h. The protocol allows both primary and secondary alcohols to be employed as alkylating agents. Apart from various oxindole derivatives, the method could also enable the C5-monoalkylation of *N*-alkyl barbituric acids, a class of heterocyclic compounds that shows a wide range of biological activities such as antimicrobial and antitumor [16]. The mechanism of alcohol dehydrogenation and vinyl amide hydrogenation proceeds similarly to that of typical Shvo-type complexes (Scheme 6) [29]. Initially, PPh_3 mediates the active catalyst **Fe1-a** generation via the CO de-coordination of the precatalyst **Fe1**. **Fe1-a** then dehydrogenates the alcohol, where the **Fe**-center accepts the hydridic part and the protic part goes to the ligand oxygen to form the intermediate **Fe1-b** and an aldehyde. Subsequent aldol condensation and hydrogenation deliver the product and regenerate the catalyst.

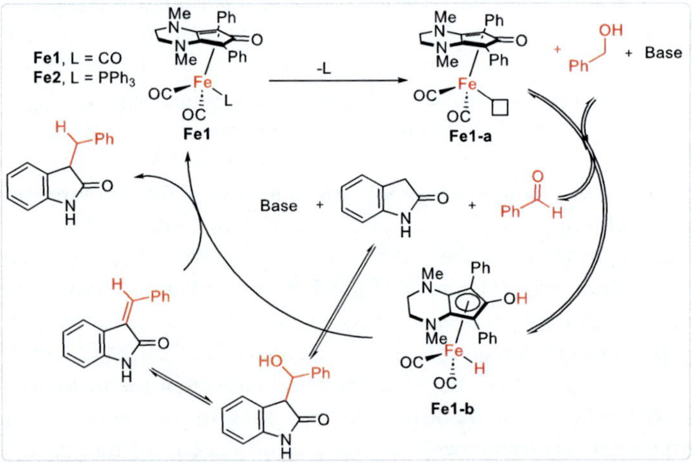

Scheme 6 Fe-catalyzed C3-alkylation of oxindole with primary alcohol

2.3 Co-Catalyzed C-Alkylation of Amide

Molecularly defined complexes of cobalt have also been utilized in hydrogen transfer reactions [30–40]. Cobalt-catalyzed C-alkylation of amides was reported in 2016 by Kempe (Scheme 3g) [12]. Optimization studies revealed that **Co1** and **Co2** could provide the desired alkylated product in 69 and 74% yield, respectively. However, due to the good solubility in THF and lower cost, **Co1** was chosen as the optimal catalyst for further investigation. Interestingly, cobalt complexes derived from structurally similar pyridine-based pincer ligands were found to be silent. Total 22 examples with up to 93% yield were demonstrated using the combination of **Co1** (2.5 mol%) with KO^tBu (1.2 equiv.) in THF (1 M) at 100°C for 24 h. No example of secondary alcohol was shown. Apart from electron-donating substituents, halogen and heteroarene-containing benzyl alcohols rendered good to excellent yields. However, a high catalyst loading (5 mol%) was necessary to observe the good conversion of 3- and 4-pyridine methanol substrates. Moreover, to avoid the formation of dechlorination products for 4-chlorobenzyl alcohol, NaO^tBu was used instead of KO^tBu.

In 2020, Sundararaju employed [Cp*Co(CO)I_2] **Co3** as the catalyst for the same reaction (Scheme 3h) [41]. Besides the challenging secondary alcohols, this protocol could accommodate huge substrate scope (36 examples) with up to 91% yield. The reaction was performed in toluene in the presence of K_2CO_3 (2 equiv.), albeit at a high 150°C temperature. This method was also amenable to the late-stage functionalization of cholesterol and the C-alkylation of barbituric acids with secondary alcohols. Unlike other amide substrates, as the acidity of the α-hydrogens of the barbituric acid ($pK_a \sim 4.8$) is higher, the α-alkylation of barbituric acids occurred in the absence of the base. The reaction was found to be first order for **Co3** and substrate (rate = k[**Co3**][Oxindole]). The deuterium labeling experiment revealed the BH pathway as 8% and 63% D incorporation took place at the α- and β-position of the amide carbonyl, respectively (Scheme 7c). The authors proposed that, due to the higher acidity of the oxindole, the deprotonated oxindole might act as a ligand leading to **Co3-g** or **Co3-h** (X is the anionic oxindole) (Scheme 7a). The β-hydride elimination from **Co3-a** (probably containing a monodentate C-bonded oxindole ligand) was proposed to provide **Co3-b** with coordinated ketone. The dissociation of ketone from **Co3-b** produces a 16-electron hydride complex **Co3-c** (Scheme 7b). The aldol condensation with the liberated ketone generates the C3-unsaturated oxindole, which coordinates to the Co-center, yielding **Co3-d**. Selective hydrogenation at the γ-position of the olefin delivered **Co3-e**. Finally, a proton exchange with another molecule of alcohol regenerated the active catalyst **Co3-a** with the release of 2-hydroxyoxindole, which then rearranges to provide C3-alkylated oxindole.

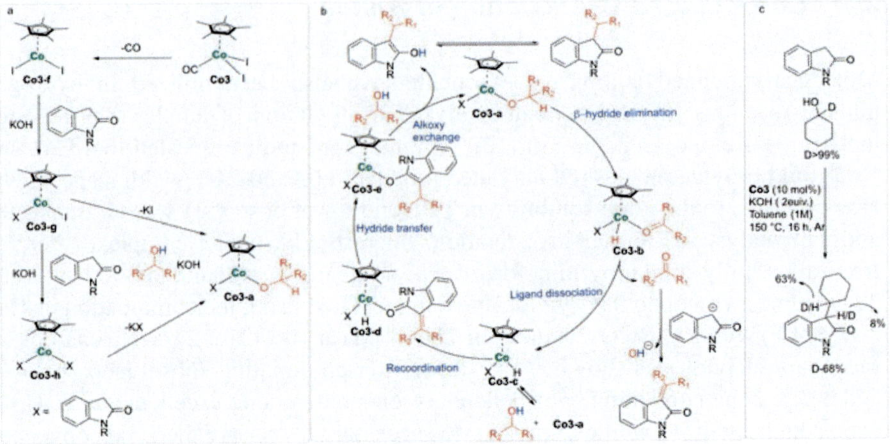

Scheme 7 (**a**) Interaction of oxindole with **Co3**-catalyst. (**b**) Co(III)-catalyzed α-alkylation of oxindole with secondary alcohols. (**c**) Deuterium labeling experiment

2.4 *Ni-Catalyzed C-Alkylation of Amide*

Nickel is an abundant and low-cost transition metal with manifold applications in the hydrogen auto-transfer reaction [42, 43], various cross-coupling [4, 44], and photoredox catalysis [45–47]. In 2018, the Balaraman group synthesized an air-stable octahedral (NNN)Ni(II)-dihalide aqua complex **Ni1** on a gram scale and used it as a catalyst for the C-alkylation reactions (Scheme 3i). Remarkably, **Ni1** in 1 mol% loading showed high reactivity toward the alkylation of cyclic or acyclic amides with differently substituted primary alcohols in the presence of KO^{t}Bu (1.1 equiv.) in *n*-octane. With 0.1 mol% catalyst loading, 82% of the product could be isolated. The calculated high turnover number, TON = 820, demonstrates the robustness of this catalytic system. The generality of the protocol was presented with 19 examples with up to 90% yield. However, the reaction was limited to the primary alcohols. Notably, cyclic amides and oxindole molecules were also diversified under these conditions. Zhao and coworkers later computationally studied the reaction [48]. The hydrogen transfer step, which could proceed via a bifunctional (involving MLC) or non-bifunctional mechanism (an inner-sphere or an outer-sphere), is one of the key steps in this process. All three possibilities were studied. After a systematic comparison, the author concluded that the bifunctional mechanism was the optimal pathway. The octahedral **Ni1**-complex is paramagnetic due to the low energy gap (Δ) between the t_{2g} and e_g levels arising from the weak field of the *N*-arm ligand, leading to a high spin complex. However, a typical Ni(II)-catalyst with d^8 electronic configuration tends to adopt a low-spin square-planar geometry due to a high energy gap between the d_{z^2} orbital and $d_{x^2-y^2}$ orbital (based on the ligand field effect). DFT calculations revealed that the high spin state of nickel is more favorable than the low

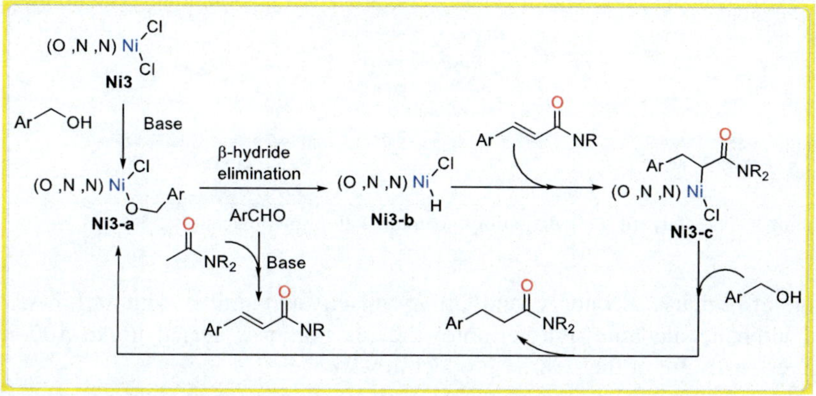

Scheme 8 **Ni3**-catalyzed α-alkylation of amides with alcohols

spin state for this reaction. The former also offers more vacant sites to promote the proton and hydride transfer steps.

Using methanol and ethanol as alkylating reagents is challenging and has not been discussed [49, 50]. Recently, Jin and coworkers demonstrated that various readily available alcohols, including methanol and ethanol, could be utilized for the C-alkylation of unactivated amides (Scheme 3j) [51]. A well-defined N,N,O-Ni-complex, **Ni3** was used as the catalyst in toluene (0.5 M) at 140°C in the presence of ^{t}AmOK (1.5 equiv.). The ligand was crucial in improving efficiency and selectivity. Compared to other bidentate ligands, an additional diethylaminomethyl group in the tetrahydroquinoline backbone improved the reactivity. The substrate scope includes aryl and heteroaryl alcohols and linear and *β*-branched aliphatic alcohols. A higher nickel and base loading was required for unactivated amides (cyclic and acyclic). The proposed mechanism involves the formation of **Ni3-a** intermediate from **Ni3** in the presence of a base, which liberates the aldehyde intermediate through β-hydride elimination and converts to Ni-H species **Ni3-b** (Scheme 8). Like previous mechanisms, the condensation between amide and aldehyde generated the α,β-unsaturated amide, which then reacts with **Ni3-b** to form a Ni-alkyl intermediate **Ni3-c**. Finally, protonolysis of **Ni3-c** by one molecule of alcohol yields the desired product and **Ni3-a**.

So far, our discussion includes the development of diverse well-defined Ni-catalysts to achieve the C-alkylation of amide scaffolds. In 2020, Tang utilized an in situ complex derived from $Ni(OAc)_2/P(t\text{-}Bu)_3$ as the catalyst to enable the selective α-alkylation of amides with alcohols via BH catalysis (Scheme 3k) [52]. The reaction was carried out using $Ni(OAc)_2$ (1 mol%), $P(t\text{-}Bu)_3$ (1.2 mol%), and KO^{t}Bu (1 equiv.) in 1,4 dioxane at 80°C for 12 h. A high temperature >80°C resulted in both *N*- and C-alkylated products in nearly equivalent amounts. Weak bases such as K_3PO_4 and K_2CO_3 produced neither the desired C- nor the *N*-alkylation product. This mild and easy-handle protocol afforded 47 examples of the alkylated amides with high yields of up to 99%. The scope includes combining a

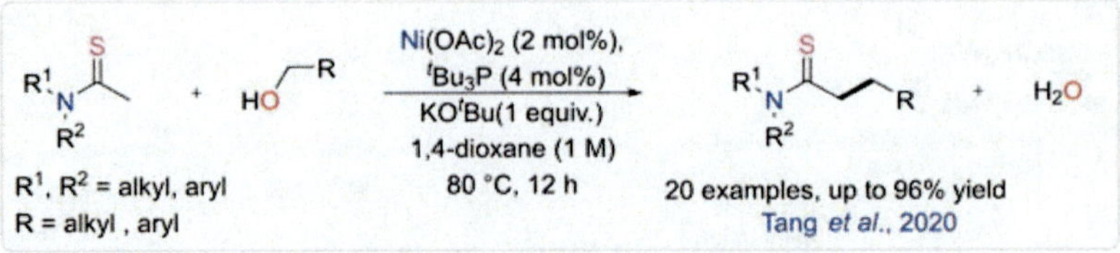

Scheme 9 Ni-catalyzed *α*-alkylation of thioamide with primary alcohol

range of primary, secondary, and tertiary unactivated amides with aryl, heteroaryl, and aliphatic alcohols. Furthermore, the reaction was scaled up to 100 mmol, delivering 16.3 g of the desired product in 92% yield.

The Tang group also extended their work for the alkylation of thioamide (Scheme 9) [52]. Thioamides have more acidic *α*-hydrogen than amides (p*K*a ~ 25.7 in DMSO vs. 35 for the latter) [53]. Notably, these compounds have several bio-activities and are also introduced as isosteres of amide which can enhance the therapeutic effect through modification at the desired position of the peptide [54, 55]. They are also an important building block for constructing many sulfur-containing heterocycles [56]. The $Ni(OAc)_2/P(t\text{-}Bu)_3$ catalyst system efficiently diversified tertiary and cyclic thioamides with various primary alcohols. The desired product was obtained with up to 96% yield.

3 α-Alkylation of Ester

Like amides, esters could also be diversified via BH catalysis employing alcohols as the carbon source [3]. Esters are more amicable substrates due to slightly higher acidity than amides. However, the early reports used noble metals as catalysts in the presence of a stoichiometric base [57]. Recently, 3d metals have been used for these reactions (Scheme 10). The following section summarizes such development.

3.1 Mn-Catalyzed α-Alkylation of Ester

Various Mn catalysts with different ligand backbone have been used for the C-alkylation of esters (Scheme 10a–c). In 2017 Milstein employed **Mn2** as a catalyst for the alkylation of *t*-butyl acetate with four primary alcohols (Scheme 10a) [24]. The reaction proceeded via BH catalysis, similar to that in Scheme 4b. The products were isolated up to 75% yield in the presence of stoichiometric KO^{t}Bu as the base. In 2018, El-Sepelgy reported a similar C-alkylation of *t*-butyl acetate with primary alcohol using **Mn3** as the catalyst (Scheme 10b) [25]. Six examples were shown with different primary alcohols that gave up to 61% yield in the presence of KO^{t}Bu. The reaction yield was low when Cs_2CO_3, Na_2CO_3, KOH, and NaO^{t}Bu

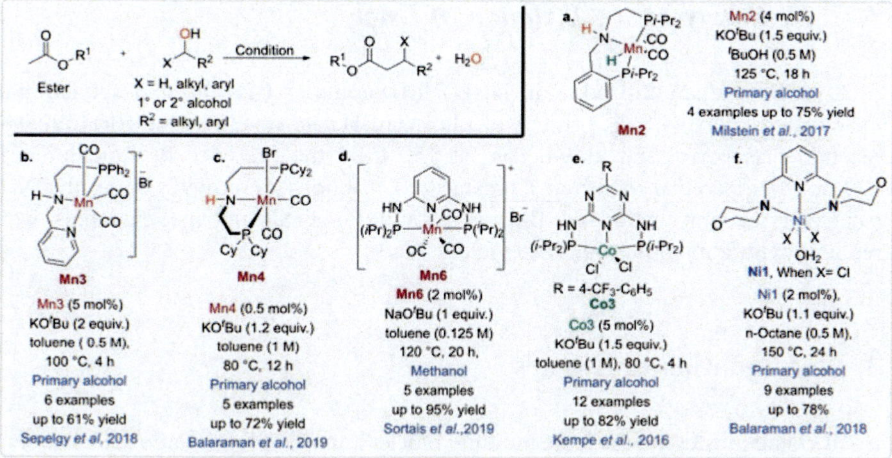

Scheme 10 Manganese-, cobalt-, and nickel-catalyzed α-alkylation of esters with alcohols

were used. Compared to **Mn2**, the reaction with **Mn3** could be performed in less time and lower temperature. Later in 2019, Balaraman performed the same reaction using **Mn4** in lower loading under similar conditions (Scheme 10c) [26]. Besides *t*-butyl acetate, other ester partners, such as *t*-butyl propionate and benzyl acetate, performed poorly. Transesterification was mainly observed for the benzyl acetate. Like previous reports, the transformation remained limited to primary alcohols. In the same year, the Sortais group reported the more challenging α-methylation of esters which was barely achieved, even with the precious metal catalysts [58]. They had shown five examples with up to 95% yield utilizing **Mn6** (3 mol%) as catalyst, and NaO^{t}Bu (1 equiv.) as the base in toluene at 120°C for 20 h (Scheme 10d). Several aryl acetic esters including brominated substrate were methylated using methanol to give the desired product in moderate to good ^{1}H NMR yield. However, due to difficulty in isolation, the isolated yields of the final products were low.

3.2 *Co-Catalyzed α-Alkylation of Ester*

In 2016, Kempe disclosed the sole report on cobalt-catalyzed C-alkylation of *t*-butyl acetate with primary alcohols (Scheme 10e) [12]. The **Co3**-catalyzed reaction required 1.5 equiv. of KO^{t}Bu. Twelve examples and up to 82% yields were demonstrated. The reaction follows a general mechanism, as shown in Scheme 1.

3.3 Ni-Catalyzed α-Alkylation of Ester

In 2018, Balaraman utilized their (NNN)Ni(II)-complex **Ni1** for the α-alkylation of unactivated esters with the primary alcohol via BH catalysis (Scheme 10f) [59]. The reaction was performed in n-octane at 150°C in the presence of stoichiometric KO^{t}Bu. Compared to Mn and Co-catalyzed reactions described above, the **Ni1**-catalyzed reaction includes t-Bu-, Me-, and Ph-acetates. Nine examples were reported with a maximum of 78% yield.

4 α-Alkylation of Nitriles

α-Alkylated nitriles are versatile building blocks important for organic and medicinal chemistry [60, 61]. Moreover, the easy transformation of the nitrile group into aldehydes, ketones, amides, carboxylic acids, and oxazolines could produce high-value chemicals [60–64]. Traditionally, the preparation of α-alkylated nitriles relied on using alkyl halides as alkylating agents and a stoichiometric amount of strong bases such as NaH and $NaNH_2$. However, the toxic nature of the organohalide reagents and the production of inevitable salt wastes often limit the utility of these protocols. In this regard, using alcohol as an alternative to organohalides in a BH strategy has gained much attention [65, 66]. Over the decades, significant efforts have been devoted toward developing this reaction utilizing catalysts based on precious noble metals such as ruthenium, osmium, rhodium, iridium, and palladium [62, 67–70]. As discussed earlier, replacing such metals with earth-abundant first-row-transition metals is highly demanding, considering the toxicity, high cost, and limited availability of the precious noble metals. However, the reactivity of those first-row-transition metal-based complexes lies in the multifunctional ligand design and choice of suitable ancillary ligands [71]. This section summarizes the advancement in developing sustainable catalysts based on manganese, iron, nickel, and cobalt for the α-alkylation of nitriles with alcohols.

4.1 Mn-Catalyzed α-Alkylation of Nitrile

In 2018, Maji first reported the α-alkylation of arylacetonitriles with primary alcohols employing an Mn(I) complex bearing a bidentate hydrazone ligand as the catalyst (Scheme 11a) [71]. This report described the synthesis of two (NNS)Mn (I)-complexes (**Mn6** and **Mn7**) for the first time. Notably, the newly developed Mn catalysts are stable under air and could be operated under low catalyst loading. The report demonstrated 35 examples with good functional group tolerance along with the best yield of 88%. The proposed mechanism is shown in Scheme 12a, b. The N-H deprotonation of the precatalyst **Mn6** in the presence of KO^{t}Bu forms a

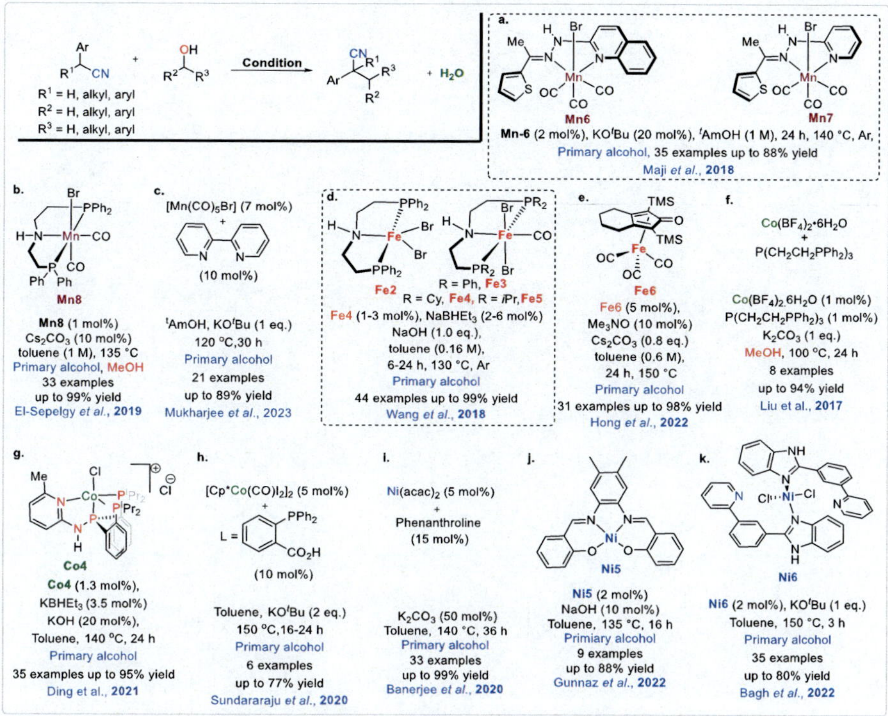

Scheme 11 Manganese-, iron-, nickel-, cobalt-catalyzed α-alkylation of nitriles

coordinatively unsaturated intermediate **Mn6-a** (dearomatized form), which is then saturated by the alcohol activation to form intermediate **Mn6-b**. Through the β-hydride elimination, alcohol undergoes oxidation, and intermediate **Mn6-b** is converted to **Mn6-c**. An aldehyde and the Mn-H complex **Mn6-c** were liberated in a moderately slow step, as depicted by the kinetic isotope effect study ($k_H/k_D = 1.54$). This MLC via protonation and deprotonation of an acidic N-H is involved in many BH and dehydrogenative coupling reactions [72]. Additionally, this bifunctional behavior of the hydrazone ligand was verified by treating an N-Me (**Me6-Me**) analog of the ligand **L1** which resulted in a poor yield of the desired product. Next, the liberated aldehyde undergoes aldol condensation with the nitrile in the presence of KO^{t}Bu to generate the vinyl nitrile. This condensation step is only base dependent and the control experiment suggested that the manganese catalyst might not be involved in this C–C bond-forming step. Finally, the outer-sphere hydrogenation of the vinyl nitrile by **Mn6-c** affords the α-alkylated nitrile.

Later, El-Sepelgy and Rueping utilized **Mn8** for the same reaction (Scheme 11b) [73]. Optimized reaction was performed in toluene at 135°C in the presence of Cs_2CO_3 (10 mol%). The use of polar solvents such as *t*-amyl alcohol, 1,4-dioxane, or 2-methyltetrahydrofuran resulted in moderate yields with the generation of undesired alkenyl nitrile. The reaction is compatible with diverse benzyl cyanides

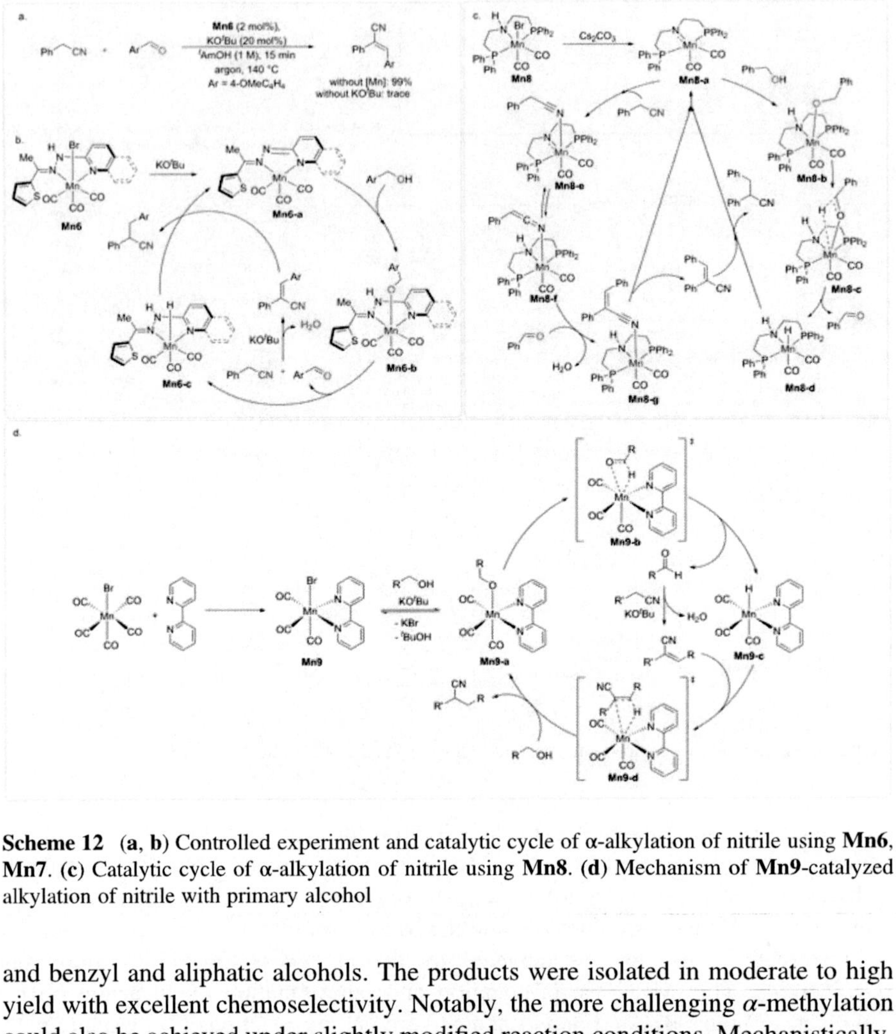

Scheme 12 (**a, b**) Controlled experiment and catalytic cycle of α-alkylation of nitrile using **Mn6**, **Mn7**. (**c**) Catalytic cycle of α-alkylation of nitrile using **Mn8**. (**d**) Mechanism of **Mn9**-catalyzed alkylation of nitrile with primary alcohol

and benzyl and aliphatic alcohols. The products were isolated in moderate to high yield with excellent chemoselectivity. Notably, the more challenging α-methylation could also be achieved under slightly modified reaction conditions. Mechanistically, the reaction was initiated by generating an imido complex **Mn8-a** from the precatalyst in the presence of the base, which was confirmed by ^{31}P NMR measurements. **Mn8-a** triggered the alcohol dehydrogenation and converted to Mn-H species **Mn8-d** via the Mn-alkoxy complex **Mn8-b**, and an aldehyde is liberated. The complex **Mn8-a** also activates the nitrile through initial N-coordination to form **Mn8-e**. Subsequent hydrogen abstraction by the ligand backbone produced a highly nucleophilic species **Mn8-f**. **Mn8-e** and **Mn8-f** remained in equilibrium. The addition of **Mn8-f** to the in situ generated aldehyde formed **Mn8-g** with elimination of one molecule of H_2O. Due to the high lability of the coordinated nitrile, α,β-unsaturated nitrile is dissociated from **Mn8-g**. Finally, the hydrogenation of the alkenyl nitrile occurs in the presence of H – Mn – N – H (**Mn8-d**) species to produce the α-alkylated nitrile and **Mn8-a**.

Recently, the Mukherjee group devised an operationally simple protocol for the α-alkylation of aryl nitriles with readily available alcohols [74]. Instead of using a pre-synthesized Mn catalyst, they explored the reaction with an Mn(I) salt and a suitable ligand, thus generating the active catalyst in situ (**Mn9**) (Scheme 11c). The best result was obtained by using $Mn(CO)_5Br$ (7 mol%), 2,2′-bipyridyl (10 mol%), KO^tBu (1 equiv.) at 120°C in ^{t}AmOH for 30 h. The reaction was sensitive to the amount of catalyst and ligand introduced, as increasing both loading led to diminished yield. The protocol worked well with diverse benzyl cyanides and primary alcohols, delivering the α-branched products in moderate to good yields. The mechanism of the reaction is depicted in Scheme 12d. At first, the reaction between $Mn(CO)_5Br$ and 2,2′-bipyridyl ligand produced Mn(I)-complex **Mn9** in situ. The next step involves the activation of alcohol in the presence of KO^tBu, which leads to the Mn-alkoxy complex **Mn9-a**. Subsequent β-hydride elimination via intermediate **Mn9-b** generated an Mn-H complex **Mn9-c** with the liberation of aldehyde. The base-catalyzed condensation between the aldehyde and nitrile formed the vinyl nitrile, which underwent hydrogenation in the presence of alcohol through intermediate **Mn9-d** to form the desired α-alkylated product and generated the active catalyst.

4.2 *Fe-Catalyzed α-Alkylation of Nitrile*

Wang and coworkers disclosed the first report on iron-catalyzed nitrile alkylation via BH catalysis (Scheme 11d) [75]. Different (PNP)Fe-complexes **Fe2-Fe5** were prepared to examine this reaction. It was found that **Fe4** complex bearing the cyclohexyl substituents on the phosphorous atoms performed the best in the presence of $NaBHEt_3$ as an activator and NaOH as a base in toluene. The proposed mechanism is described in Scheme 13. At first, **Fe4** gets activated by $NaBHEt_3$ to generate the

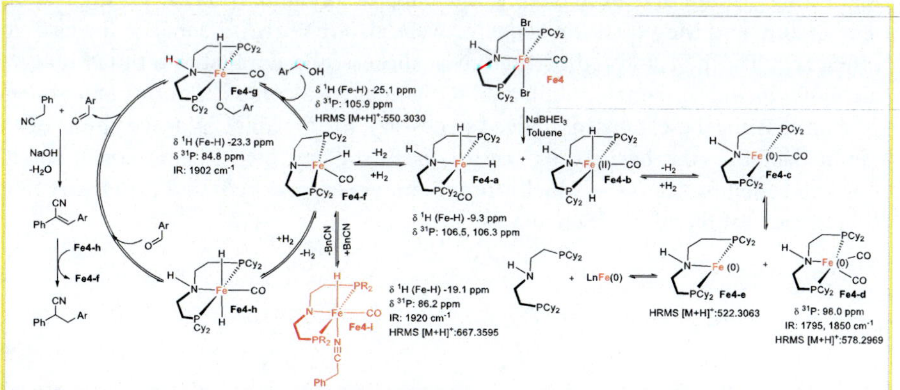

Scheme 13 Mechanistic cycle of **Fe4**-catalyzed α-alkylation of nitrile using primary alcohol

Scheme 14 Mechanistic cycle for **Fe6**-catalyzed α-alkylation of nitrile with primary alcohol

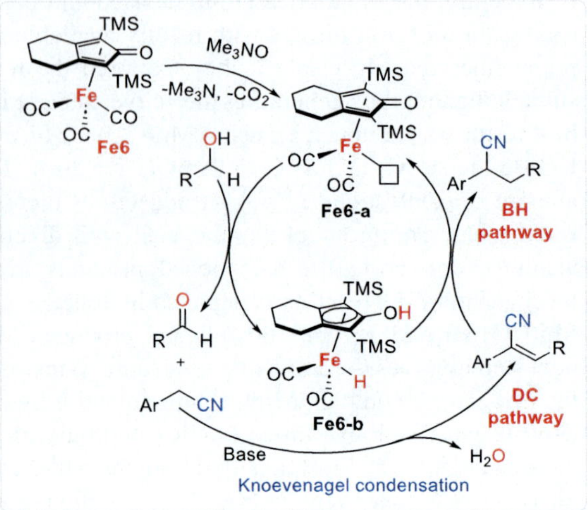

Fe-imido complex **Fe4-f**. **Fe4-i**, which could be the catalyst resting state, is formed reversibly in the presence of nitrile. **Fe4-f** performed the alcohol dehydrogenation and produced a Fe-dihydride **Fe4-h** by eliminating an aldehyde via the intermediacy of a Fe-alkoxy complex **Fe4-g**. The produced aldehyde undergoes aldol condensation, and **Fe4-h** hydrogenated the resulting vinyl nitrile to yield the product and close the cycle. Extensive experimental studies involving control experiments, NMR, IR, and HRMS analysis were performed to identify the proposed reaction intermediates and support the mechanistic hypothesis.

Later, Hong devised a highly chemoselective Fe-catalyzed α-alkylation and α-olefination of nitriles using alcohols (Scheme 11e) [76]. The chemoselectivity between BH and dehydrogenative coupling was achieved by simply changing the base under a (cyclopentadienone)iron(0) carbonyl (CIC) catalyst **Fe6**. The α-olefination strategy is discussed in Sect. 6. A thorough screening of bases revealed that Cs_2CO_3 is the most effective toward this α-alkylation process. A high 150°C temperature and Me_3NO as an activator were also required to enhance the catalytic efficiency. The Fe-catalyzed protocol was successfully applied to a broad range of alcohols, including benzyl, aliphatic, and cinnamyl alcohols. Control experiments proved that the Fe-catalyst is indeed necessary for the final hydrogenation of the olefin. Notably, the basicity or size of the metal ion could be reasoned for the observed chemoselectivity, which strongly influences the **Fe6**-mediated hydrogenation process of the olefin (Scheme 14).

4.3 Co-Catalyzed α-Alkylation of Nitrile

Liu and coworkers first utilized a cobalt salt $Co(BF_4)_2 \cdot 6H_2O$ with a tetradentate phosphine ligand $P(CH_2CH_2PPh_2)_3$ as the catalyst for the α-methylation of nitriles with methanol (Scheme 11f) [77]. Later, Ding reinvestigated this reaction in detail by employing a tetradentate PPPN-Co catalyst, **Co4** (Scheme 11g) [78]. The reaction operates under a low 1.3 mol% catalyst loading with $KHBEt_3$ (3.5 mol%) as the catalyst activator and KOH (20 mol%) as the base. The reaction with the N-Me analog of **Co4** rendered an excellent yield of the α-alkylation product, suggesting that N-H on the ligand backbone did not show MLC. Various nitriles could be efficiently coupled with a wide range of alcohols to afford the alkylated nitriles in moderate to excellent yields. Control experiments revealed that *α,β*-unsaturated nitrile is a possible intermediate. Further investigation revealed that base could mediate the hydrogenation of the *α,β*-unsaturated nitrile in the absence of Co-catalyst/$KHBEt_3$. This result supported a base-mediated Meerwein-Ponndorf-Verley (MPV) hydrogenation pathway (Scheme 15). In all the above examples, secondary alcohols could not be utilized as alkylating agents in *α*-alkylations of nitriles. To fill this gap, in 2020, Sundararaju utilized $[Cp^*Co(CO)I_2]$/2-(diphenylphosphino)benzoic acid as the catalyst for the *α*-alkylation of nitriles with secondary alcohols (Scheme 11h) [41]. Unlike previous protocols, a stoichiometric amount of KO^tBu as the base was needed to obtain the *α*-alkylated nitriles in moderate to good yields. However, the scope was limited to cyclic secondary alcohols, and the protocol was inefficient toward linear and aryl secondary alcohols.

Scheme 15 Mechanistic cycle for **Co4**-catalyzed α-alkylation of nitrile

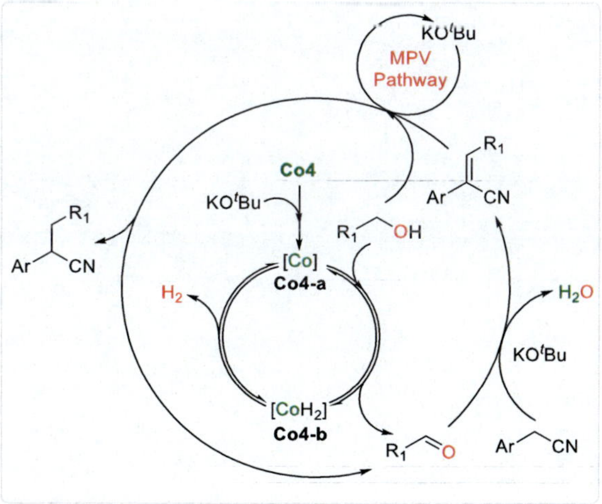

4.4 Ni-Catalyzed α-Alkylation of Nitrile

In 2020, the Banerjee group reported the first nickel-catalyzed α-alkylation of nitriles with alcohols via BH catalysis (Scheme 10i) [79]. The authors found that the combination of $Ni(acac)_2$ (5 mol%), 1,10-phenanthroline (10 mol%), and K_2CO_3 (50 mol%) in toluene was optimal in providing the α-alkylated nitrile in high yields. Other ligands, such as PCy_3 or the derivatives of bipyridine and phenanthroline, were ineffective under these conditions. The combination of $NiCl_2 \cdot$dme with NaOH improved the reactivity of electron-poor and heteroaryl alcohols. However, the alkyl alcohols reacted poorly under standard reaction conditions. Again, the combination of $NiCl_2 \cdot$dme with NaOH helped to achieve the desired products in moderate to good yields. The chemoselective transformation of cinnamyl alcohol and citronellol delineated the synthetic utility of this protocol. The reaction of the olefin intermediate with benzyl alcohol resulted in the α-alkylated product in 65% yield. This experiment, along with ^{1}H NMR and GC-MS analysis of the crude reaction mixture, proved the intermediacy of the olefin intermediate during the catalysis. The Ni-H-mediated hydrogenation step was verified by the stoichiometric reactions of **Ni7** with the olefin intermediate (Scheme 16a). Furthermore, the KIE experiment suggested that the oxidation of alcohol to aldehyde is the rate-limiting step of the overall process.

Later in 2022, Günnaz and coworkers demonstrated that a Ni-salen complex **Ni5** could catalyze the same reaction (Scheme 11j) [80]. Salen ligands are widely explored in homogeneous catalysis and could be easily prepared by condensation of salicylaldehyde with diamine [81–83]. The authors synthesize three different

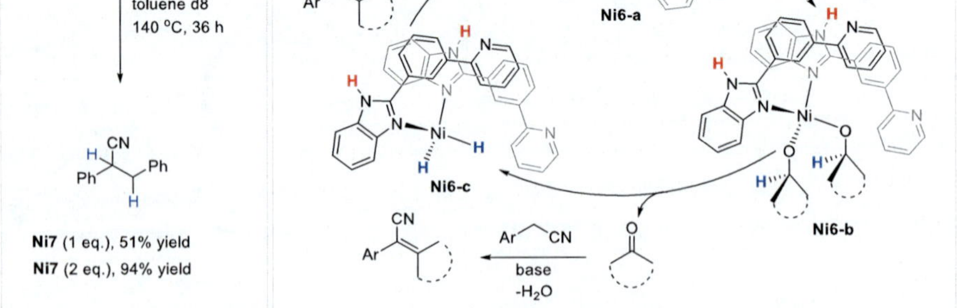

Scheme 16 (**a**) Hydrogenation of intermediate olefine using **Ni7** hydride complex. (**b**) **Ni6**-catalyzed α-alkylation of nitrile with alcohols

Ni-salen complexes; among them, **Ni5** showed high reactivity for this reaction. Using **Ni5** (2 mol%) and NaOH (10 mol%) in toluene at 135°C as standard conditions, a few examples of α-alkylation of nitriles with alcohols were performed. Around the same time, the Bagh group focused on utilizing the more challenging secondary alcohols as alkylating agents for this reaction (Scheme 11k) [84]. This was achieved by employing a Ni(II)-catalyst, **Ni6**, prepared by treating an imidazole-based ligand with $NiCl_2$(DME) in a 2:1 stoichiometric ratio. Despite having two coordinating sites in a single ligand, the complexation occurred with two imidazole moieties of two ligands, as confirmed by single-crystal X-ray analysis. The α-alkylation of nitriles with secondary alcohols proceeded chemoselectively in the presence of **Ni6** (2 mol%), KO^{t}Bu (1 equiv) in toluene at 150°C in a short reaction time. However, prolonged heating was required for certain substrates to improve the yields. Apart from good functional group tolerance of various arylacetonitriles, this process was also amenable to cyclic and acyclic secondary alcohols. However, a few secondary alcohols, such as isopropanol, 1-phenyl ethanol, diphenyl methanol, and 1-pyridyl ethanol, remained challenging substrates. A plausible reaction mechanism is described in Scheme 16b. The mechanism includes the generation of the active catalyst via base-promoted dehydrochlorination and secondary alcohol activation via the intermediacy of a metal-coordinated alkoxy intermediate **Ni6-b**. The β-hydride elimination from **Ni6-b** releases the ketone and the Ni-hydride species **Ni6-c**. The ketone reacts with the nitrile to produce α,β-unsaturated nitrile intermediate with the elimination of water. Finally, the hydrogenation of the vinyl nitrile occurs in the presence of **Ni6-c** to afford the desired alkylated product, and the active catalyst is regenerated. However, the author proposed that the reaction might also proceed via forming a Ni-monoalkoxide intermediate followed by the generation of Ni-monohydride species.

5 Mn-Catalyzed α-Alkylation of Nitrile with Allyl Alcohols

δ-Hydroxynitriles are essential intermediates for synthesizing lactones, diols, and amino alcohols [85–89]. They are ideally prepared by the Michael addition of nucleophilic nitrile to α,β-unsaturated carbonyl followed by reduction of the carbonyl group [90–92]. However, the requirement of multistep procedures and tedious reaction conditions producing inevitable waste limit the utility of the conventional method. Hence, a direct formal conjugate addition of nitriles to allylic alcohols to access δ-hydroxynitriles would be appealing and was previously disclosed by Gunanathan using a ruthenium pincer complex [93]. The use of base-metal-based metal complexes adds another dimension to this protocol toward sustainable development. In this regard, Zhong, Ling, and coworkers exploited the **Mn8** complex for the formal conjugate addition of nitriles to allylic alcohols to synthesize δ-hydroxynitriles (Scheme 17a) [94]. Among various bases, K_2CO_3 was optimal to provide the desired product in a short reaction time. Notably, α-alkylation of nitrile occurred through condensation at the C3-position of allylic alcohol. A drop in

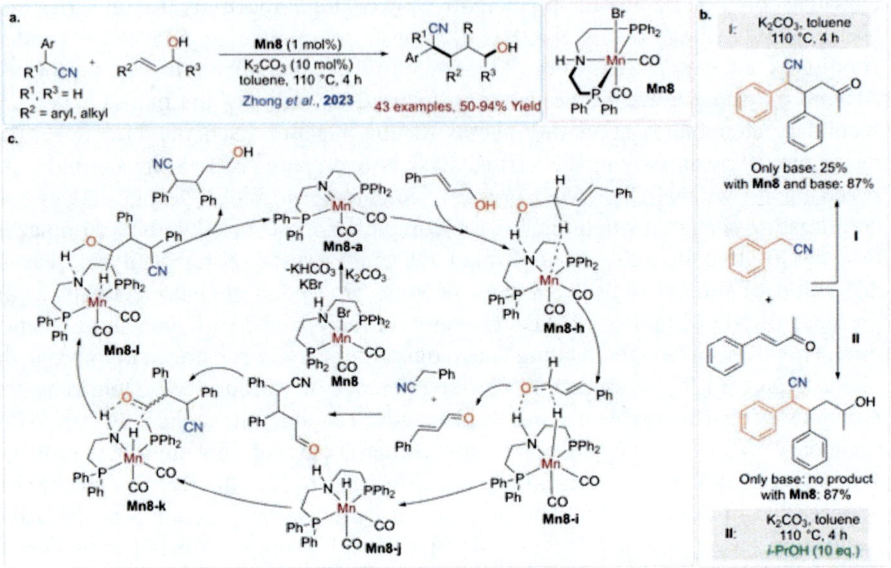

Scheme 17 **Mn8**-catalyzed α-alkylation of nitrile with allylic alcohol and mechanism of the reaction

the yield was found when the reaction medium was changed from toluene to *t*-amyl alcohol or 1-butanol. The electron-rich benzyl nitriles were observed to have relatively higher yields than those bearing an electron-withdrawing substituent. Various allyl alcohols, including internal and terminal allyl alcohols and propylene alcohol, are applied to this transformation. Mechanistic investigations revealed that **Mn8** complex might accelerate the 1,4-conjugate addition reaction as the reaction of benzyl cyanide with benzalacetone poorly yielded the product in its absence (Scheme 17b, condition **I**). A similar experiment with ^{i}PrOH revealed that the hydrogen transfer process could not be achieved without the Mn catalyst (Scheme 17b, condition **II**). The plausible mechanism of this reaction is similar to the above-mentioned reports (Scheme 12 b-d). At first, base-assisted active catalyst generation occurs, followed by dehydrogenation of allylic alcohol to form corresponding cinnamaldehyde and Mn-H species **Mn8-d** (Scheme 17c). Next, the 1,4-conjugate addition of nucleophilic nitrile to cinnamaldehyde led to γ-cyanoaldehyde, which was finally hydrogenated to give δ-hydroxynitrile and regenerated the active catalyst.

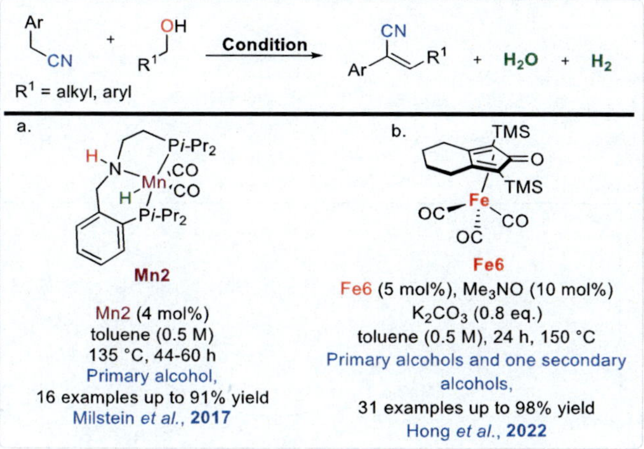

Scheme 18 Manganese and iron-catalyzed α-alkenylation of nitrile with primary alcohol via ADC

6 α-Olefination of Nitriles with Alcohols

In 2017, Milstein reported the ADC of nitriles with alcohols to synthesize α-olefinated nitriles (Scheme 18a) [90]. The previously reported pincer complex **Mn2** (4 mol%) catalyzed the reaction in toluene at 135°C without a base. The **Mn2** complex is involved in activating the α-C-H bond of nitriles, which was further confirmed by detecting a cationic complex **Mn2-d** through ^{31}P, ^{19}F, and ^{1}H NMR measurements. A sharp decrease in the yield was observed when the temperature was lowered to 110°C. A diverse range of benzyl alcohols bearing electron-donating and electron-withdrawing substituents were successfully coupled with various benzyl nitriles rendering the olefinated products in good to excellent yields. In addition, aliphatic alcohols and cinnamyl alcohol could also undergo this ADC reaction smoothly under these conditions. The ^{1}H NMR analysis of the alkenylated product confirmed the formation of Z-isomer. The utility of this protocol was further showcased by a large-scale reaction providing a 66% yield of the isolated olefin. The plausible mechanism of this reaction is depicted in Scheme 19. Initially, H_2 liberation from **Mn2** generated the amido complex **Mn2-a**, which stays in equilibrium with a thermodynamically more stable complex **Mn2-b** through intramolecular C-H activation. Next, activation of the O-H bond of alcohol by **Mn2-a** or **Mn2-b** leads to an alkoxy intermediate **Mn2-c**, which releases the aldehyde following a *β*-hydride elimination step. At the same time, ligand-based *α*-C-H deprotonation of the nitrile generates complex **Mn2-d** and a nitrile carbanion. Subsequently, condensation of the carbanion and the aldehyde resulted in *α,β*-unsaturated nitrile with the regeneration of amido complex **Mn2-a**.

In 2022, Hong utilized **Fe6** for the chemoselective α-olefination of nitriles with alcohols (Scheme 18b) [76]. As previously discussed, the authors observed that different bases could alter the chemoselectivity. **Fe6** (5 mol%), along with Me_3NO

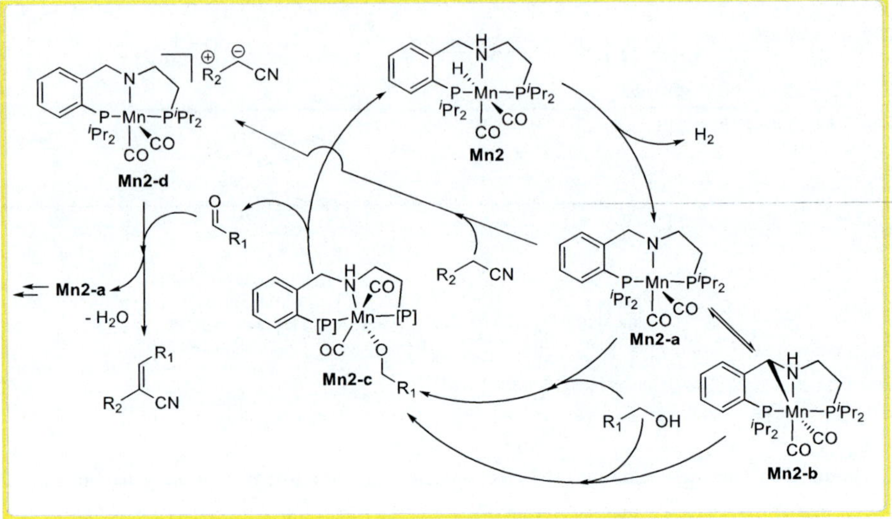

Scheme 19 Reaction mechanisms of alkylation of nitrile using Mn catalysts

(10 mol%) as an activator and K_2CO_3 (1.0 equiv.) as the base, could promote the ADC in toluene at 150°C. Although alcohol acts as a hydrogen atom donor, a high stoichiometry (3 equiv.) of the alcohol partner was necessary to obtain the α-olefination product in high yield. Lowering the temperature to 140°C also resulted in lower conversion. The protocol was not only amenable to diversifying benzylic, heteroaryl, and aliphatic alcohols but also tolerated diverse aryl acetonitriles bearing electron-donating and electron-withdrawing halogen substituents. The mechanism of this reaction is shown in Scheme 14.

7 Conclusion

In conclusion, we have summarized the alkylation of unactivated carboxylate derivatives with primary and secondary alcohol via the hydrogen transfer-mediated coupling reactions (BH and ADC). Molecularly defined manganese, iron, cobalt, and nickel metals-based complexes demonstrated excellent activities in catalyzing these C–C bond-forming reactions employing abundant alcohols as carbon sources. The utilization of diverse phosphine and non-phosphine-based catalysts has been shown to give an elaborative idea about the reactivity and selectivity of the catalysts based on the ligand backbone. Alkylation of nitriles has also been carried out with allylic alcohols using a manganese complex, providing synthetically useful δ-hydroxynitriles scaffolds. Furthermore, α-alkenylation of nitrile with primary alcohol via acceptorless dehydrogenation coupling (ADC) has been successfully exploited using manganese and iron catalysts. The alkenylation of amide, ester, or thioamide with alcohols as carbon sources is yet to be developed. Another challenge is the requirement of high reaction temperatures to facilitate the dehydrogenation process. Reaching an optimum condition by suitably tuning different reaction parameters is essential to achieve the desired α-functionalization sustainably. Due to the lower acidity of the α-hydrogen atoms of the unactivated amides, thioamides, esters, and nitriles, deprotonation by the base remained a challenge. Consequently, stoichiometric strong bases were needed to populate nucleophilic enolate in sustained concentration for the condensation reactions. Besides, the catalyst loading is high in most cases. As a result, large-scale applications are limited to date.

References

1. Li C-J (2009) Cross-dehydrogenative coupling (CDC): exploring C–C bond formations beyond functional group transformations. Acc Chem Res 42(2):335–344
2. Scheuermann CJ (2010) Beyond traditional cross couplings: the scope of the cross dehydrogenative coupling reaction. Chem Asian J 5(3):436–451
3. Donthireddy SNR, Tiwari CS, Kumar S, Rit A (2021) Atom-economic alk(en)ylations of esters, amides, and methyl heteroarenes utilizing alcohols following dehydrogenative strategies. Asian J Chem 10(3):464–484
4. Mesganaw T, Garg NK (2013) Ni- and Fe-catalyzed cross-coupling reactions of phenol derivatives. Org Process Res Dev 17(1):29–39
5. Corma A, Navas J, Sabater MJ (2018) Advances in one-pot synthesis through borrowing hydrogen catalysis. Chem Rev 118(4):1410–1459
6. Baráth E (2018) Hydrogen transfer reactions of carbonyls, alkynes, and alkenes with noble metals in the presence of alcohols/ethers and amines as hydrogen donors. Catalysts 8(12):671
7. Das K, Barman MK, Maji B (2021) Advancements in multifunctional manganese complexes for catalytic hydrogen transfer reactions. Chem Commun 57(69):8534–8549
8. Evans DH (2008) One-electron and two-electron transfers in electrochemistry and homogeneous solution reactions. Chem Rev 108(7):2113–2144
9. Reed-Berendt BG, Polidano K, Morrill LC (2019) Recent advances in homogeneous borrowing hydrogen catalysis using earth-abundant first row transition metals. Org Biomol Chem 17(7): 1595–1607
10. Li F, Ma J, Wang N (2014) α-Alkylation of ketones with primary alcohols catalyzed by a Cp*Ir complex bearing a functional bipyridonate ligand. J Org Chem 79(21):10447–10455
11. Guo L, Ma X, Fang H, Jia X, Huang Z (2015) A general and mild catalytic α-alkylation of unactivated esters using alcohols. Angew Chem Int Ed 54(13):4023–4027
12. Deibl N, Kempe R (2016) General and mild cobalt-catalyzed C-alkylation of unactivated amides and esters with alcohols. J Am Chem Soc 138(34):10786–10789
13. Kuwahara T, Fukuyama T, Ryu I (2013) Ruthenium hydride/nitrogen tridentate ligand-catalyzed α-alkylation of acetamides with primary alcohols. RSC Adv 3(33):13702–13704
14. Grigg R, Whitney S, Sridharan V, Keep A, Derrick A (2009) Iridium catalysed alkylation of 4-hydroxy coumarin, 4-hydroxy-2-quinolones and quinolin-4(1H)-one with alcohols under solvent free thermal conditions. Tetrahedron 65(36):7468–7473
15. Jensen T, Madsen R (2009) Ruthenium-catalyzed alkylation of oxindole with alcohols. J Org Chem 74(10):3990–3992
16. Irrgang T, Kempe R (2019) 3d-metal catalyzed N- and C-alkylation reactions via borrowing hydrogen or hydrogen autotransfer. Chem Rev 119(4):2524–2549

17. Peña-López M, Piehl P, Elangovan S, Neumann H, Beller M (2016) Manganese-catalyzed hydrogen-autotransfer C–C bond formation: α-alkylation of ketones with primary alcohols. Angew Chem Int Ed 55(48):14967–14971
18. Chaudhari, C.; Siddiki, S. M. A. H.; Kon, K.; Tomita, A.; Tai, Y.; Shimizu, K.-i., C-3 alkylation of oxindole with alcohols by Pt/CeO2 catalyst in additive-free conditions. Cat Sci Technol 2014, 4 (4), 1064–1069
19. Dambatta MB, Polidano K, Northey AD, Williams JMJ, Morrill LC (2019) Iron-catalyzed borrowing hydrogen C-alkylation of oxindoles with alcohols. ChemSusChem 12(11): 2345–2349
20. Kagata T, Saito S, Shigemori H, Ohsaki A, Ishiyama H, Kubota T, Kobayashi JI (2006) Paratunamides A–D, oxindole alkaloids from cinnamodendron axillare. J Nat Prod 69(10): 1517–1521
21. Kogure N, Ishii N, Kitajima M, Wongseripipatana S, Takayama H (2006) Four novel gelsenicine-related oxindole alkaloids from the leaves of Gelsemium elegans Benth. Org Lett 8(14):3085–3088
22. Galliford CV, Scheidt KA (2007) Pyrrolidinyl-spirooxindole natural products as inspirations for the development of potential therapeutic agents. Angew Chem Int Ed 46(46):8748–8758
23. Ziarani GM, Gholamzadeh P, Lashgari N, Hajiabbasi P (2013) Oxindole as starting material in organic synthesis. ARKIVOC 2013(1):470–535
24. Chakraborty S, Daw P, Ben David Y, Milstein D (2018) Manganese-catalyzed α-alkylation of ketones, esters, and amides using alcohols. ACS Catal 8(11):10300–10305
25. Jang YK, Krückel T, Rueping M, El-Sepelgy O (2018) Sustainable alkylation of unactivated esters and amides with alcohols enabled by manganese catalysis. Org Lett 20(24):7779–7783
26. Rana J, Gupta V, Balaraman E (2019) Manganese-catalyzed direct C–C coupling of α-C–H bonds of amides and esters with alcohols via hydrogen autotransfer. Dalton Trans 48(21): 7094–7099
27. Nondek L, Sedláček J (1975) Mechanism of dehydrogenation of secondary alcohols on chromia. J Catal 40(1):34–39
28. Rana J, Nagarasu P, Subaramanian M, Mondal A, Madhu V, Balaraman E (2021) Manganese-catalyzed C(α)-alkylation of oxindoles with secondary alcohols via borrowing hydrogen. Organometallics 40(6):627–634
29. Conley BL, Pennington-Boggio MK, Boz E, Williams TJ (2010) Discovery, applications, and catalytic mechanisms of Shvo's catalyst. Chem Rev 110(4):2294–2312
30. Friedfeld MR, Zhong H, Ruck RT, Shevlin M, Chirik PJ (2018) Cobalt-catalyzed asymmetric hydrogenation of enamides enabled by single-electron reduction. Science 360(6391):888–893
31. Chirik PJ (2015) Iron- and cobalt-catalyzed alkene hydrogenation: catalysis with both redox-active and strong field ligands. Acc Chem Res 48(6):1687–1695
32. Friedfeld MR, Margulieux GW, Schaefer BA, Chirik PJ (2014) Bis(phosphine)cobalt dialkyl complexes for directed catalytic alkene hydrogenation. J Am Chem Soc 136(38):13178–13181
33. Lin TP, Peters JC (2014) Boryl-metal bonds facilitate cobalt/nickel-catalyzed olefin hydrogenation. J Am Chem Soc 136(39):13672–13683
34. Lin T-P, Peters JC (2013) Boryl-mediated reversible H2 activation at cobalt: catalytic hydrogenation, dehydrogenation, and transfer hydrogenation. J Am Chem Soc 135(41):15310–15313
35. Korstanje TJ, van der Vlugt JI, Elsevier CJ, de Bruin B (2015) Hydrogenation of carboxylic acids with a homogeneous cobalt catalyst. Science 350(6258):298–302
36. Mukherjee A, Srimani D, Chakraborty S, Ben-David Y, Milstein D (2015) Selective hydrogenation of nitriles to primary amines catalyzed by a cobalt pincer complex. J Am Chem Soc 137(28):8888–8891
37. Zhang G, Hanson SK (2013) Cobalt-catalyzed acceptorless alcohol dehydrogenation: synthesis of imines from alcohols and amines. Org Lett 15(3):650–653
38. Jeletic MS, Mock MT, Appel AM, Linehan JC (2013) A cobalt-based catalyst for the hydrogenation of CO2 under ambient conditions. J Am Chem Soc 135(31):11533–11536

39. Srimani D, Mukherjee A, Goldberg AF, Leitus G, Diskin-Posner Y, Shimon LJ, Ben David Y, Milstein D (2015) Cobalt-catalyzed hydrogenation of esters to alcohols: unexpected reactivity trend indicates ester enolate intermediacy. Angew Chem Int Ed 54(42):12357–12360
40. Zhang G, Scott BL, Hanson SK (2012) Mild and homogeneous cobalt-catalyzed hydrogenation of C=C, C=O, and C=N bonds. Angew Chem Int Ed 51(48):12102–12106
41. Chakraborty P, Garg N, Manoury E, Poli R, Sundararaju B (2020) C-alkylation of various carbonucleophiles with secondary alcohols under CoIII-catalysis. ACS Catal 10(14): 8023–8031
42. Vellakkaran M, Singh K, Banerjee D (2017) An efficient and selective nickel-catalyzed direct N-alkylation of anilines with alcohols. ACS Catal 7(12):8152–8158
43. Das S, Maiti D, De Sarkar S (2018) Synthesis of polysubstituted quinolines from α-2-Aminoaryl alcohols via nickel-catalyzed dehydrogenative coupling. J Org Chem 83(4): 2309–2316
44. Rosen BM, Quasdorf KW, Wilson DA, Zhang N, Resmerita A-M, Garg NK, Percec V (2011) Nickel-catalyzed cross-couplings involving carbon−oxygen bonds. Chem Rev 111(3): 1346–1416
45. Terrett JA, Cuthbertson JD, Shurtleff VW, MacMillan DW (2015) Switching on elusive organometallic mechanisms with photoredox catalysis. Nature 524(7565):330–334
46. Hwang SJ, Powers DC, Maher AG, Anderson BL, Hadt RG, Zheng S-L, Chen Y-S, Nocera DG (2015) Trap-free halogen photoelimination from mononuclear Ni(III) complexes. J Am Chem Soc 137(20):6472–6475
47. Welin ER, Le C, Arias-Rotondo DM, McCusker JK, MacMillan DW (2017) Photosensitized, energy transfer-mediated organometallic catalysis through electronically excited nickel(II). Science 355(6323):380–385
48. Du C, Zhou X, Li W, Wen X, Ke Z, Zhao C (2021) Unusual mechanism of paramagnetic nickel-catalysed alpha-alkylation of amides. Dalton Trans 50(20):6923–6932
49. Jin J, MacMillan DWC (2015) Alcohols as alkylating agents in heteroarene C–H functionalization. Nature 525(7567):87–90
50. Couch DB, Baker RC (2002) Ethanol-enhanced cytotoxicity of alkylating agents. Alcohol Clin Exp Res 26(3):381–385
51. Yang X, Tian X, Sun N, Hu B, Shen Z, Hu X, Jin L (2023) Geometry-constrained N,N,O-nickel catalyzed α-alkylation of unactivated amides via a borrowing hydrogen strategy. Organometallics 42(1):38–47
52. Yang P, Wang X, Ma Y, Sun Y, Zhang L, Yue J, Fu K, Zhou JS, Tang B (2020) Nickel-catalyzed C-alkylation of thioamide, amides and esters by primary alcohols through a hydrogen autotransfer strategy. Chem Commun 56(90):14083–14086
53. Reich's H. Collection. Bordwell pKa table. Collection. Bordwell pKa Table https://organicchemistrydata.org/hansreich/resources/pka/#kaamid
54. Yu K-L, Torri AF, Luo G, Cianci C, Grant-Young K, Danetz S, Tiley L, Krystal M, Meanwell NA (2002) Structure–activity relationships for a series of thiobenzamide influenza fusion inhibitors derived from 1,3,3-Trimethyl-5-hydroxy-cyclohexylmethylamine. Biorg Med Chem Lett 12(23):3379–3382
55. Culik RM, Jo H, DeGrado WF, Gai F (2012) Using thioamides to site-specifically interrogate the dynamics of hydrogen bond formation in β-sheet folding. J Am Chem Soc 134(19): 8026–8029
56. Jagodziński TS (2003) Thioamides as useful synthons in the synthesis of heterocycles. Chem Rev 103(1):197–228
57. Yeung CS, Dong VM (2011) Catalytic dehydrogenative cross-coupling: forming carbon−carbon bonds by oxidizing two carbon−hydrogen bonds. Chem Rev 111(3):1215–1292
58. Bruneau-Voisine A, Pallova L, Bastin S, César V, Sortais J-B (2019) Manganese catalyzed α-methylation of ketones with methanol as a C1 source. Chem Commun 55(3):314–317

59. Midya SP, Rana J, Pitchaimani J, Nandakumar A, Madhu V, Balaraman E (2018) Ni-catalyzed alpha-alkylation of unactivated amides and esters with alcohols by hydrogen auto-transfer strategy. ChemSusChem 11(22):3911–3916
60. Kulp SS, McGee MJ (1983) Oxidative decyanation of benzyl and benzhydryl cyanides. A simplified procedure. J Org Chem 48(22):4097–4098
61. Fleming FF, Yao L, Ravikumar PC, Funk L, Shook BC (2010) Nitrile-containing pharmaceuticals: efficacious roles of the nitrile pharmacophore. J Med Chem 53(22):7902–7917
62. Motokura K, Nishimura D, Mori K, Mizugaki T, Ebitani K, Kaneda K (2004) A ruthenium-grafted hydrotalcite as a multifunctional catalyst for direct α-alkylation of nitriles with primary alcohols. J Am Chem Soc 126(18):5662–5663
63. Hartmann RW, Batzl C (1986) Aromatase inhibitors. Synthesis and evaluation of mammary tumor inhibiting activity of 3-alkylated 3-(4-aminophenyl)piperidine-2,6-diones. J Med Chem 29(8):1362–1369
64. Takaya H, Yoshida K, Isozaki K, Terai H, Murahashi S (2003) Transition-metal-based Lewis acid and base ambiphilic catalysts of iridium hydride complexes: multicomponent synthesis of glutarimides. Angew Chem Int Ed 42(28):3302–3304
65. Hamid MHSA, Slatford PA, Williams JMJ (2007) Borrowing hydrogen in the activation of alcohols. Adv Synth Catal 349(10):1555–1575
66. Dobereiner GE, Crabtree RH (2010) Dehydrogenation as a substrate-activating strategy in homogeneous transition-metal catalysis. Chem Rev 110(2):681–703
67. Löfberg C, Grigg R, Whittaker MA, Keep A, Derrick A (2006) Efficient solvent-free selective monoalkylation of arylacetonitriles with mono-, bis-, and tris-primary alcohols catalyzed by a Cp*Ir complex. J Org Chem 71(21):8023–8027
68. Anxionnat B, Pardo DG, Ricci G, Cossy J (2011) Monoalkylation of acetonitrile by primary alcohols catalyzed by iridium complexes. Org Lett 13(15):4084–4087
69. Panda S, Saha R, Sethi S, Ghosh R, Bagh B (2020) Efficient α-alkylation of arylacetonitriles with secondary alcohols catalyzed by a phosphine-free air-stable iridium(III) complex. J Org Chem 85(23):15610–15621
70. Li F, Zou X, Wang N (2015) Direct coupling of arylacetonitriles and primary alcohols to α-alkylated arylacetamides with complete atom economy catalyzed by a rhodium complex–triphenylphosphine–potassium hydroxide system. Adv Synth Catal 357(7):1405–1415
71. Jana A, Reddy CB, Maji B (2018) Manganese catalyzed α-alkylation of nitriles with primary alcohols. ACS Catal 8(10):9226–9231
72. Elsby MR, Baker RT (2020) Strategies and mechanisms of metal–ligand cooperativity in first-row transition metal complex catalysts. Chem Soc Rev 49(24):8933–8987
73. Borghs JC, Tran MA, Sklyaruk J, Rueping M, El-Sepelgy O (2019) Sustainable alkylation of nitriles with alcohols by manganese catalysis. J Org Chem 84(12):7927–7935
74. Bera K, Mukherjee A (2023) Chemoselective α-alkylation of nitriles with primary alcohols by manganese(I)-catalysis. Chem Asian J 18(13):e202300157
75. Ma W, Cui S, Sun H, Tang W, Xue D, Li C, Fan J, Xiao J, Wang C (2018) Iron-catalyzed alkylation of nitriles with alcohols. Chem A Eur J 24(50):13118–13123
76. Putta RR, Chun S, Lee SB, Hong J, Choi SH, Oh D-C, Hong S (2022) Chemoselective α-alkylation and α-olefination of arylacetonitriles with alcohols via iron-catalyzed borrowing hydrogen and dehydrogenative coupling. J Org Chem 87(24):16378–16389
77. Liu Z, Yang Z, Yu X, Zhang H, Yu B, Zhao Y, Liu Z (2017) Methylation of C(sp3)–H/C(sp2)–H bonds with methanol catalyzed by cobalt system. Org Lett 19(19):5228–5231
78. Paudel K, Xu S, Ding K (2020) α-Alkylation of nitriles with primary alcohols by a well-defined molecular cobalt catalyst. J Org Chem 85(23):14980–14988
79. Bera S, Bera A, Banerjee D (2020) Nickel-catalyzed hydrogen-borrowing strategy: chemo-selective alkylation of nitriles with alcohols. Chem Commun 56(50):6850–6853
80. Genç S, Arslan B, Gülcemal D, Gülcemal S, Günnaz S (2022) Nickel-catalyzed alkylation of ketones and nitriles with primary alcohols. Org Biomol Chem 20(48):9753–9762

81. Cozzi PG (2004) Metal–Salen Schiff base complexes in catalysis: practical aspects. Chem Soc Rev 33(7):410–421
82. Gupta KC, Sutar AK (2008) Catalytic activities of Schiff base transition metal complexes. Coord Chem Rev 252(12):1420–1450
83. Pessoa JC, Correia I (2019) Salan vs. salen metal complexes in catalysis and medicinal applications: virtues and pitfalls. Coord Chem Rev 388:227–247
84. Saha R, Panda S, Nanda A, Bagh B (2023) Nickel-catalyzed α-alkylation of arylacetonitriles with challenging secondary alcohols. J Org Chem
85. Bao X, Wang Q, Zhu J (2019) Dual photoredox/copper catalysis for the remote C(sp3)–H functionalization of alcohols and alkyl halides by N-alkoxypyridinium salts. Angew Chem Int Ed 58(7):2139–2143
86. Dorsey AD, Barbarow JE, Trauner D (2003) Reductive cyclization of delta-hydroxy nitriles: a new synthesis of glycosylamines. Org Lett 5(18):3237–3239
87. Justribó V, Pellegrinet SC, Colombo MI (2007) Studies on the intramolecular cyclizations of bicyclic delta-hydroxynitriles promoted by triflic anhydride. J Org Chem 72(10):3702–3712
88. Taylor SK, Chmiel NH, Simons LJ, Vyvyan JR (1996) Conversion of hydroxy nitriles to lactones using Rhodococcus rhodochrous whole cells. J Org Chem 61(26):9084–9085
89. Darko LL, Cannon JG (1967) Acid catalyzed-rearrangement of cyclopropylphenylglycolic acid. J Org Chem 32(7):2352–2354
90. Chakraborty S, Das UK, Ben-David Y, Milstein D (2017) Manganese catalyzed α-olefination of nitriles by primary alcohols. J Am Chem Soc 139(34):11710–11713
91. Percino MJ, Cerón M, Castro ME, Soriano-Moro G, Chapela VM, Meléndez FJ (2014) Michael addition of phenylacetonitrile to the acrylonitrile group leading to diphenylpentanedinitrile. Structural data and theoretical calculations. Chem Pap 68(5):668–680
92. Arseniyadis S, Kyler KS, Watt DS (2005) Addition and substitution reactions of nitrile-stabilized carbanions. In. Org React:1–364
93. Thiyagarajan S, Sankar RV, Anjalikrishna PK, Suresh CH, Gunanathan C (2022) Catalytic formal conjugate addition: direct synthesis of δ-hydroxynitriles from nitriles and allylic alcohols. ACS Catal 12(4):2191–2204
94. Wang S, Song D, Shen F, Chen R, Cheng Y, Zhao C, Shen Q, Yin S, Ling F, Zhong W (2023) Manganese catalyzed cross-coupling of allylic alcohols and nitriles: an elegant route for access to δ-hydroxynitriles. Green Chem 25(1):357–364

Top Organomet Chem (2024) 73: 93–128
https://doi.org/10.1007/3418_2023_112

Published online: 11 November 2023

Current State-of-Art in the Guerbet-Type *β*-Alkylation of Secondary Alcohols with Primary Alcohols Catalyzed by Complexes Based on 3d Metals

Himani Narjinari, Akshara Bisarya, Vinay Arora, Pran Gobinda Nandi, Kanu Das, and Akshai Kumar

Contents

Abstract Development of greener, sustainable, and atom-economical methods for the formation of Carbon–Carbon bonds leading to versatile fuels and value-added chemicals has been an area of recent global research. For example, the methodology for C–C bond formation proves to be useful for the valorization of ethanol to butanol which acts as an alternative fuel. This methodology can be further applied in natural

H. Narjinari, A. Bisarya, V. Arora, P. G. Nandi, and K. Das
Department of Chemistry, Indian Institute of Technology Guwahati, Guwahati, Assam, India

A. Kumar (✉)
Department of Chemistry, Indian Institute of Technology Guwahati, Guwahati, Assam, India

Center for Nanotechnology, Indian Institute of Technology Guwahati, Guwahati, Assam, India

Jyoti and Bhupat Mehta School of Health Science and Technology, Indian Institute of Technology Guwahati, Guwahati, Assam, India
e-mail: akshaikumar@iitg.ac.in

product synthesis including but not limited to synthesis of several cholesterol and flavan derivatives. This greatly negates the typical disadvantages of classical coupling of alkyl halides that generally are accompanied by the formation of hazardous waste. Transition metal-catalyzed Guerbet-type coupling of alcohols is a promising route to new C–C bonds as they are greener and atom-economical with water as the sole by-product. Typically, these reactions follow the sequence of catalytic dehydrogenation to carbonyl compounds, aldol condensation and a tandem catalytic hydrogenation of the resulting α,β-unsaturated carbonyl compound. Barring the large library of precious metal (Ru, Ir, Rh, and Pd) based homogeneous catalytic systems that accomplish the β-alkylation of alcohols, there are only a few reports on corresponding homogeneous catalytic systems derived from 3d metals and are mainly based on Mn, Fe, Co, Cr, and Ni. In this chapter, an attempt has been made to shed light on the current state-of-art in the 3d-metal catalyzed β-alkylation of secondary alcohols with primary alcohols from a synthetic and mechanistic point-of-view.

1 Introduction

Alcohols have widespread application in diverse field of organic transformations and chemical enterprises [1, 2]. The new C–C bond formations can be deployed as a substitute of carbon source in hydrocarbon-based chemical industries and other petrochemical industries. The traditional process of synthesizing β-alkylated alcohols involves a multistep synthetic pathway (Scheme 1) [5–8]. In this method, the ratio of waste production is very high [5–7], along with the requirement of highly expensive as well as toxic reagents and strong bases [5–8]. Synthesis of alcohols via an alternative methodology is a current global challenge.

Among the various approaches, methods where β-alkylation of alcohols have been accomplished via transition metal-catalyzed C–C bond formation have gathered huge attention. Formation of higher β-alkylated alcohols via acceptorless dehydrogenative coupling of alcohols proves to be an environmentally benign process and this typically employs the hydrogen-borrowing strategy. In this pathway, the transition metal catalyst brings about the dehydrogenation of primary and

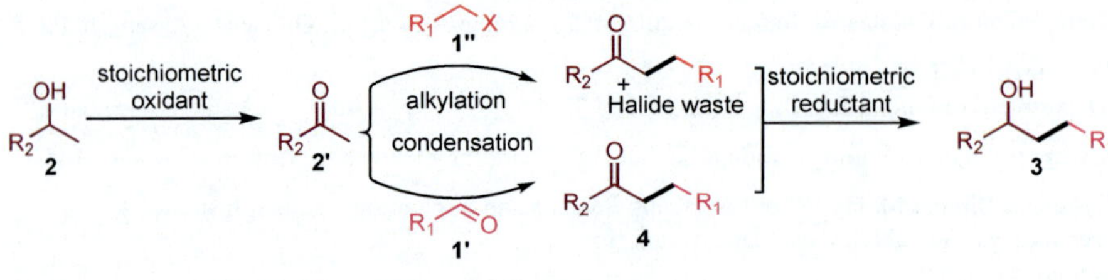

Scheme 1 Traditional routes to β-alkylation of alcohols [3, 4]

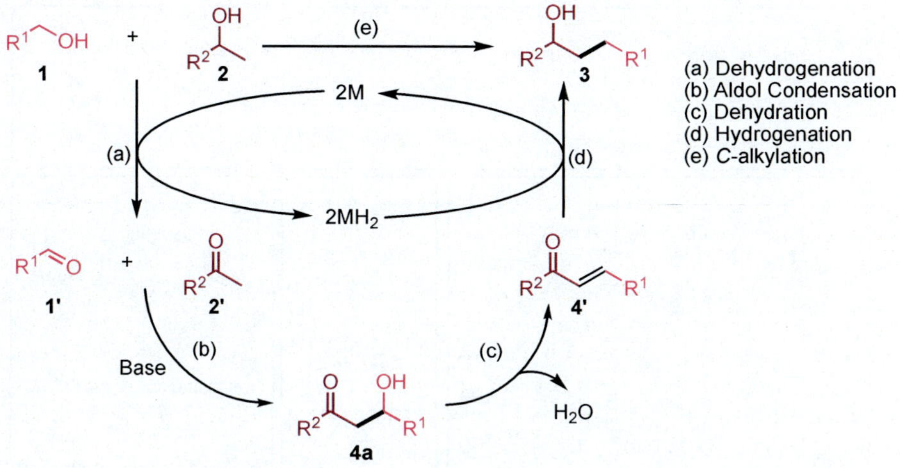

Scheme 2 A greener approach towards β-alkylation of alcohols [4]

secondary alcohols to their corresponding carbonyl compounds. This is accompanied by an uncatalyzed base-mediated aldol condensation of the resulting carbonyl compounds to give rise to an α,β-unsaturated ketone, which on further catalytic hydrogenation affords the β-alkylated alcohols (Scheme 2) giving rise to water as the sole by-product.

Thus, in organic synthesis, hydrogen-borrowing methodology has become more appealing, as it employs alcohols as alkylating agents and takes place in milder reaction conditions. On this ground, transition metal-catalyzed dehydrogenation of alcohols have been widely explored in anaerobic conditions [9–11]. This is mainly owing to the fact that, under aerobic conditions, dehydrogenation of alcohols are severely limited by the air sensitive nature of the key metal-hydride species (i.e., MH_2; Scheme 2) that either gets deactivated or not readily formed under the non-inert reaction conditions [12, 13].

A plethora of transition metal catalysts based on Ir [14–23], Rh [24], Ru [3, 13, 25–30], Pd [31], Mn [32, 33], Co [34], Fe [35], Ni [36], and Cu [37] have been successfully employed in achieving the β-alkylation of primary alcohols with secondary alcohols (*see* Fig. 1). Sun et al. have demonstrated the acceptorless β-alkylation of primary alcohols with secondary alcohols that are catalyzed by inexpensive, environmentally friendly, and phosphine-free iron complex **7** in presence of catalytic amount of NaOH [35]. Li and Lang have studied the utility of the nickel-thiolate-based cluster complex **9** in the C–C cross-coupling of primary and secondary alcohols which resulted in either α,β-unsaturated ketones, α-alkylated ketones or β-alkylated secondary alcohols with multiple chemoselectivities under milder conditions [36]. Xu et al. have described the superior catalytic activity of the ligand-free copper catalyst under air towards the β-alkylation of alcohols (*see* Fig. 1) [37]. Rather than the involvement of a copper-hydride species, they proposed a new mechanism involving a Meerwein-Pondorf-Verley type transition state for the β-alkylation of

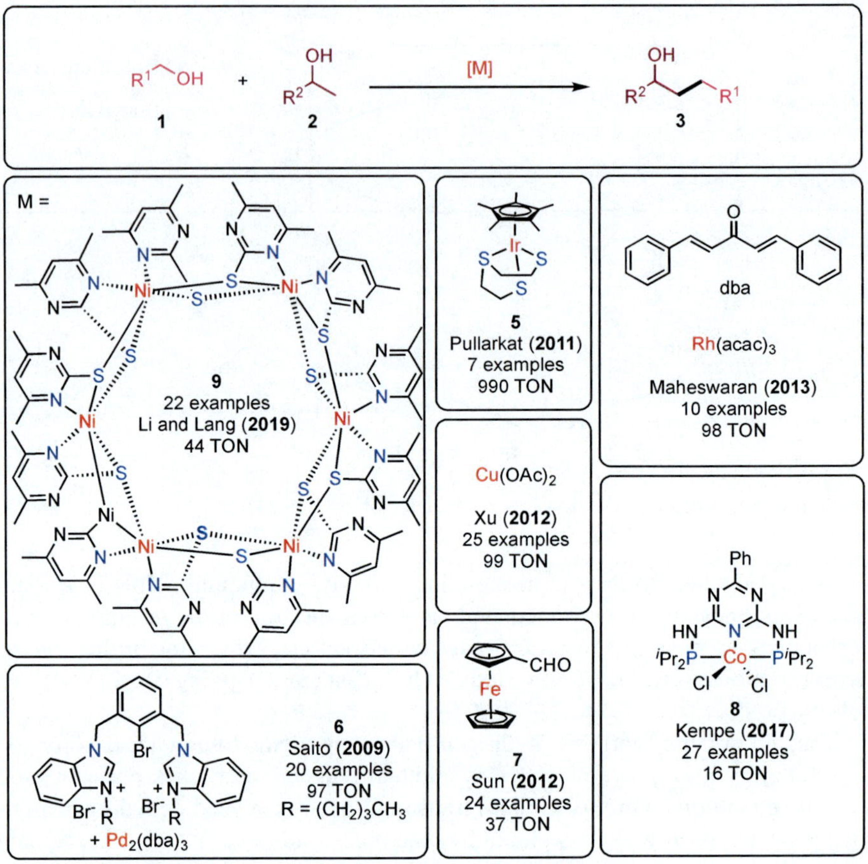

Fig. 1 Representative examples of catalytic systems for *C*-alkylation of alcohols

secondary alcohols with primary alcohols [38]. There have also been several influential reports by the groups of Kempe [34], Pullarkat [16], Saito [31], and Maheswaran [24] who have demonstrated the hydrogen-borrowing methodology in the *C*-alkylation reaction of primary alcohols with secondary alcohols. However, in general, higher reactivity and selectivity have been exhibited by ruthenium and iridium-based complexes in comparison to other transition metals. In particular, there are several reports on the pincer-Ru catalyzed β-alkylation of secondary alcohols.

1.1 β-Alkylation of Secondary Alcohols with Primary Alcohols Catalyzed by Noble Metal Complexes

In 2017, Achard et al. reported the catalytic alcohol dehydrogenation leading to esters by utilizing metal catalysts based on bi/tri-dentate phosphine pyridine ligands. Interestingly, complex **10** afforded the α-alkylated ketone via dehydrogenative cross-coupling of secondary alcohol with primary alcohol with the concomitant formation of trace amount of β-alkylated alcohol (Scheme 3) [12].

The α-alkylated ketones can be synthesized either from coupling of primary alcohols with a ketone or from secondary alcohols with an aldehyde in the presence of a base and catalyst. However, the sequential de/hydrogenation strategy demonstrated in Archad's report is an effective alternative pathway to synthesize the α-alkylated ketone directly from alcohols. This also shows that the reaction of secondary alcohol with primary alcohols in the presence of transition metal catalysts leads to either the β-alkylated alcohol or results in further dehydrogenation of the β-alkylated alcohol resulting in the α-alkylated ketone. In the proposed mechanism (Scheme 3), the first step is the base mediated dissociation of PPh_3 which gives rise to the catalytically active intermediate L_nRu **10a**. Then, L_nRuH_2 **10b** a dihydride intermediate is generated via hydride transfer from alcohol to the ruthenium center. Base mediated condensation of aldehyde and ketone generates the α, β-unsaturated ketone **4′**. The hydrogenation of **4** and **4′** by ruthenium dihydride complex **10b** leads to the generation of the β-alkylated alcohol **3**. The final product α-alkylated ketone **4** is obtained by the dehydrogenation of **3** which further leads to the regeneration of L_nRu **10a** along with the release of a H_2 molecule [12].

Later in 2022, Shim et al. employed $RuCl_2(PPh_3)_3$ for the catalytic β-alkylation of secondary alcohols with primary alcohols in the presence of 5 equivalents of 1-dodcene (Scheme 4) [13]. This protocol followed a selective pathway leading to the β-alkylated product with a variety of substrate scope (ranging from alkyl methyl, aryl methyl and cyclic carbinols, apart from alkyl methyl carbinols).

Similarly, Gülcemal et al. studied the imidazol-2-ylidene ligand-based iridium (I) catalyst **11** for the β-alkylation of secondary alcohols. At a very low loading of the catalyst **11**, this simple catalytic system furnished the β-alkylated alcohols (including but not limited to cholesterol derivatives) with good yields (Scheme 5) [39]. The iridium (I) complex gave up to 940,000 TON in the presence of catalytic amounts of NaOH or KOH under air via the hydrogen-borrowing strategy.

Recently, Bera et al. investigated the catalytic β-alkylation of secondary alcohols with primary alcohols that is catalyzed by Ir(III)-*N*-heterocyclic carbene **12** as catalyst. The complex **12** achieved remarkable activity owing to its pyridyl (benzamide)-functionalized NHC backbone, which takes part in dearomatization/aromatization of its pyridine backbone during the course of the catalysis. Within a short span of reaction time, these catalytic systems exhibited a broad range of substrate scope at a very low base and catalyst loading (Scheme 6) [40]. Apart from the synthesis of quinoline and lactone derivatives, the same methodology, was extended to the α-alkylation of ketones utilizing primary alcohols [40].

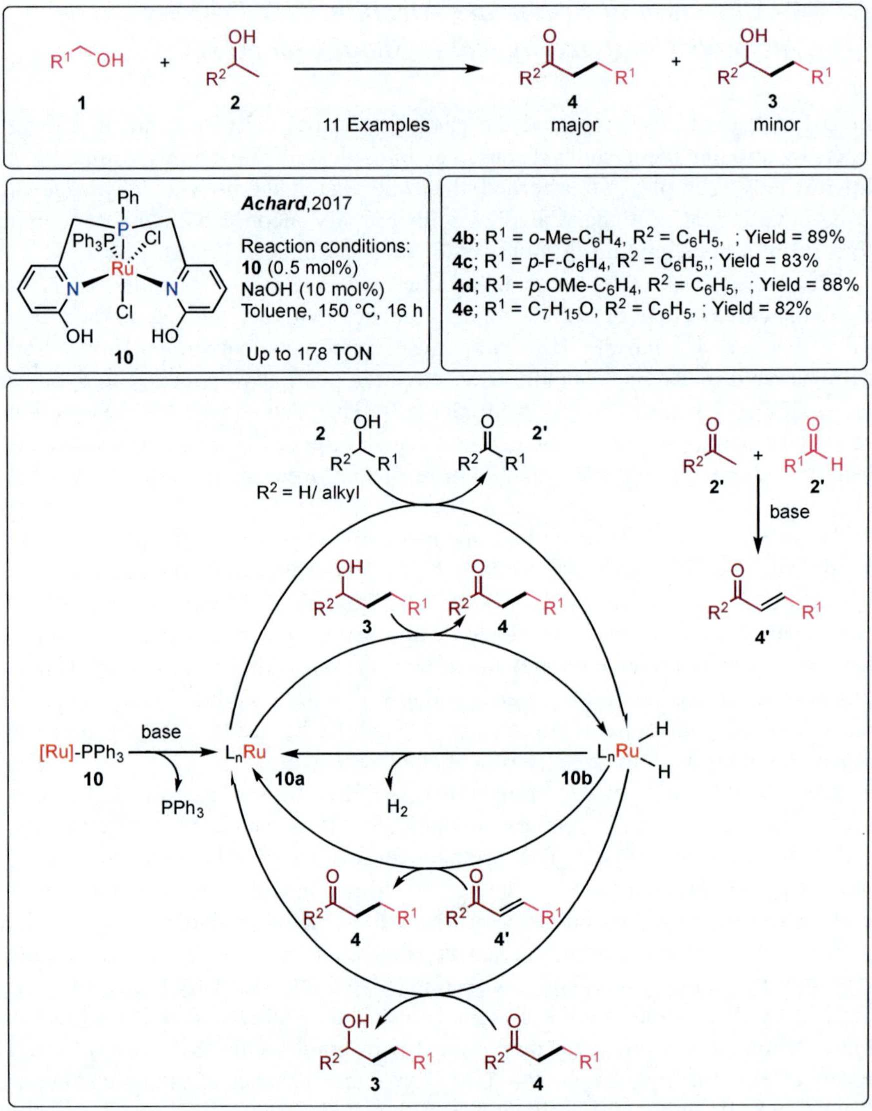

Scheme 3 Synthesis of α-alkylated ketone obtained from further dehydrogenation of β-alkylated alcohol in the catalytic alkylation of alcohols using pincer-Ru complexes [12]

Single-pot acceptorless catalytic coupling of a mixture of low-molecular weight alcohols (primary and secondary) thus leads to value-added higher-molecular weight alcohols and ketones. In this context, Gelman et al. studied the pincer-ruthenium catalyzed cross-coupling of secondary alcohols with primary/secondary alcohols that efficiently led to the selective synthesis of β-alkylated ketones. The non-innocent nature of the ligand comes into play during the mechanistic cycle

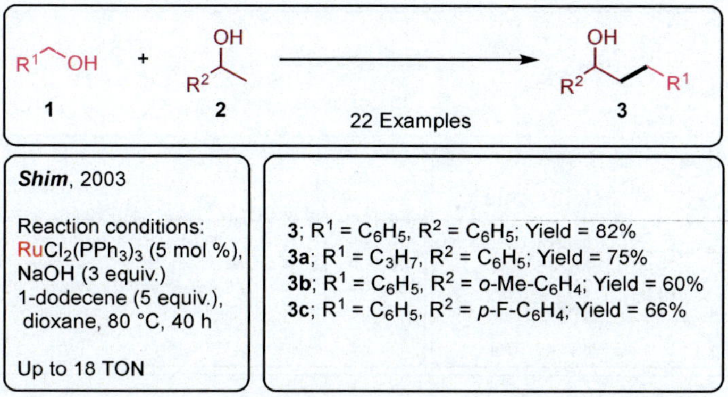

Scheme 4 $RuCl_2(PPh_3)_3$-catalyzed β-alkylation of alcohols [13]

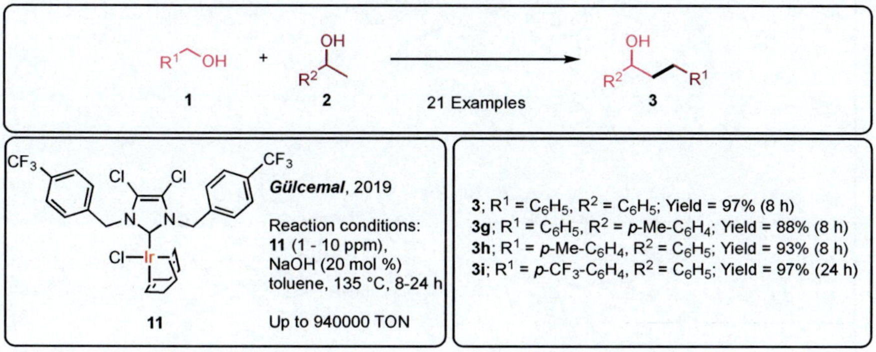

Scheme 5 The β-alkylation of secondary alcohols with primary alcohols employing a Ir(I)-catalyst [39]

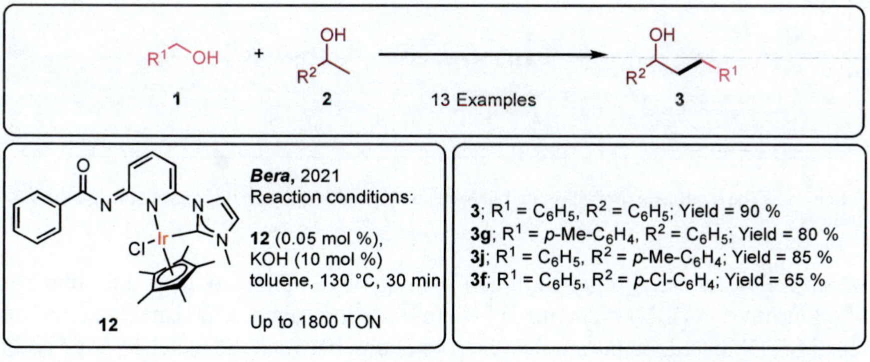

Scheme 6 The Ir(III)-catalyzed β-alkylation of secondary alcohols with primary alcohols [40]

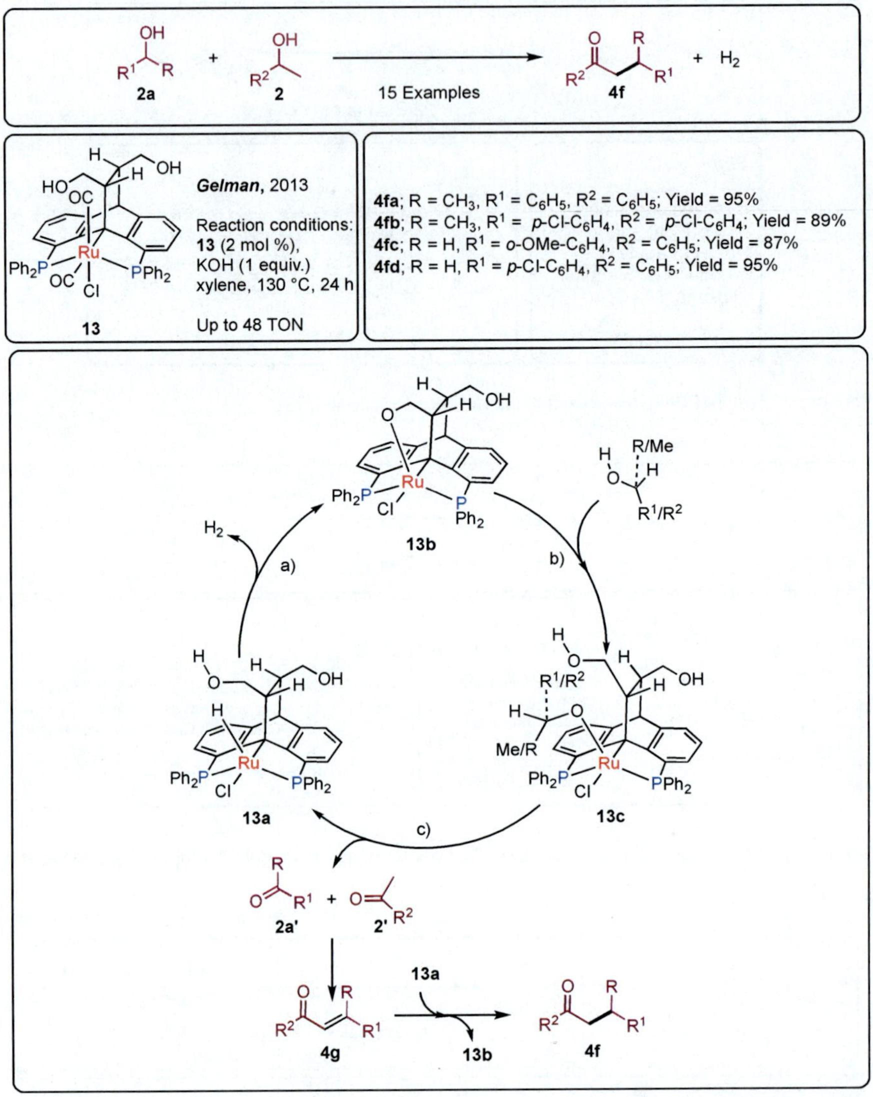

Scheme 7 The synthesis of β-alkylated ketones catalyzed by pincer-ruthenium complex **13** [41]

which follows a dehydrogenation and hydrogenation pathway to yield a maximum of 48 turnovers (TON) (Scheme 7) [41]. In the proposed mechanism (Scheme 7) the reaction is initiated via intramolecular interaction between the metal-hydride and the acidic side-arm of the ligand (**13a**) giving rise to the arm-closed species (**13b**) along with the extrusion of a H_2 molecule. This is followed by the formation of the arm-open iridium alkoxide species (**13c**) resulting from the ligand exchange step. The subsequent β-hydride elimination gives rise to the carbonyl product **2a′** and **2′**

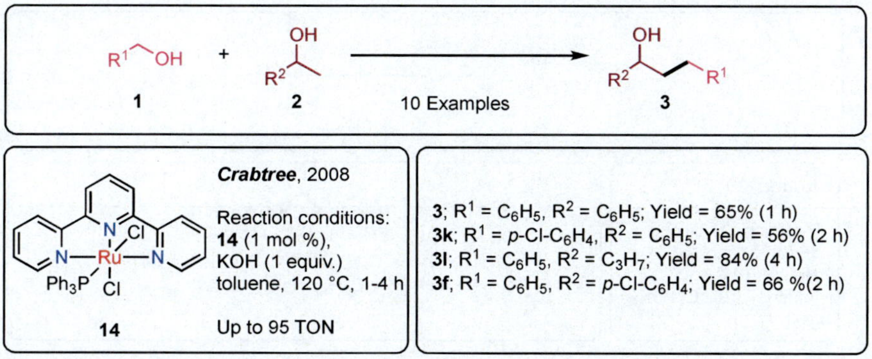

Scheme 8 Pincer-ruthenium catalyzed β-alkylation of secondary alcohols with primary alcohols [22]

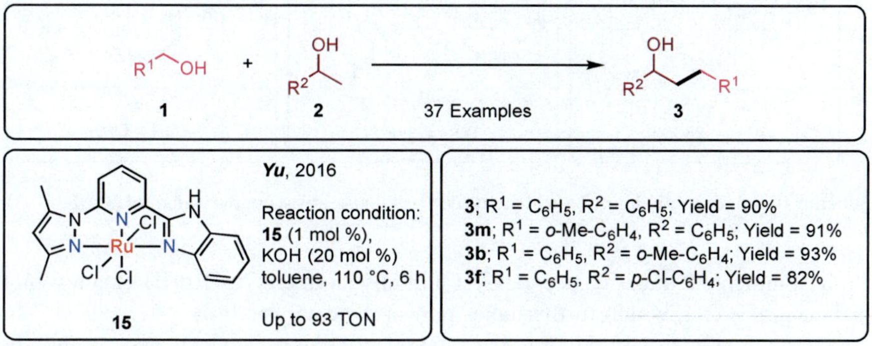

Scheme 9 The β-alkylation of secondary alcohols with primary alcohols catalyzed by the pincer ruthenium catalyst **15** [3]

along with the regeneration of the hydride species **13a**. This step is followed by base-mediated cross-aldol condensation of **2a′** and **2′** to give **4g** which on further hydrogenation gives rise to the β-alkylated ketone **4f** along with the species **13b**.

The NNN 2,2′:6′,2″-terpyridine (terpy) ligand is not only a stronger π-acceptor ligand in comparison with other *N*-donor ligands [42], but also demonstrates oxidatively and thermally robust properties [43]. Regardless of the incompatible nature of these ligands in transition metal complex systems, they mediate a variety of very important catalytic organic transformations [44]. In this context, Crabtree et al. have investigated the terpy based pincer-ruthenium **14** catalyzed β-alkylation of secondary alcohols with primary alcohols (Scheme 8) [22]. The **14** (1 mol%) catalyzed reaction in the presence of one equivalent KOH led to the quantitative yields of the alkylated products with a maximum of 95 TON with the concomitant formation of H_2O as a by-product (Scheme 8) [22].

Using Ru(III) **15**, Yu et al. have achieved up to 93 TON in the β-alkylation of secondary alcohols with primary alcohols (Scheme 9) [3, 45]. Detailed mechanistic

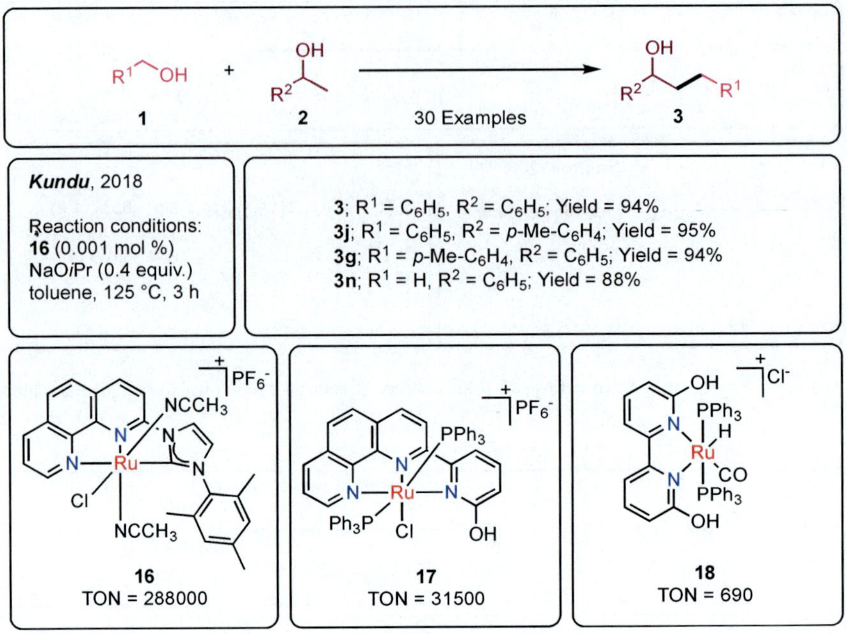

Scheme 10 Pincer-ruthenium **16–18** catalyzed *β*-alkylation of secondary alcohols [26]

studies indicated that the pre-catalyst initially converts to Ru(II) species in the presence of KOH, which further takes part in the catalysis [46].

Kundu et al. synthesized a series of pincer-ruthenium catalysts and applied them towards the *β*-alkylation of secondary alcohols with primary alcohols. They observed that the NHC ligand-based NNC pincer-ruthenium complex **16** resulted in very high turnover of 288,000 for the *C*-alkylation reaction upon significantly decreasing the catalyst loading (Scheme 10) [28]. Moreover, the catalytic system works efficiently under solvent-free conditions and demonstrated a vast substrate scope that includes aliphatic, aromatic, and heterocyclic alcohols. On similar lines, Kundu et al. demonstrated the role of metal-ligand cooperativity in the bifunctional catalyst **17** (0.1 mol%) in bringing about the *β*-alkylation of various secondary alcohols with very high selectivity while achieving 31,500 TON in the presence of 0.5 equivalents NaOH in refluxing toluene within 90 min of reaction (Scheme 10) [29]. Complex **18** has shown a moderate reactivity (690 TON) at a catalyst loading of 0.1 mol% in presence of 0.5 equivalents of KO^tBu in refluxing toluene within 45 min of reaction [27].

In 2021, Ghosh et al. synthesized an amido-functionalized ruthenium carbene complex **19** and have utilized it towards the catalytic *β*-alkylation of alcohols (Scheme 11) [47]. They also stated that a similar cationic picolyl functionalized NHC ruthenium complex displayed a comparable reactivity towards *C*-alkylation [47]. The catalyst **19** led to the formation of a hydride intermediate **19d** and was highly active up to about 120 h.

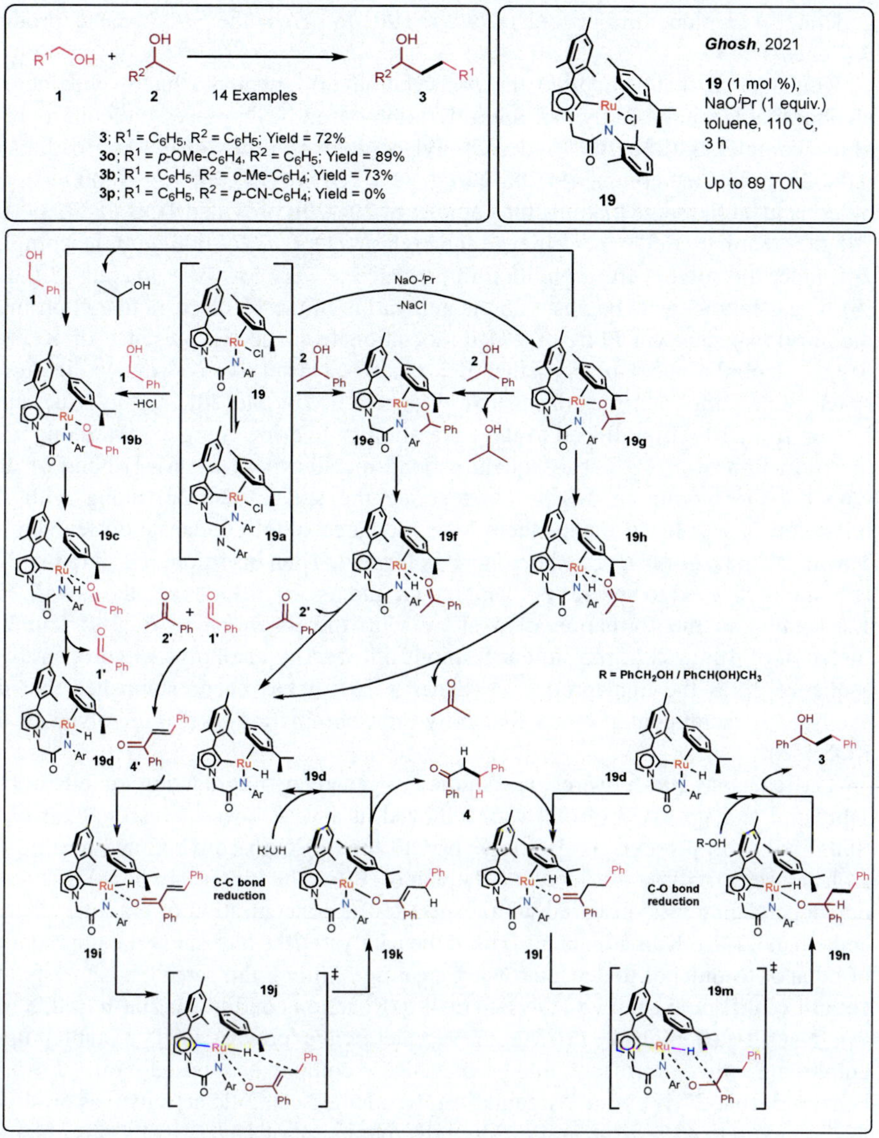

Scheme 11 The *C*-alkylation reaction catalyzed by ruthenium-carbene complex **19** [47]

The mechanism initiates with the formation of the Ru-alkoxide species **19b** and **19e** on the reaction of **19** with **1** and **2** respectively. The active species **19b** and **19e** undergoes the β-hydride elimination giving rise to **19c** and **19f** followed by the formation of the hydride species **19d** via extrusion of aldehyde **1′** and ketone **2′** respectively. The cross-aldol product **4′** reacts with the hydride species **19d** to generate **4** which further reacts with **19d** to give the *β*-alkylated product **3** (Scheme 11).

Kumar et al. have reported the *β*-alkylation of 1-phenyl ethanol with benzyl alcohol that is catalyzed by *bis*(imino)pyridine-based NNN pincer-ruthenium phosphine complexes (**20a-20d**) and carbonyl complexes (**20e-20h**). At a loading of 0.00025 mol%, the complexes **20a-20d** gave a TON of 356,000 to 320,000. On the other hand, at the same loading, the complexes **20e-20h** provided TON of 372,000 to 248,000. Kumar et al. have achieved the highest TON of 372,000 with the complex **20f** under the solvent-free condition in presence of very low base loading (2.5 mol %). The catalytic cycle begins with the generation of the 16-electron five-coordinate dichloride species via PPh_3 or CO dissociation from **20**. In presence of KO^tBu, benzyl alcohol **1** and 1-phenyl ethanol **2**, the 16-electron five-coordinate dichloride species undergoes salt-metathesis and gives rise to **20i** and **20j**. A *β*-hydride elimination from **20i/20j** will lead to the generation of hydride complex **20k** along with the formation of **1′/2′**. A subsequent σ-bond metathesis of the M–H bond of **20k** with the O–H bond of **1/2** will regenerate the species **20i/20j** along with the liberation of H_2. In an independent base mediated aldol condensation reaction of **1′** with **2′**, the *α*-*β*-unsaturated ketone **4′** is formed. Then insertion of **4′** into the Ru–H bond in **20 k** led to species **20l** which then undergoes σ-bond metathesis with **H_2**/**1/2** leading to the formation of **4** along with the regeneration of **20k /20i/20j**. Insertion of **4** into **20k** in a similar fashion followed by alcoholysis/hydrogenolysis sequence gives the final product **3** (Scheme 12). It has been established that the catalysis is metal-centered and that *β*-hydride elimination step is involved in the RDS [48].

The catalyst **20a-h** were also studied to catalyze upgradation of ethanol to 1-butanol and up to 335 TON was achieved at a very low catalyst loading of **2f** (0.05 mol%) in presence of 10 mol% NaOEt, at 140°C in conventional heating for 24 h at a selectivity of 90% towards 1-butanol. From the kinetic studies, a first order rate dependency was observed with respect to the concentration of both the catalyst and ethanol [48]. Kumar et al. also used the catalysts **20a-h** to study the upgradation of ethanol to butanol under microwave heating. Among the screened catalysts, the complex **20d** could achieve TONs up to 490 (Reaction conditions: **20a-h** (0.025 mol %), NaOEt (10 mol%), 110°C, 75 W microwave power) [49]. Notably, upon employing the corresponding pincer-Ru complexes based on 2,6-*bis*(-benzimidazole-2-yl) pyridine ligands, unprecedented catalytic activity was observed leading to up to 42% *n*-butanol yields and a maximum of 13,022 TON was obtained under identical conditions [49]. Attempts were also made to immobilize these complexes on various solid supports leading to catalytic systems that resulted in turnovers as high as 15,510 TON [49].

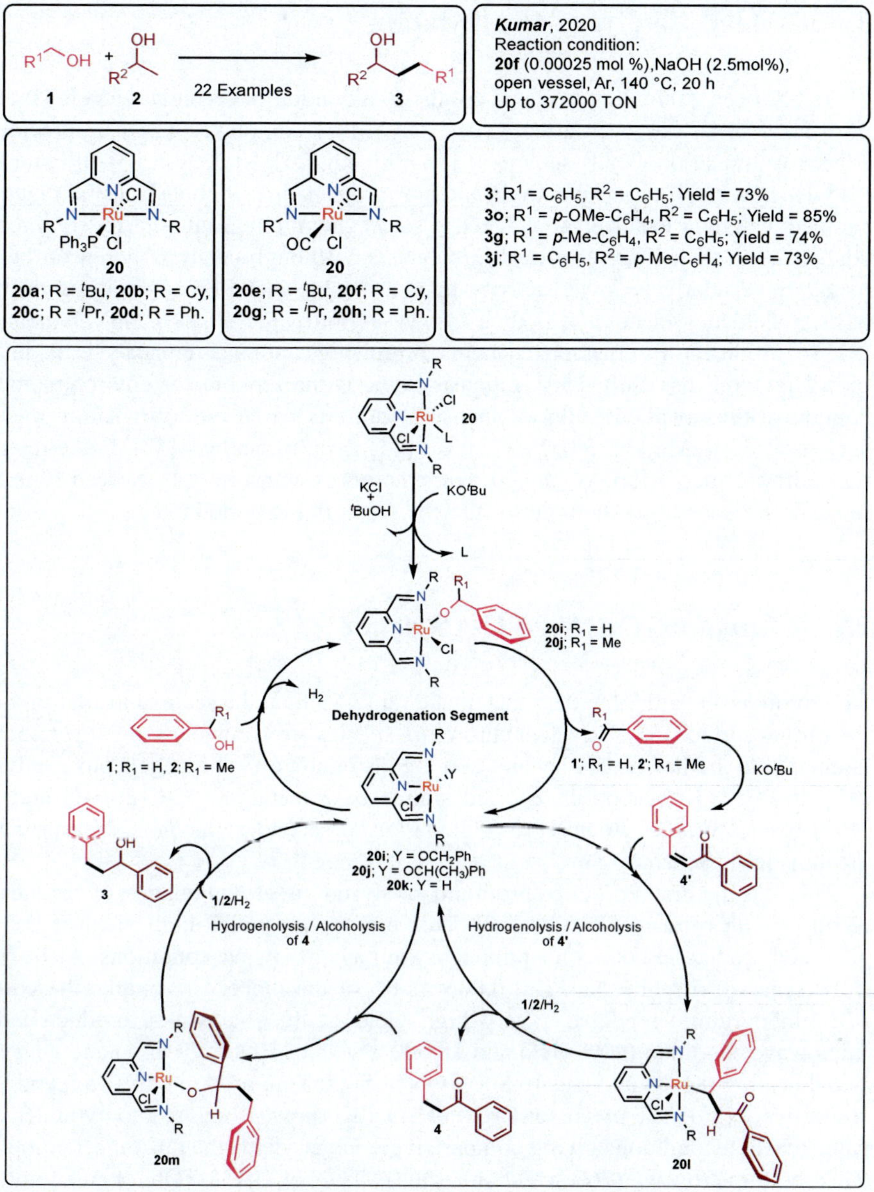

Scheme 12 The β-alkylation of secondary alcohols with primary alcohols catalyzed by the NNN pincer-ruthenium complexes **20a-h** [48, 49]

2 3d-Metal Catalyzed *C*-Alkylation

There has been exemplary cases where the β-alkylation of secondary alcohols with primary alcohols has been effectively catalyzed by complexes based on precious, expensive transition metals such as Ir [14–23], Rh [24], Ru [3, 13, 25–30], and Pd [31]. However, the expensive, toxic and hazardous nature of noble transition metals coupled with their localized abundance in select geographical areas poses significant challenges to their long-term uses in catalysis. Although catalysts based on base-metals have relatively lower activity than the noble metal-based catalysts on the basis of yield and selectivity [8], their easy accessibility [32–36, 50], high abundance [32–36, 50] in earth's crust makes them a promising candidate for catalysis. Rightly, global research has shifted its emphasis towards the lookout for environmentally benign, readily available, and inexpensive catalysts based on 3d-transition metals such as Cr [50], Mn [32, 33], Fe [35], Co [34], Ni [36], and Cu [37]. In the review that follows, such efforts on the current state-of-art with catalysts derived from 3d metals are discussed in the order of their position in the periodic table.

2.1 Chromium-Catalyzed C-*Alkylation*

Chromium is an earth-abundant metal, and ranks 13th on the scale of availability on the earth's surface [51]. Trivalent chromium species are known to show low toxicity compared to the hexavalent species and are also known for homogeneous catalysis [51]. In 2022, Kumar et al. reported simple base metal salt $CrCl_3.6H_2O$ and its NNN-pincer complex to catalyze β-alkylation of secondary alcohols with primary alcohols [50] (Scheme 13).

This was the first report on chromium-catalyzed direct β-alkylation of secondary alcohols with primary alcohols [50]. This protocol provided high yield of the β-alkylated product in both conventional as well as microwave conditions. At 140°C, in presence of 5 mol% NaO^{t}Bu, 0.005 mol% of the pincer-Cr complex based on *bis*(iminopyridine) rendered a very high yield of the β-alkylated product under microwave condition (90% yield and 18,000 TON at 12000 TOF h^{-1}) and conventional heating (84% yield and 16,800 TON at 5600 TOF h^{-1}) in only 1.5 h and 3 h respectively [50]. The precursor $CrCl_3.6H_2O$ also showed high reactivity under the same reaction conditions giving comparatively lower yields than its pincer complex **21** in both microwave (76% yield and 15,200 TON at 10133 TOh^{-1}) and conventional heating (79% yield and 15,800 TON at 5267 TOh^{-1}) [50].

To shed light on the mechanistic insight, various control experiments were performed. From EPR analysis, it was found that with both $CrCl_3.6H_2O$ and its pincer-Cr(III) complex **21**, the active species had Cr in its +2 oxidation state. The homogenous and well-defined molecular nature of the catalyst in the reaction mixture was evident from the inferences obtained from HRMS analysis, hot filtration test and mercury drop test. Deuterium labelling studies provided a KIE of 7.33 from

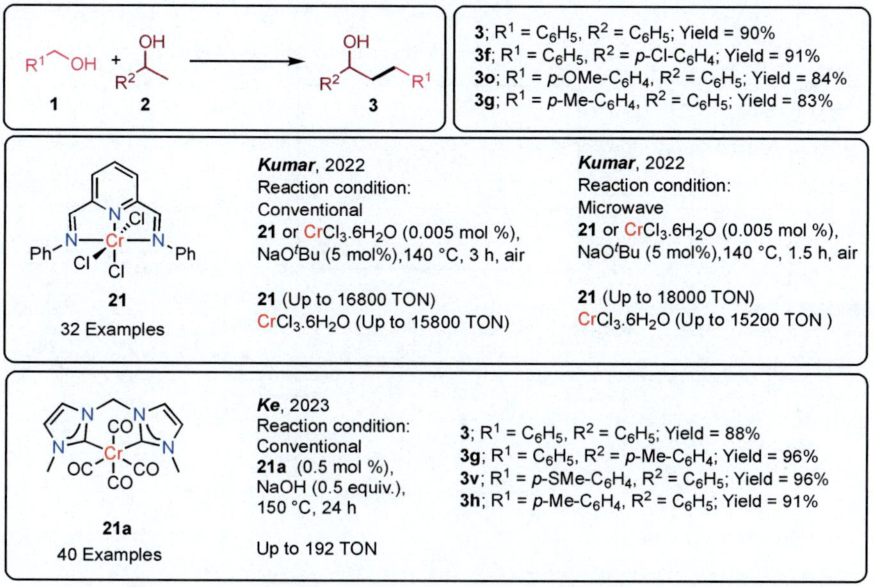

Scheme 13 Chromium-catalyzed β-alkylation of secondary alcohols with primary alcohols [50]

which the authors have inferred that the insertion of the α-alkylated ketone into the Cr–H bond is the RDS of the reaction. These observations formed the basis for the proposed mechanism.

The reaction initiates with the conversion of both the Cr(III) precursors **21** and **22**, to a Cr(II) active species **23** and **24** respectively, in the presence of NaO^{t}Bu (Scheme 14**).** Next step is the generation of alkoxides **25** and **26** as a result of salt metathesis of **23** and **24** with NaO^{t}Bu in the presence of **1** and **2** (Scheme 14). Salt metathesis is followed by a β-hydride elimination from **25/26** that gives the Cr-H species **27** along with the extrusion of corresponding benzaldehyde (**1′**) or acetophenone (**2′**). Alcoholysis of **27** with **1/2** would result in evolution of hydrogen along with the regeneration of **25/ 26**. The β-hydride elimination is followed by a base mediated uncatalyzed condensation of **1′** and **2′** which results in the formation of **4′**. Then insertion of **4′** into the Cr–H bond in **27** led to species **28** which then undergoes σ-bond metathesis with **H_2/1/2** leading to the formation of **4** along with the regeneration of **27/25/26**. Insertion of **4** into **27** in a similar fashion followed by alcoholysis/hydrogenolysis sequence gives the final product **3** [50].

In 2023, Ke have reported a *bis*-(*N*-heterocyclic carbene)-based Cr(0) **21a** (Scheme 13) catalyzed C-alkylation of primary and secondary alcohols, which followed a hydrogen auto-transfer or hydrogen-borrowing strategy. They could achieve up to 96% yield of the β-alkylated product at a catalyst loading of 0.5 mol % in presence of 0.5 equivalents of NaOH, at 150°C after 24 h. From both experimental and computational studies, they have explained the fact that Cr

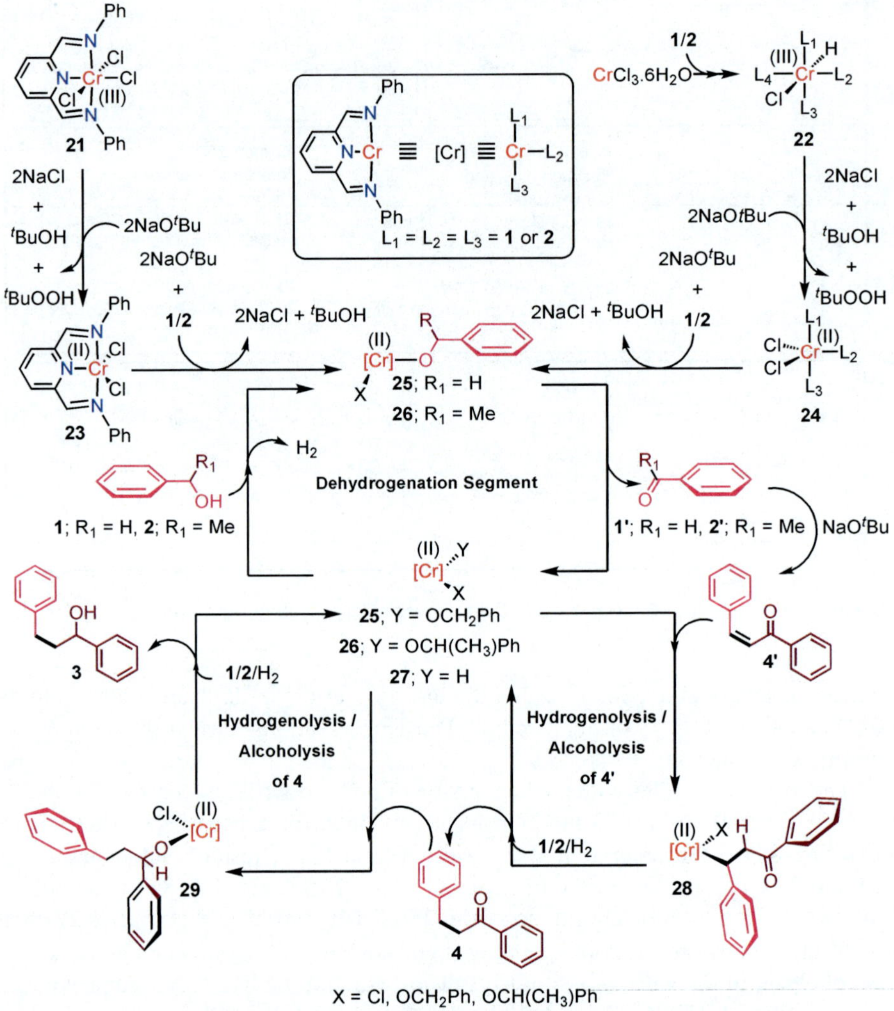

Scheme 14 Mechanism of Cr-catalyzed β-alkylation proposed by Kumar et al. [50]

(0) can show better reactivity than the Cr(III) /Cr(II) species due to the absence of dπ-pπ interaction in the metal-alkoxy species [52].

2.2 *Manganese-Catalyzed* C-*Alkylation*

The third most abundant metal, manganese, has proven to be a great alternative for several (de)hydrogenation and related reactions. In the context of β-alkylation of secondary alcohols by primary alcohols, Yu et al. in 2018 first reported the

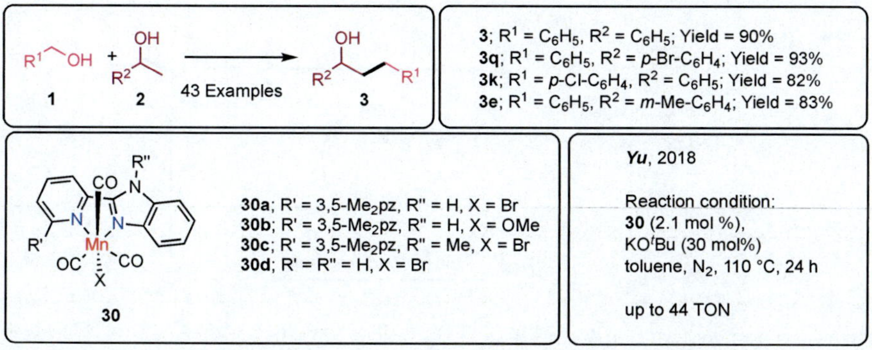

Scheme 15 Manganese-catalyzed β-alkylation of secondary alcohols with primary alcohols [32]

Mn-catalyzed β-alkylation reaction, where they have synthesized a panel of phosphine-free Mn(I) complexes bearing a pyridyl-supported pyrazolyl-imidazolyl ligand (pz) [32]. The synthesized complexes **30a-d** were then employed for direct β-alkylation of secondary alcohols by primary alcohols (43 examples) and 92% yield of the β-alkylated product was obtained using 2.1 mol% of catalyst loading in presence of KO^{t}Bu (30 mol%) at 110°C with water as the sole by product. This protocol was further extended for one pot synthesis of flavan derivatives from simple alcohols and for β-alkylation of cholesterol and its derivatives (Scheme 15).

Transition metal-hydride species are known to participate in the hydrogen-borrowing approach. To gain the mechanistic insight, an independent reaction was performed where complex **30b** was treated with KO^{t}Bu or KO^{i}Pr in refluxing ethanol or isopropanol to investigate the formation of corresponding Mn(I)-Hydride species but unfortunately no formation of hydride species was detected. A simplified reaction mechanism was proposed based on the reaction pathway where Mn (I) species oxidizes the primary **1** and secondary alcohols **2** to their corresponding aldehyde **1′** and ketone **2′** along with the generation of Mn-hydride species. The in-situ generated aldehyde and ketone will then undergo base-mediated cross aldol condensation and leads to the formation of α,β-unsaturated ketone as an intermediate which on subsequent transfer hydrogenation from Mn-hydride species will form the targeted β-alkylated product with water as the sole by-product [32] (Scheme 16).

In a similar parallel investigation, Rueping et al. have synthesized a series of new Mn(I) complexes (**31–33**) bearing PNN and PNP pincer ligands and employed them for successfully catalyzing the β-alkylation of secondary alcohols by primary alcohols through double hydrogen auto-transfer [33]. They have obtained 99% conversion using 1 mol% of PNN pincer-Mn(I) catalyst **31a** in presence of KO^{t}Bu (25 mol %) at 135°C and a broad range of substrates (29 examples) were converted into higher-value alcohols in good yields (Scheme 17). Mechanistic investigations pointed out the involvement of hydrogen auto-transfer mechanism in which the non-innocent ligand plays a significant role (Scheme 18).

In 2023, Kumar et al. have developed another series of Mn(I) complexes **36** based on *bis*(imino) pyridine ligands of the type ^{R2}NNN (R = Cy, ^{i}Pr, Ph, *p*-FC$_6$H$_4$)

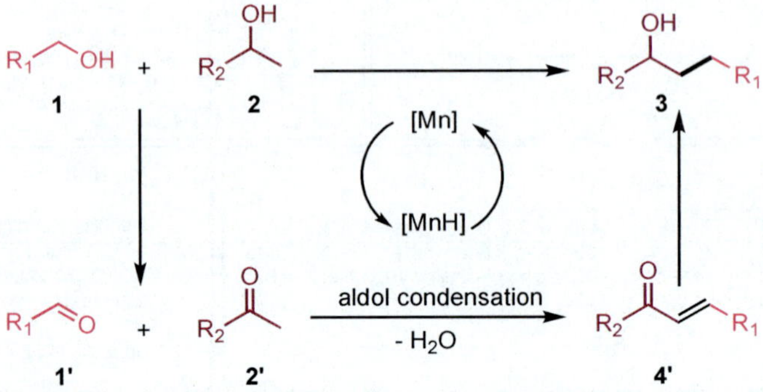

Scheme 16 Mechanism of Mn-catalyzed β-alkylation proposed by Yu [32]

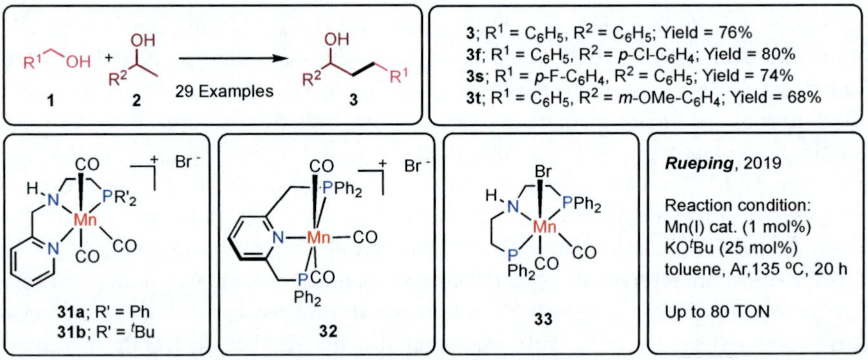

Scheme 17 Manganese-catalyzed β-alkylation of secondary alcohols with primary alcohols [33]

[53]. The complexes in the solid form were found to be present as an NN bidentate-Mn(I) *tris* carbonyl species while in solution, there is a gradual loss of one CO molecule that resulted in the formation of NNN tridentate-Mn(I) *bis* carbonyl species which eventually disproportionates to an NNN pincer-Mn(II) dibromide and an NNN pincer-Mn(0) complex (Scheme 19). All the complexes were then employed for catalyzing the β-alkylation of secondary alcohols by primary alcohols (35 examples) at 0.05 mol% catalyst loading in presence of 2.5 mol% of NaO^{t}Bu under both conventional (12 h) and microwave heating (2 h). Among the considered complexes, $^{p\text{-FC6H4}}$NNNMn$(CO)_2$Br has given the maximum yield of 92% (1840 TON) under conventional heating while yield of 85% (1700 TON) was obtained under microwave conditions.

Kumar et al. demonstrated that both Mn(I) species **36** and Mn(II) species **37** are capable of catalyzing the β-alkylation, with the latter being less active. Deuterium labelling studies indicates the involvement of C–H bond activation in the rate determining step (RDS) with a KIE of 9.0. Further, based on IR, EPR, and HRMS analyses, two catalytic cycles were proposed involving NNN pincer-Mn(I) *bis* carbonyl species and a NNN pincer-Mn(II)/Mn(0) couple [53].

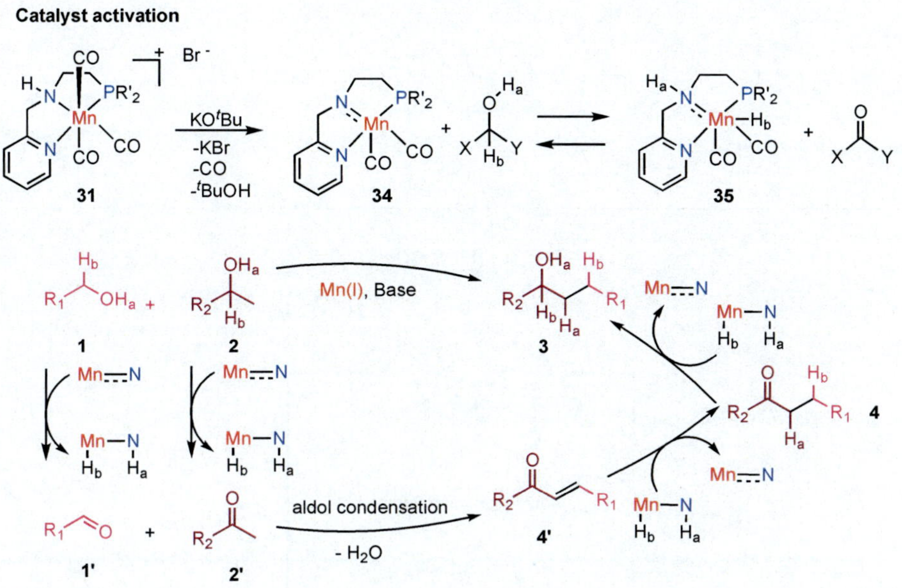

Scheme 18 Mechanism of Mn-catalyzed β-alkylation proposed by Rueping [33]

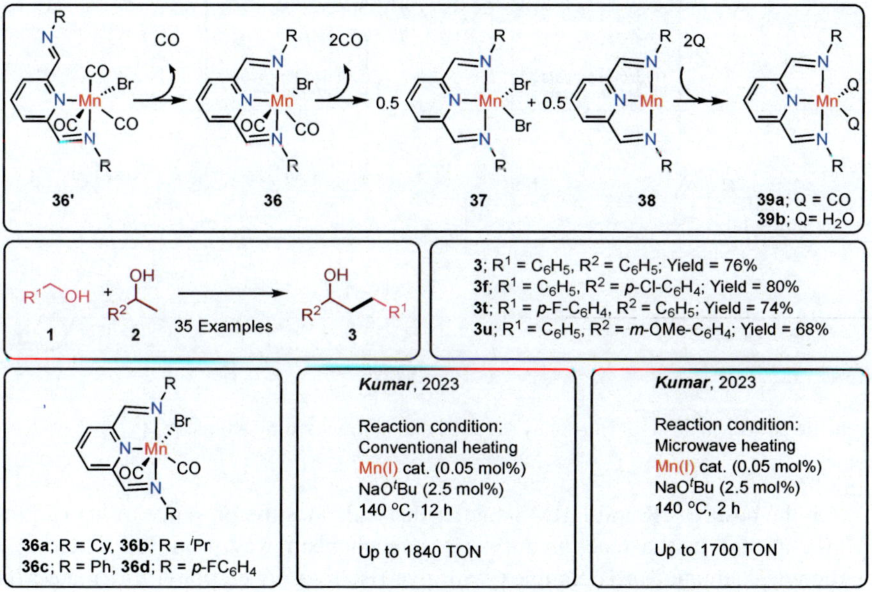

Scheme 19 Manganese-catalyzed β-alkylation of secondary alcohols with primary alcohols [53]

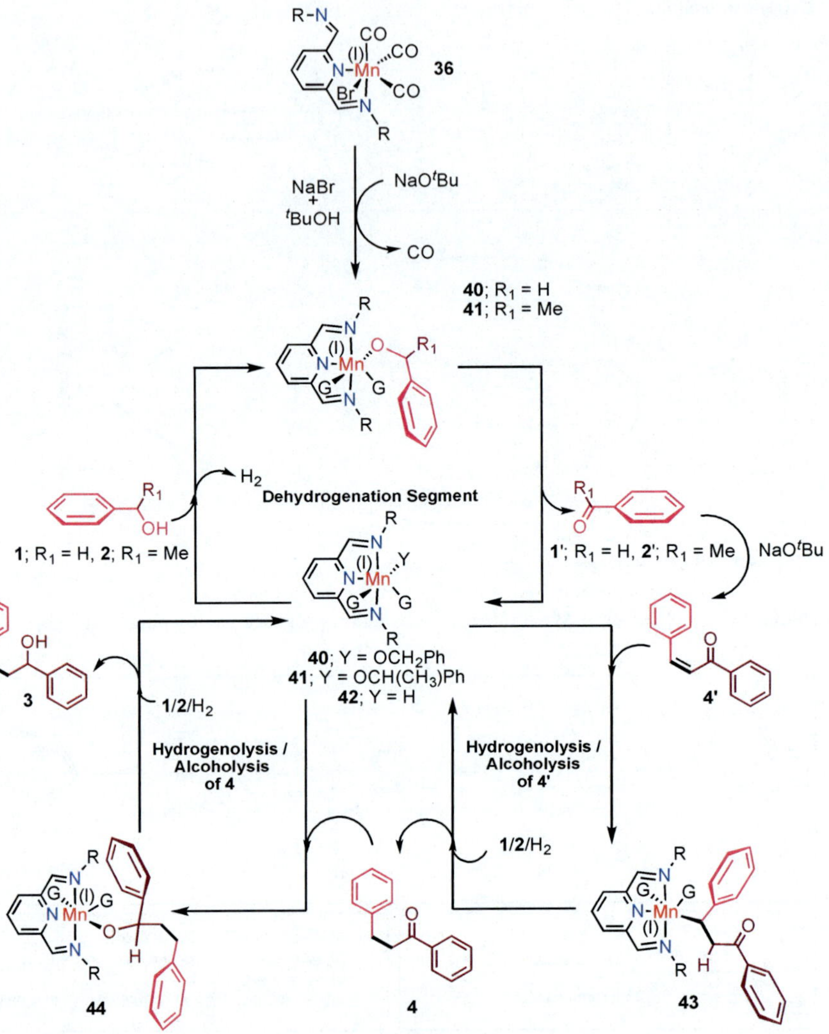

Scheme 20 Mechanism of Mn-catalyzed β-alkylation proposed by Kumar [53]

On the basis of IR and HRMS studies that indicates the presence of *bis*-carbonyl Mn(I) species in the reaction mixture, a mechanism was proposed where a *tris* carbonyl bidentate-Mn(I) complex will give rise to a *bis* carbonyl tridentate-Mn(I) species in presence of NaO^{t}Bu and **1/2** along with the generation of NaBr (Scheme 20). A β-hydride elimination from **40/41** will lead to the generation of hydride complex **42** along with the formation of **1′/2′**, which are both identified by NMR in a separate dehydrogenation of **1** and **2**. A subsequent σ-bond metathesis of the M–H bond of **42** with the O–H bond of **1/2** will regenerate the species **40/41** along with

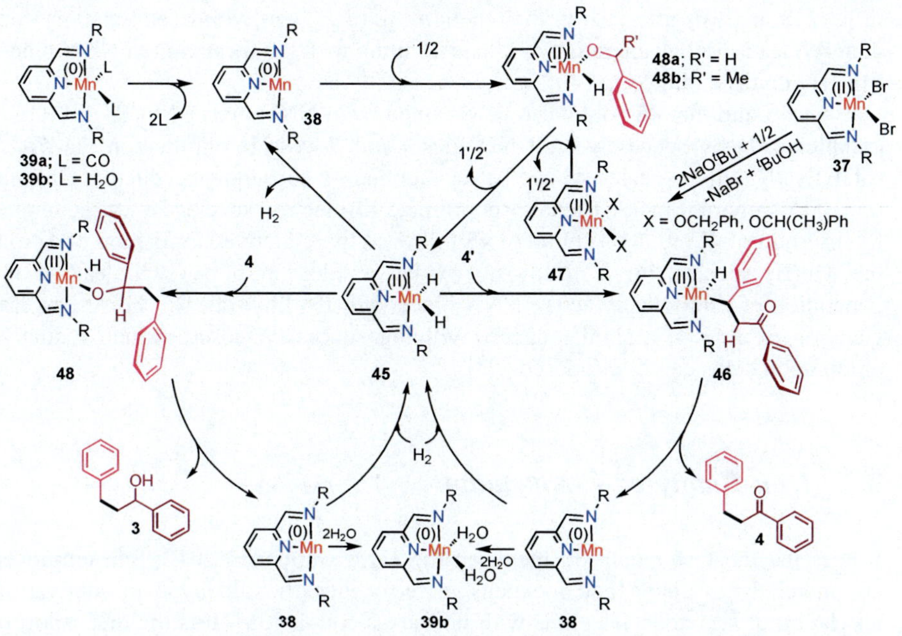

Scheme 21 Alternative mechanism of Mn-catalyzed β-alkylation proposed by Kumar [53]

the liberation of H_2. (detected by GC). In an independent base mediated aldol condensation reaction of **1′** and **2′**, the α-β-unsaturated ketone (**4′**) is formed (detected by HRMS analysis). This is followed by the insertion of the double bond of α-β-unsaturated ketone **4′** into M H bond of **42** to generate **43** which is then followed by another σ-bond metathesis of Mn–C bond of **43** with either H_2 (hydrogenolysis) or O–H bond of **1/2** (alcoholysis) to yield the corresponding ketone **4** while regenerating **42/40/41**. A similar cycle involving insertion and hydrogenolysis/alcoholysis was proposed for the transformation of **3** to **4**.

While the EPR analysis of the reaction mixture after heating it to 1 h indicates its EPR silent nature, the magnetic moment measurement using Evans method indicates the presence of paramagnetic species that corresponds to Mn(0) species with $\mu_{eff} = 3.80\ \mu_B$ (theoretical value $\mu_{eff} = 3.87\ \mu_B$). This along with the observation of lower yields of **3** obtained in independent experiments catalyzed by **37** under the optimized conditions allowed Kumar et al. to infer that, apart from a significant contribution from the Mn(I) species **36** (Scheme 20), there is also a slight contribution of a Mn(II)/Mn(0) couple (Scheme 21) in the catalytic β-alkylation with Mn (0) species **38** serving as the resting state. Not surprisingly, adducts corresponding to both Mn (II) and Mn(0) species are found to be present in the reaction mixture as indicated by HRMS analysis. The NNN pincer-Mn(0) diaquo complex **39b** will lead to the generation of catalytically active NNN pincer-Mn(II) alkoxide−hydride species **48a/48b** via formation of three-coordinate 13- electron NNN pincer-Mn(0) species **38**. Interestingly, this three-coordinated Mn(0) species was found to appear in HRMS analysis after 1 h in the reaction mixture. A β-hydride elimination from **48a/48b** leads to the generation of **1′** and **2′** along with the formation of NNN pincer Mn(II) dihydride **45**.

It was found that **45** could also be obtained from NNN pincer-Mn(II) dibromide complex (**37**) by successive salt metathesis and β-hydride elimination via **47**. A catalytically active species **48** can be regenerated by subsequent reductive elimination of hydrogen from **45**. The three coordinate 13-electron species **38** was generated by insertion of **4′** into M–H bond of **45** followed by reductive elimination of **4** from the Mn(II) species **46**. A follow-up oxidative addition of H_2 will lead to the generation of catalytically active NNN pincer-Mn(II) dihydride **45**. For the hydrogenation of **3** to **4,** a similar cycle involving successive insertion and reductive elimination steps can be proposed [53].

2.3 *Iron-Catalyzed* C-*Alkylation*

Iron is the most abundant 3d metal on the earth's crust. In 2012, Sun employed commercially available and inexpensive ferrocenecarboxaldehyde to catalyze β-alkylation of secondary alcohols with primary alcohols [35]. It is the first report of iron mediated β-alkylation of secondary alcohols with primary alcohols. In this protocol, 99% yield of β-alkylated product with 97% selectivity was observed when 5 mol% of ferrocenecarboxaldehyde was allowed to react with benzyl alcohol and 1-phenyl ethanol in presence of 20 mol% NaOH, in *p*-xylene, at 130°C for 12 h. The developed synthetic protocol is generic in nature and the ferrocenecarboxaldehyde-catalyzed β-alkylation of various substrates were conveniently achieved (Scheme 22).

A mechanism (Scheme 23) for the β-alkylation of secondary alcohols with primary alcohols was proposed that commences with the base-mediated nucleophilic attack of alkoxy anions to the aldehyde group of **7b**. It is followed by the generation

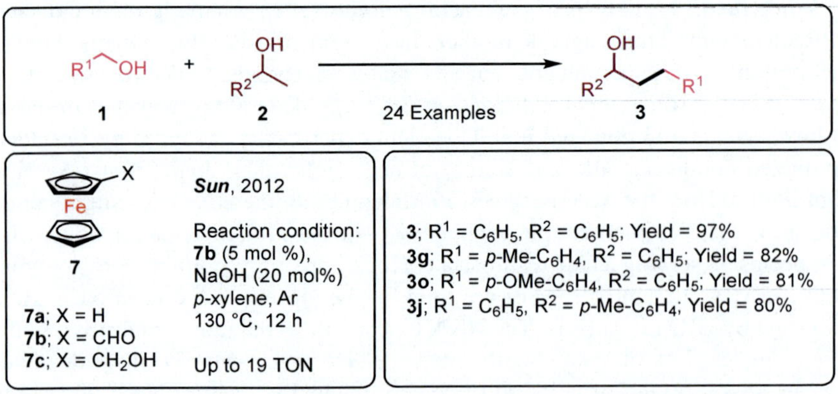

Scheme 22 Ferrocene-catalyzed β-alkylation of secondary alcohols with primary alcohols [35]

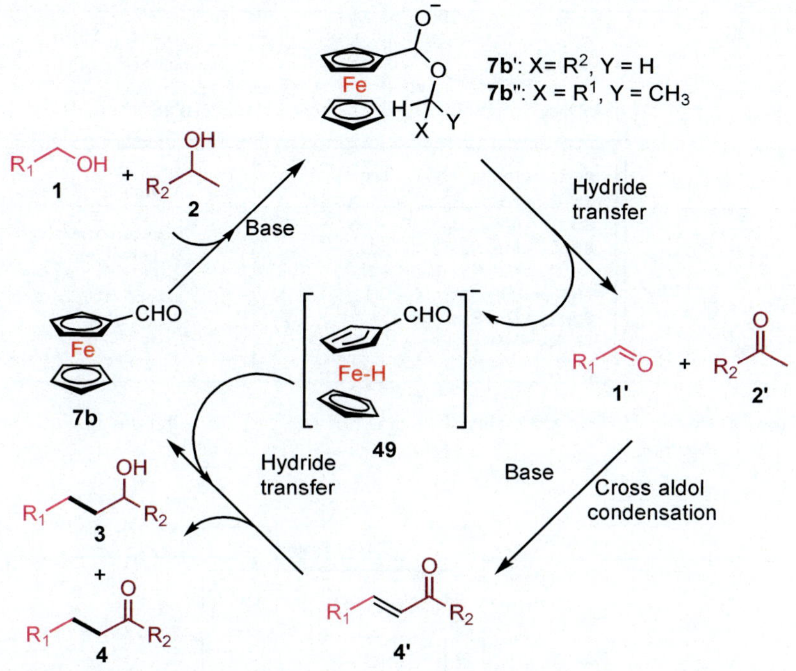

Scheme 23 Mechanism of Fe-catalyzed β-alkylation proposed by Sun [35]

of iron hydride species along with the oxidation of primary alcohol to aldehyde and secondary alcohol to ketone. In the next step, base-mediated cross-aldol condensation gives rise to an α,β-unsaturated ketone, where transfer hydrogenation from the hydride species generates the desired product along with the regeneration of the catalyst.

2.4 *Cobalt-Catalyzed C-Alkylation*

Cobalt is found in earth's crust only in a chemically combined form. Among the halide salts four cobalt halide salts (CoF_2, $CoCl_2$, $CoBr_2$, CoI_2) are known. $CoCl_2$ is easily available and it is hydrated and anhydrous form shows red and blue color, respectively. In 2017, Kempe et al. first reported the Co-catalyzed β-alkylation of alcohols using a PN_5P pincer-Co complex **8** (5 mol%) in the presence of 1.1 equiv of KHMDS (potassium hexamethyldisilazane) to get up to 90% yield of the β-alkylated product. The catalyst system based on **8** was applicable to a broad range of substrates. Notably, the β-alkylation was driven by **8** based on a hydrogen-borrowing strategy [34] (Scheme 24).

In 2021, Ding et al. reported a Co catalyst **50** (0.7 mol%) supported by a $^{iPr}PPPN^{H}Py^{Me}$ terdentate ligand for the β-alkylation of alcohols in a sealed vessel

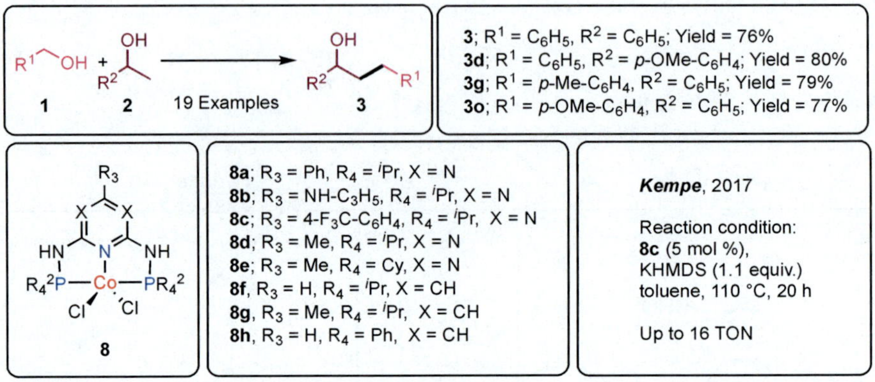

Scheme 24 Pincer-Co catalyzed β-alkylation of secondary alcohols with primary alcohols [34]

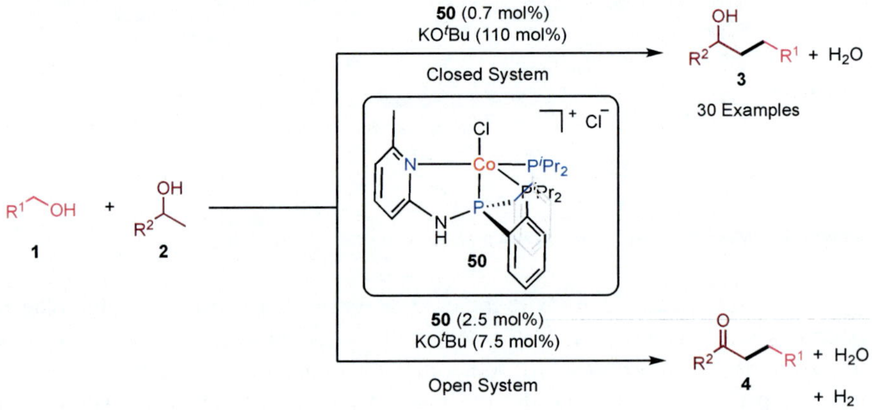

Scheme 25 Pincer-Co catalyzed β-alkylation of secondary alcohols with primary alcohols [57]

which gave up to 80% yield at 110°C at a very high loading (110 mol%) of KO^{t}Bu [54]. When the same reaction was carried out with 7.5 mol% KO^{t}Bu and 2.5 mol% of the catalyst **50** in an open vessel, the ketone product **4** was obtained [55] (Scheme 25).

Kumar et al. recently reported that cobaltous chloride (0.01 mol%) efficiently achieves the catalytic β-alkylation of alcohols with high yields (up to 87%) and unprecedented turnovers (ca. 8,700) in the presence of only 2.5 mol% of NaO^{t}Bu [56] (Scheme 26). Instant formation of nano-particles was observed at higher loadings of cobaltous chloride (1 mol%) which were confirmed by SEM and TEM analysis. The nano-particles formed were found to be sensitive in air and resulted in poor yields (25%) of β-alkylated products when utilized to catalyze the reaction in air. Hot-filtration experiments and mercury-drop tests were indicative of the molecular nature of the reaction mixture. EPR silent nature of the reaction mixture suggested the involvement of Co(II) species in an octahedral environment. Deuterium labelling studies demonstrated a KIE of 1.61 which indicated the involvement

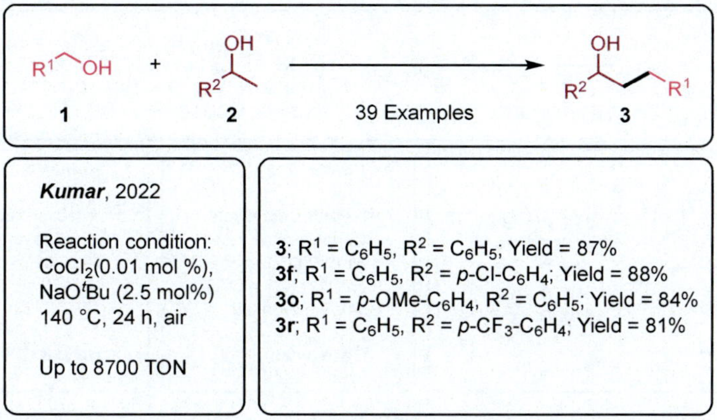

Scheme 26 Cobaltous chloride-catalyzed β-alkylation of secondary alcohols with primary alcohols [56]

of C-H activation in the cobaltous chloride-catalyzed β-alkylation of alcohols. Kinetic studies were performed which suggested a first order dependency of rate on the concentration of cobaltous chloride and NaO^{t}Bu. On the other hand, a nonlinear dependency of rate on the concentration of benzyl alcohol and 1-phenyl ethanol was observed. Based on HRMS analysis, GC studies and NMR experiments, a mechanism similar to that proposed for the corresponding reaction catalyzed by pincer-Co complex **51** as shown in Scheme 28 has been proposed.

Kumar et al. reported the synthesis of a series of NNN pincer-Co complexes based on *bis*(imino)pyridine ligands to yield complexes of the type (^{R2}NNN)$CoCl_2$ (**51a**, R = ^{t}Bu; **51b**, R = Cy; **51c**, R = ^{i}Pr; **51d**, R = Ph; **51e**, R = *p*-F-C_6H_4). While cobaltous chloride at a 0.0025 mol% loading in the presence of 2.5 mol% of NaO^{t}Bu at 140°C gives about a 66% yield (26,400 TON at 1100 TOF h^{-1}) in the β-alkylation of 1-phenyl ethanol with benzyl alcohol, its corresponding pincer complex (^{iPr2}NNN)$CoCl_2$ (0.0025 mol%) is highly productive (ca. 1.3-fold vs $CoCl_2$) and results in up to 85% yield (ca. 34,000 TON at 1417 TOF h^{-1}) of the β-alkylated product under otherwise identical conditions (Scheme 27) [57]. All of the considered pincer-Co complexes have been found to efficiently catalyze β-alkylation of 1-phenylethanol with benzyl alcohol. In particular, (^{iPr2}NNN)$CoCl_2$ has demonstrated the best activity in air at 140°C. The corresponding reaction under microwave conditions (140°C at 75 W) was complete in only 2 h with a comparable yield (83%, 33,200 TON) and far better TOF (16,600 TOF h^{-1}). Under similar conditions, the precursor $CoCl_2$ (0.0025 mol%) gave relatively poor results under microwave heating (ca. 61%, 24,400 TON at 12200 TOF h^{-1}) [57].

The ESR analysis and magnetic moment studies via Evans method suggested the presence of octahedral Co(II) species in the reaction mixture [56]. The NNN pincer complex **51/52** upon treatment with NaO^{t}Bu in the presence of **1** or **2** produces octahedral complex **53/54** via a salt metathesis reaction as confirmed from HRMS study. The first step of the catalytic cycle involves the β- hydride elimination from

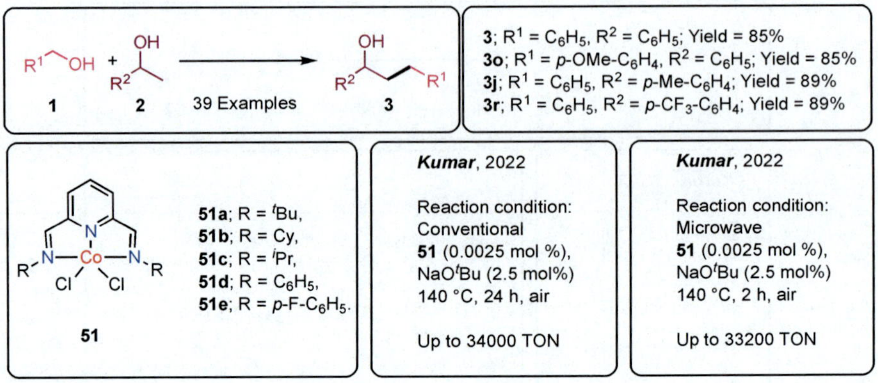

Scheme 27 Pincer-cobalt catalyzed β-alkylation of secondary alcohols [57]

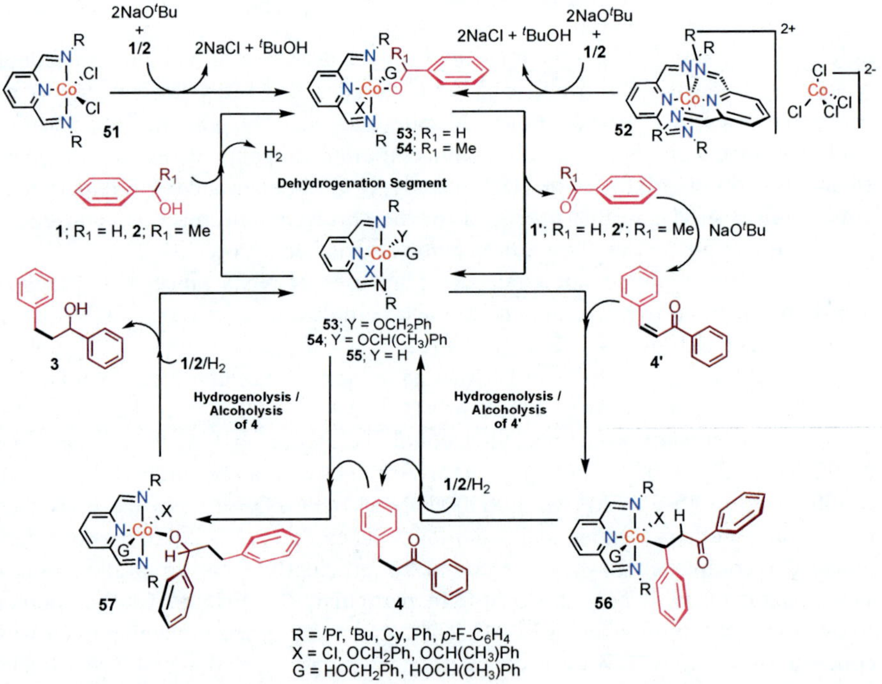

Scheme 28 Mechanism of pincer-cobalt catalyzed β-alkylation proposed by Kumar et al. [57]

53/54 to produce the carbonyl species **1′** and **2′**. The formation of **1′** and **2′** was confirmed by independent NMR experiments (Scheme 28). The non-appearance of Co adducts with either **4** or **3** in the HRMS study leads to the conclusion that the alcohol dehydrogenation fragment and more specifically the β-hydride elimination is the rate determining step, which is in good agreement with the reported primary KIE of 6.14.

The species **55** further interact with alcohols (**1/2**) to get back the active species **53/54** through the σ-bond metathesis of O–H bond of alcohols with the Co–H bond of **55** along with the elimination of H_2. The evolution of H_2 in these reactions was confirmed by gas chromatography (GC) studies.

The insertion of aldol product **4′** to Co–H bond of octahedral complex **55** yielded **56** which on further alcoholysis with **1/2** or hydrogenolysis with H_2 gives rise to **4** (Scheme 28). A similar cycle involving insertion of **4** and alcoholysis/hydrogenolysis via **55** → **57** → **53/54/55** was proposed for the transformation of **4** to **3** (Scheme 28). The reaction was unaffected by either a closed-vessel or an open-vessel condition which indicates that alcoholysis is a major contributor to product formation.

2.5 *Nickel-Catalyzed C-Alkylation*

Nickel ranks 22nd among the earth's most abundant element and is the seventh most abundant transition metal [58]. It is a silver white crystalline metal that occurs in meteors or combined with other elements in ores. The salts of nickel are cheap and easily available, with $NiCl_2$ being 58 times cheaper than $PdCl_2$ [59]. The first report on nickel-catalyzed alkylation of alcohols was described by Lang et al. in 2019, where they studied the utility of hexanickel cluster embedded with a 4,6-dimethylpyrimidine-2-thione ligand (**9**) in catalyzing the C–C bond formation based on hydrogen borrowing methodology (Scheme 29) [36]. The Ni(II) catalyst (**9**) displayed good activity towards synthesis of α-alkylated ketones, α,β-unsaturated ketones, and quinolines under slightly different optimized conditions. Performing the reaction at higher temperatures and under sealed conditions afforded complete hydrogenation of α,β-unsaturated ketones and resulted in corresponding alcohol.

Later, Balaraman demonstrated the β-alkylation of alcohols using $NiBr_2$/TMEDA (1:1) in the presence of 1 equiv. of KOH at 130°C using *n*-octane as solvent (Scheme 30) [60]. The $NiBr_2$/TMEDA (1:1) tolerated a broad range of substrates including aromatic, cyclic, acyclic, aliphatic alcohols and the scope was extended towards the double alkylation of cyclopentanol with various alcohols. The preliminary mechanistic studies and isotopic labelling experiments pointed towards the involvement of borrowing hydrogenation in the C–C coupling with the formation of water as a by-product.

In 2021, Kumar et al. reported a synthesis of a series of NNN pincer-Ni(II) complexes (**60a-e**) based on *bis*(imino)pyridine ligands ((^{R2}NNN)$NiCl_2(CH_3CN)$; R = ^{i}Pr, ^{t}Bu, Cy, Ph and *p*-F-C_6H_4) and employed them for the catalytic β-alkylation of various primary alcohols with secondary alcohols at 140°C [61]. Among all the complexes studied, very high TONs (up to 18,400) were been observed for the alkylation of 1-phenyl ethanol with benzyl alcohol catalyzed by 0.005 mol% of

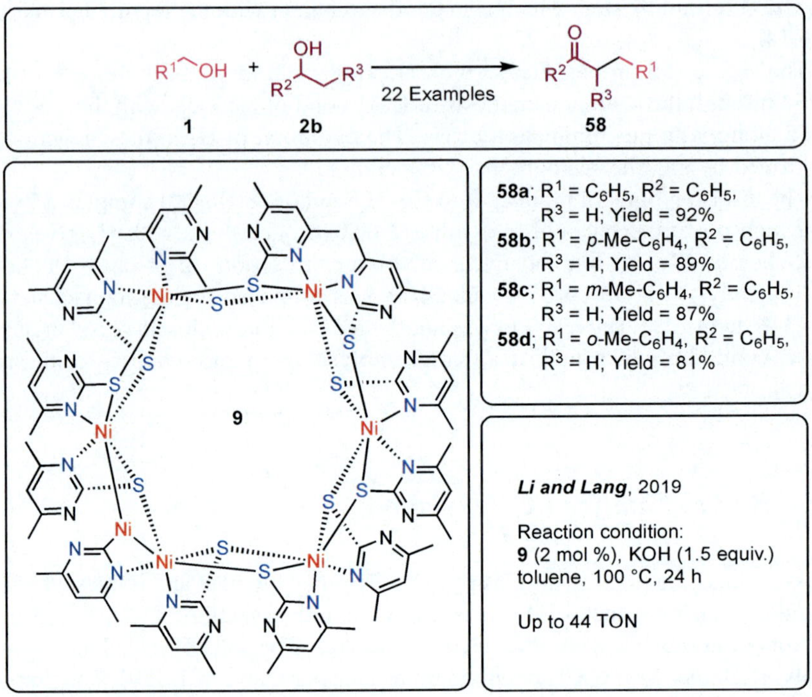

Scheme 29 Cluster nickel-catalyzed β-alkylation of secondary alcohols with primary alcohols [36]

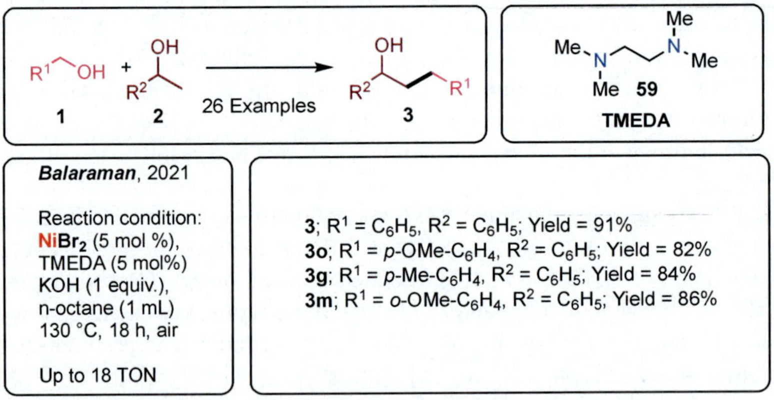

Scheme 30 Nickel bromide-catalyzed β-alkylation of secondary alcohols [60]

(^{Ph2}NNN)NiCl$_2$(CH$_3$CN) (**60d**) in the presence of 5 mol% NaO^{t}Bu at 140°C after 24 h (Scheme 31).

HRMS studies pointed towards the involvement of α,β-unsaturated ketone as a key intermediate. The HRMS analysis of the reaction mixture consisted of peaks which were assigned to various adducts of **60d** with **1/2** and on the other hand EPR

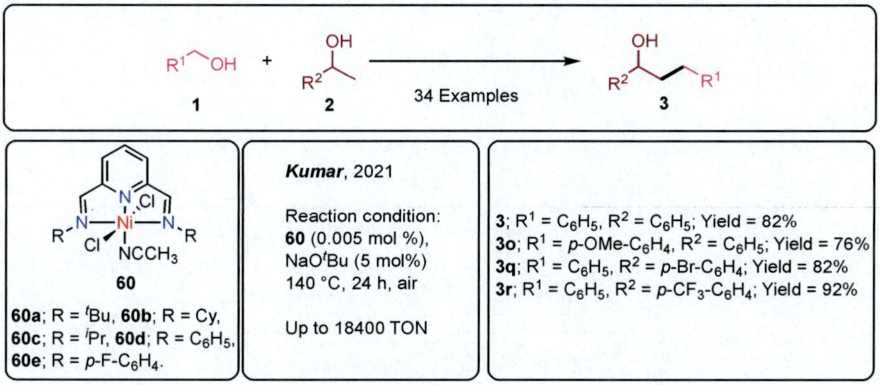

Scheme 31 Pincer-nickel catalyzed β-alkylation of secondary alcohols with primary alcohols [61]

studies were indicative of the existence of octahedral Ni(II) species in the mechanistic cycle. The aldol condensation being the rate determining step was apparent from the kinetic studies with the overall reaction showing zero-order dependence of rate on the catalyst concentration and first-order rate dependence on concentration of base and substrates.

Based on above observations, Kumar et al. described a plausible mechanism (Scheme 32) where the first step is the dissociation of acetonitrile from **60** or the ligand from **61**, which on reaction with NaO^{t}Bu, in the presence of benzyl alcohol (**1**) and 1-phenyl ethanol (**2**) yields **62** and **63** respectively (Scheme 32). The Ni-H species **64** is obtained via the β-hydride elimination of **62** and **63** resulting in the formation of benzaldehyde (**1′**) and acetophenone (**2′**) (Scheme 32). The aldol condensation of **1′** and **2′** affords the α,β-unsaturated ketone (**4′**). A subsequent insertion of the C-C double bond present in **4′** into the Ni–H bond of **64** generates the intermediate **66** which on further on hydrogenolysis with H_2 or alcoholysis with **1/2** results in ketone **4** along with the regeneration of **64/62/63.** Similar to this pathway, the insertion of **4** into **65** and subsequent hydrogenolysis/alcoholysis of **65** with **1/2/** **H_2** results in the formation of desired product **3** with the regeneration of **62/63/64** (Scheme 32) [61]. The fact that **3** is observed as the major product in these reactions that are performed in an open-vessel, it is likely that alcoholysis is the preferred pathway here.

Similar to Lang, Adhikari et al. recently reported the dehydrogenative cross-coupling of primary and secondary alcohols to yield α-alkylated ketones using nickel azo-phenolate catalyst **67** [62]. The Ni(II) catalyst demonstrated a wide substrate scope of primary alcohols including challenging aliphatic alcohols where they obtained up to 68% yield (Scheme 33). The process undergoes a hydrogen borrowing pathway, as proved by deuterium labelling experiments. The involvement of a radical during the reaction is proved by the addition of radical scavenger to the reaction mixture.

Very recently, Rasappan et al. established the multicomponent reaction employing readily available nickel chloride hexahydrate (Scheme 34). They were

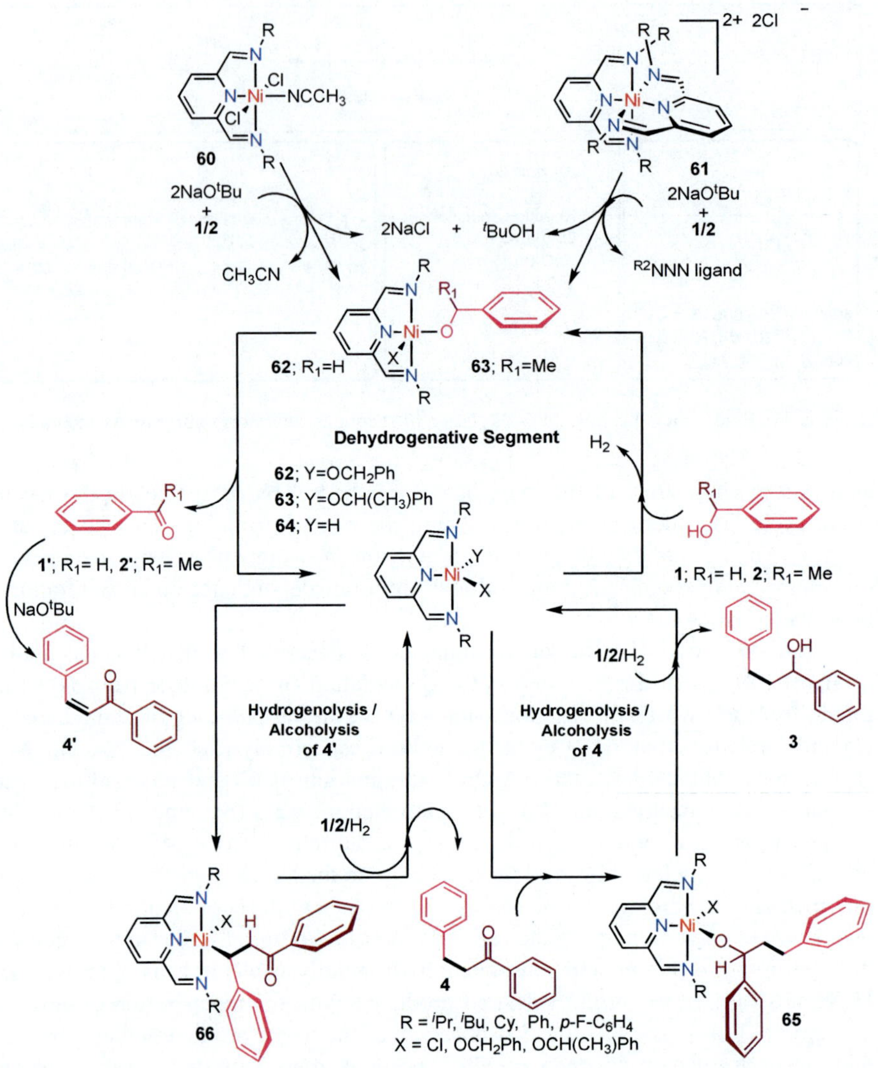

Scheme 32 Mechanism of pincer-nickel catalyzed β-alkylation proposed by Kumar [61]

successful in synthesizing the β,β-disubstituted ketones from alkylation of ketones with primary alcohols and phenyl boronic acid, in the presence of 1 mol% of catalyst loading and two equivalents of KO^{t}Bu. The catalytic system was tolerant to a broad spectrum of alcohols, ketones and boronic acids and resulted in moderate to good yields of β,β-disubstituted ketones (Scheme 34) [63].

For the detection of Ni-H species formed during the reaction, they independently synthesized Ni(II)HCl(PCy$_3$)$_2$ and ^{31}P NMR analysis of this complex revealed a peak at 33.42 ppm. The same peak was observed in the ^{31}P NMR analysis of the

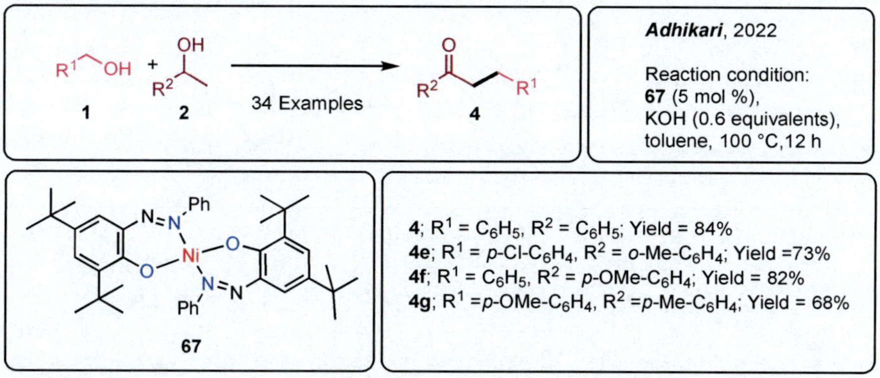

Scheme 33 Ni(II)-catalyzed dehydrogenative coupling of secondary alcohols with primary alcohols [62]

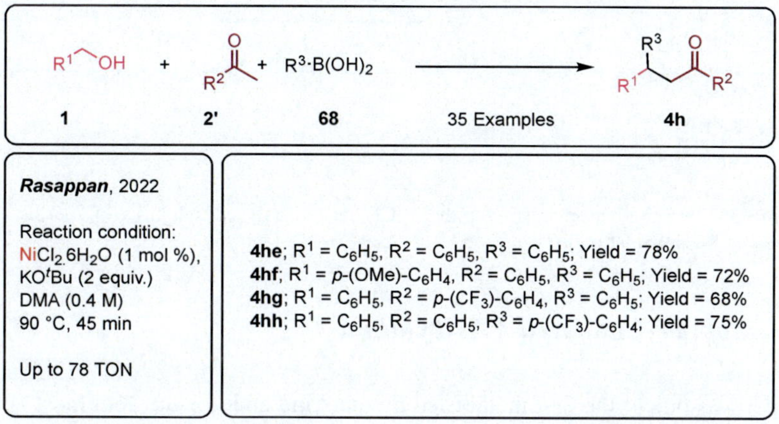

Scheme 34 $NiCl_2.6H_2O$ mediated synthesis of β,β-disubstituted ketones from alkylation of ketones with primary alcohols and phenyl boronic acid reported by Rasappan et al. [63]

reaction mixture, thereby proving the existence of N-H intermediate during the reaction. Deuterium labelling experiments indicated the reversible reaction between aldehyde intermediate and Ni-H species. The kinetic studies revealed the reaction to be first-order with respect to all the reactants and catalyst.

Based on above findings, they proposed a mechanism as shown in Scheme 35. The first step is the formation of Ni-H intermediate **71** and aldehyde **1′** starting from $NiCl_2.6H_2O$ and alcohol **1**. The aldehyde then undergoes aldol condensation with the ketone **2′**, leading to the formation of enone **4′**. Parallelly, a base mediated transmetalation of boronic acid results in the nickel intermediate **72**, which further undergoes 1,4-addition to afford the final product **4h**.

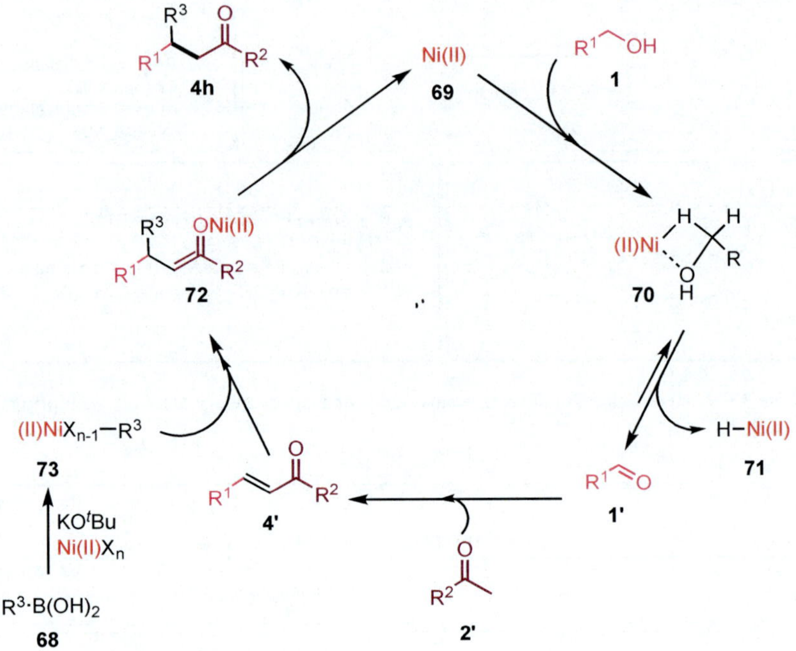

Scheme 35 Proposed mechanism for the $NiCl_2.6H_2O$-catalyzed synthesis of β,β-disubstituted ketones by Rasappan et al. [63]

2.6 *Copper-Catalyzed* C-*Alkylation*

Copper was one of the first metal used by mankind and it is the 26th most abundant metal in the earth's crust. In 2012, Xu et al. investigated the β-alkylation of secondary alcohols and α-alkylation of ketones using a ligand-free copper catalyst that holds wide scope to provide an alternative route in synthetic chemistry. They could achieve up to 99% yield of the alkylated product when they have treated benzyl alcohol and 1-phenyl ethanol with $Cu(OAc)_2.H_2O$ (1 mol%) and KOH (30 mol%) at 120°C (Scheme 36) [37]. According to the proposed mechanism discussed in Scheme 37, the reaction is initiated either with the oxidation of primary alcohols or secondary alcohols by Cu-catalyzed aerobic oxidation under air or via Cu (II)-catalyzed oxidation of primary alcohols or secondary alcohols under nitrogen (step i). It is followed by the base-mediated dehydrogenative aldol condensation of the ketone and aldehyde giving rise to an α,β-unsaturated ketone (step ii). Then the transfer hydrogenation step gives rise to the intermediates **1′** and **2′** as well as the β-alkylated product (step iii and iv). The intermediates **1′** and **2′** once formed in step iv again undergoes condensation giving rise to the α,β-unsaturated ketone (step v). From the experimental studies it was found that the reactivity of the reaction was better in aerobic conditions rather than the anaerobic condition.

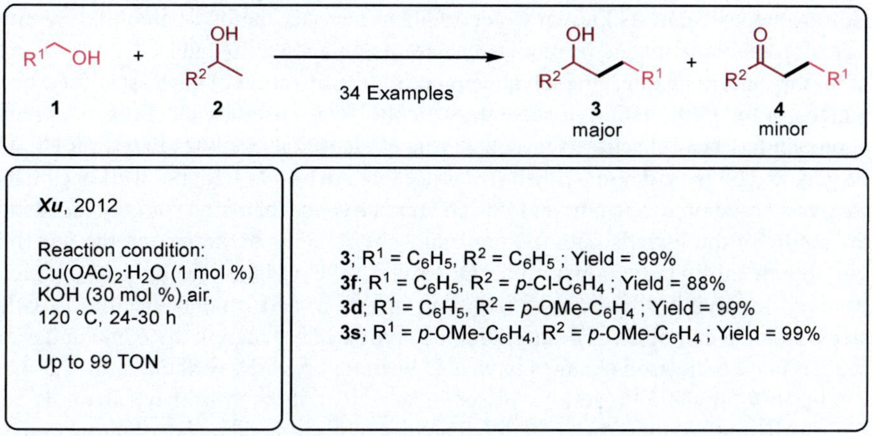

Scheme 36 Copper-catalyzed β-alkylation of secondary alcohols with primary alcohols [37]

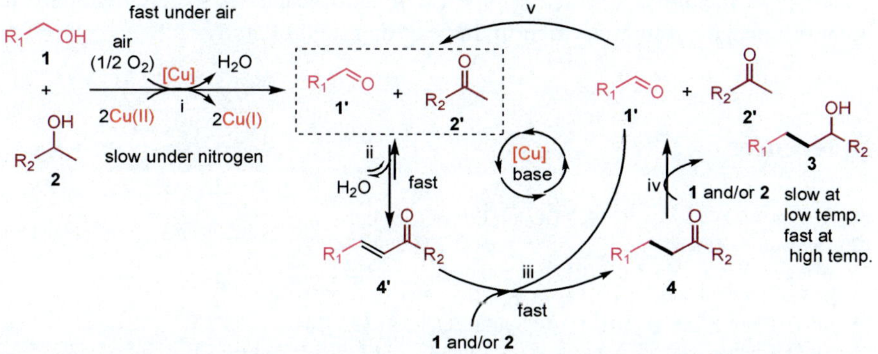

Scheme 37 Proposed mechanism for copper-catalyzed β-alkylation of secondary alcohols with primary alcohols [37]

3 Conclusion

Methods that form Carbon–Carbon bonds are key to the generation of versatile precursors to fuels and value-added chemicals. Traditional methods involving the coupling of alkyl halides tend to be non-sustainable and are typically accompanied by the formation of hazardous waste. Transition metal-catalyzed coupling of alcohols via the sequence of catalytic dehydrogenation to carbonyl compounds, aldol condensation followed by tandem catalytic hydrogenation of the resulting α,β-unsaturated carbonyl compound is a convenient way to generate new C–C bonds with water as the sole by-product. These "Guerbet-type reactions" not only provide access to a variety of α-alkylated ketones and β-alkylated alcohols of pharmaceutical importance but also are straightforward methods to obtain fuel-grade alcohols.

Homogeneous catalysts known to accomplish the β-alkylation of alcohols are either based on precious metals or require hazardous phosphine ligands.

In the current chapter, the advancement in the utilization of catalysts based on 3d metals for the β-alkylation of secondary alcohols with primary alcohols in a greener approach has been discussed in detail. It is evident that catalysts based on Cr, Mn, Fe, Co, and Ni have demonstrated reactivity that are on-par with the well-established catalysts based on Ru, Rh, Ir, and Pd. The operative mechanism is largely dictated by the nature of the ligands with the non-innocent versions orchestrating the reactivity via the metal-ligand cooperation. Otherwise, the typical mechanism is metal-centered involving β-hydride elimination, insertion, and σ-bond metathesis as key steps. Isolated examples of β-alkylation catalytic cycles wherein the metal undergoes sequential two-electron changes have also been reported. Nevertheless, the 3d-metal catalyzed β-alkylation reactions of secondary alcohols with primary alcohols have been well-studied not only to establish an efficient synthetic protocol under conventional/microwave heating but also to obtain systematic mechanistic understanding. This has led to significant advances in the global pursuit for the identification of cost-effective 3d metal-based Guerbet-type β-alkylation catalytic systems which offers great promise for future research in this exciting field of study.

References

1. DeJong W, Wallack L (1992) Health. Educ Q 19:429–442
2. Casswell S, Thamarangsi T (2009) Lancet 373:2247–2257
3. Wang Q, Wu K, Yu Z (2016) Organometallics 35:1251–1256
4. Makarov IS, Madsen R (2013) J Org Chem 78:6593–6598
5. Xu Q, Chen J, Liu Q (2013) Adv Synth Catal 355:697–704
6. Xu Q, Chen J, Tian H, Yuan X, Li S, Zhou C, Liu J (2014) Angew Chem Int Ed 53:225–229
7. Allen LJ, Crabtree RH (2010) Green Chem 12:1362
8. Irrgang T, Kempe R (2019) Chem Rev 119:2524–2549
9. Li Q, Fan S, Sun Q, Tian H, Yu X, Xu Q (2012) Org Biomol Chem 10:2966–2972
10. Feng SL, Liu CZ, Li Q, Yu XC, Xu Q (2011) Chin Chem Lett 22:1021–1024
11. Liu C, Liao S, Li Q, Feng S, Sun Q, Yu X, Xu Q (2011) J Org Chem 76:5759–5773
12. Sahoo AR, Lalitha G, Murugesh V, Bruneau C, Sharma GVM, Suresh S, Achard M (2017) J Org Chem 82:10727–10731
13. Cho CS, Kim BT, Kim H-S, Kim T-J, Shim SC (2003) Organometallics 22:3608–3610
14. Ruiz-Botella S, Peris E (2015) Chem Eur J 21:15263–15271
15. Jiménez MV, Fernández-Tornos J, Modrego FJ, Pérez-Torrente JJ, Oro LA (2015) Chem Eur J 21:17877–17889
16. Xu C, Goh LY, Pullarkat SA (2011) Organometallics 30:6499–6502
17. Segarra C, Mas-Marzá E, Mata JA, Peris E (2011) Adv Synth Catal 353:2078–2084
18. Gong X, Zhang H, Li X (2011) Tetrahedron Lett 52:5596–5600
19. Gnanamgari D, Sauer ELO, Schley ND, Butler C, Incarvito CD, Crabtree RH (2009) Organometallics 28:321–325
20. da Costa AP, Sanaú M, Peris E, Royo B (2009) Dalton Trans:6960–6966. https://doi.org/10.1039/B901195A
21. Pontes da Costa A, Viciano M, Sanaú M, Merino S, Tejeda J, Peris E, Royo B (2008) Organometallics 27:1305–1309

22. Gnanamgari D, Leung CH, Schley ND, Hilton ST, Crabtree RH (2008) Org Biomol Chem 6: 4442–4445
23. K.-i. Fujita, C. Asai, T. Yamaguchi, F. Hanasaka, R. Yamaguchi, Org Lett 7 (2005) 4017–4019
24. Satyanarayana P, Reddy GM, Maheswaran H, Kantam ML (2013) Adv Synth Catal 355:1859–1867
25. Zhang C, Zhao J-P, Hu B, Shi J, Chen D (2019) Organometallics 38:654–664
26. Roy BC, Debnath S, Chakrabarti K, Paul B, Maji M, Kundu S (2018) Org Chem Front 5:1008–1018
27. Roy BC, Chakrabarti K, Shee S, Paul S, Kundu S (2016) Chem Eur J 22:18147–18155
28. Shee S, Paul B, Panja D, Roy BC, Chakrabarti K, Ganguli K, Das A, Das GK, Kundu S (2017) Adv Synth Catal 359:3888–3893
29. Chakrabarti K, Paul B, Maji M, Roy BC, Shee S, Kundu S (2016) Org Biomol Chem 14:10988–10997
30. Prades A, Viciano M, Sanaú M, Peris E (2008) Organometallics 27:4254–4259
31. Kose O, Saito S (2010) Org Biomol Chem 8:896–900
32. Liu T, Wang L, Wu K, Yu Z (2018) ACS Catal 8:7201–7207
33. El-Sepelgy O, Matador E, Brzozowska A, Rueping M (2019) ChemSusChem 12:3099–3102
34. Freitag F, Irrgang T, Kempe R (2017) Chem Eur J 23:12110–12113
35. Yang J, Liu X, Meng DL, Chen HY, Zong ZH, Feng TT, Sun K (2012) Adv Synth Catal 354: 328
36. Zhang M-J, Li H-X, Young DJ, Li H-Y, Lang J-P (2019) Org Biomol Chem 17:3567–3574
37. Liao S, Yu K, Li Q, Tian H, Zhang Z, Yu X, Xu Q (2012) Org Biomol Chem 10:2973–2978
38. Gladiali S, Alberico E (2006) Chem SocRev 35:226–236
39. Genç S, Arslan B, Gülcemal S, Günnaz S, Çetinkaya B, Gülcemal D (2019) J Org Chem 84: 6286–6297
40. Kaur M, Din Reshi NU, Patra K, Bhattacherya A, Kunnikuruvan S, Bera JK (2021) Chem Eur J 27:10737–10748
41. Musa S, Ackermann L, Gelman D (2013) Adv Synth Catal 355:3077–3080
42. Czap A, Heinemann FW, van Eldik R (2004) Inorg Chem 43:7832–7843
43. Che C-M, Ho C, Lau T-C (1991) J Chem Soc Dalton Trans:1901–1907. https://doi.org/10.1039/DT9910001901
44. Wei C, He Y, Shi X, Song Z (2019) Coord Chem Rev 385:1–19
45. Busschaert N, Wenzel M, Light ME, Iglesias-Hernández P, Pérez-Tomás R, Gale PA (2011) J Am Chem Soc 133:14136–14148
46. Du W, Wang Q, Wang L, Yu Z (2014) Organometallics 33:974–982
47. Prakasham AP, Ta S, Dey S, Ghosh P (2021) Dalton Trans 50:15640–15654
48. Das K, Yasmin E, Das B, Srivastava HK, Kumar A (2020) Catal. Sci Technol 10:8347–8358
49. Das K, Kathuria L, Jasra RV, Dhole S, Kumar A (2023) Catal Sci Technol 13:1763–1776
50. Narjinari H, Tanwar N, Kathuria L, Jasra RV, Kumar A (2022) Catal Sci Technol 12:4753–4762
51. Miao Y, Samuelsen SV, Madsen R (2021) Organometallics 40:1328–1335
52. Su P, Chen Z, Ni J, Yang Z, Li Y, Ke Z (2023) ACS Catal 13:12481–12493
53. Bisarya A, Jasra RV, Kumar A (2023) Organometallics 42:1818–1831
54. Pandey B, Xu S, Ding K (2021) Organometallics 40:1207–1212
55. Pandey B, Xu S, Ding K (2019) Org Lett 21:7420–7423
56. Nandi PG, Kumar P, Kumar A (2022) Catal. Sci Technol 12:1100–1108
57. Nandi PG, Thombare P, Prathapa SJ, Kumar A (2022) Organometallics 41:3387–3398
58. Greenwood NN, Earnshaw A (2012) Chemistry of the elements. Elsevier
59. Ananikov VP (2015) ACS Catal 5:1964–1971
60. Babu R, Subaramanian M, Midya SP, Balaraman E (2021) Org Lett 23:3320–3325
61. Arora V, Narjinari H, Kumar A (2021) Organometallics 40:2870–2880
62. Bains AK, Biswas A, Adhikari D (2022) Adv Synth Catal 364:47–52
63. Balakrishnan V, Ganguly A, Rasappan R (2022) Org Lett 24:4804–4809

Top Organomet Chem (2024) 73: 129–172
https://doi.org/10.1007/3418_2023_108

Published online: 21 November 2023

Tandem Multicomponent Reactions for Diverse Heterocycles Synthesis Under 3d-Transition Metal Catalysis

Ganesan Sivakumar, Abhijith Karattil Suresh, and Ekambaram Balaraman

Contents

Abstract The synthesis of heterocycles from renewable starting materials is a desirable goal for chemical research, as heterocycles have many applications in pharmaceuticals, material chemistry, and natural products. Recently, there has been a notable focus on utilizing earth-abundant 3d-transition-metal catalysts in contemporary catalysis, serving as a viable alternative to noble metals. This chapter provides an in-depth discussion of the recent advancements in 3d-transition metal-catalyzed acceptorless dehydrogenative coupling (ADC) reactions for the construction of diverse heterocyclic compounds. These reactions offer an efficient and environmentally friendly approach to the synthesis of valuable heterocyclic compounds.

Keywords 3d-Metal catalysis · Alcohols · Dehydrogenative coupling · Heterocycles · Sustainable synthesis

G. Sivakumar, A. K. Suresh, and E. Balaraman (✉)
Department of Chemistry, Indian Institute of Science Education and Research (IISER), Tirupati, India
e-mail: eb.raman@iisertirupati.ac.in

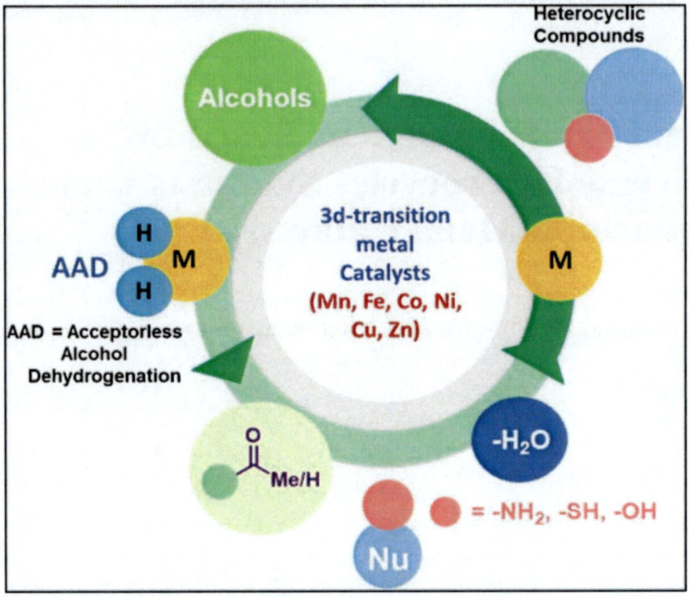

This book chapter reviews the latest developments in 3d-transition metal-catalyzed tandem multicomponent reactions for the synthesis of various heterocycles.

1 Introduction

The fundamental core of most drug molecules and natural products consists of heterocycles, making them of tremendous importance in pharmaceuticals, material chemistry, and natural product synthesis [1–5]. Interestingly, they can alter the lipophilicity, polarity, solubility, hydrogen bonding capacity, bioavailability, potency, selectivity, toxicity, etc. of drug candidates to suit their needs. The chemical market for heterocycles is a large and growing sector. The global market for heterocyclic compounds was valued at USD 19.8 billion in 2019 and is expected to reach USD 26.6 billion by 2025, with a compound annual growth rate (CAGR) of 5.1%. The major factors driving the market growth are the increasing demand for heterocyclic compounds in the pharmaceuticals, agrochemicals, dyes, and plastics industries. Consequently, their synthesis has garnered significant attention and emerged as a prominent area of chemical research. Numerous innovative methods have been developed over the past few decades to synthesize various heterocyclic compounds. However, these protocols (a) often rely on expensive and highly reactive starting materials, (b) require harsh reaction conditions, (c) involve laborious multi-step procedures, and (d) generate excess toxic waste. The challenge lies in proficiently applying these methodologies to synthesize complex molecules

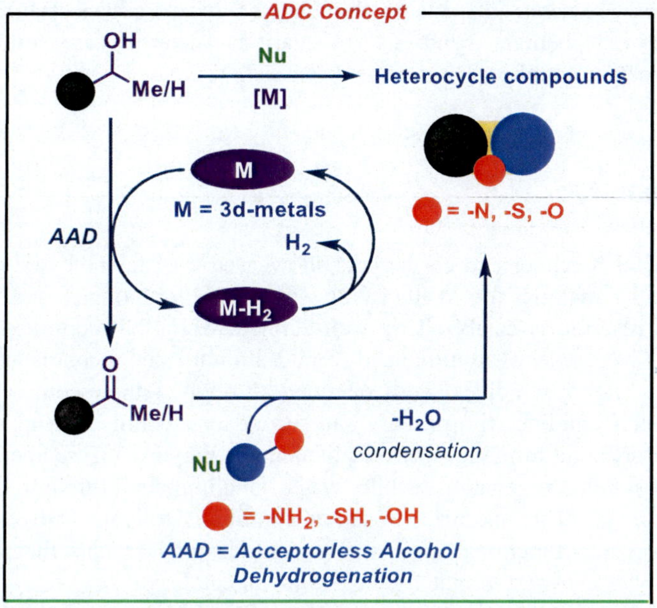

Fig. 1 Catalytic acceptorless dehydrogenative coupling (ADC) reaction

containing reactive functional groups [6, 7]. Therefore, there is a high demand for sustainable and affordable methodologies that utilize simple, abundant feedstocks and operate under benign conditions.

Alcohols are abundant and simple feedstocks that can be obtained from renewable biomass derivatives [8, 9]. They are important starting materials for making various fine and specialty chemicals through tandem C–C and C–E (E = heteroatom) bond-forming reactions [10–12]. These chemicals are synthesized by the "Acceptorless Dehydrogenative Coupling" (ADC) strategy with transition metal complex catalysts. The ADC approach is sustainable and benign and only generates water and hydrogen as by-products [13–16]. In this regard, there is a constant and focused effort on the development of new C–X (X = C and N) bond-forming reactions [17–27]. The ADC strategy has enabled the successful synthesis of various heterocycles from alcohols, which react with different kinds of nucleophiles (Fig. 1) [28–34]. However, most catalytic systems rely on precious 4d- and 5d-transition metals such as ruthenium, palladium, rhodium, and iridium [35–36].

The extensive use of these noble transition metals is accompanied by significant drawbacks, including their limited availability in terrestrial sources, high cost, and toxicity. Moreover, the stringent regulations regarding permissible metal impurities in marketed drugs and agrochemicals further exacerbate these limitations [37–38]. Consequently, the replacement of precious metals with environmentally friendly, abundantly available, and cost-effective 3d-transition metals like manganese, iron, cobalt, nickel, and copper has garnered tremendous interest in contemporary research [39–49]. In this chapter, recent advances in 3d-transition metal-catalyzed benign synthesis of various heterocyclic compounds via dehydrogenative coupling reactions is discussed.

2 Manganese

In 2016, Karl Kirchner and co-workers showcased a sustainable and environmentally friendly method for synthesizing substituted quinolines and pyrimidines [50]. This reaction is catalyzed by well-defined Mn(I)PNP complexes (Chart 1). Quinolines were selectively produced from 2-aminobenzyl alcohols and secondary alcohols, while pyrimidines were obtained through a three-component process involving benzamidine with primary and secondary alcohols. During the selective C–C and C–N bond formations, two equivalents of dihydrogen are liberated through acceptorless dehydrogenation, while water is eliminated through condensation (Scheme 1a, b). The optimized reaction conditions tolerate a diverse range of common organic functional groups. This initial study reveals that 3d-transition metal catalysts can start to match precious-metal catalysts.

Further, Kempe and co-workers reported a PN_5P-Mn-pincer complex catalyzed multicomponent synthesis of pyrimidines from amidines with secondary and primary alcohols (Scheme 2) [51]. This reaction involves condensation and dehydrogenation steps, which enable selective tandem formation of C–C and C–N bonds. Indeed, to synthesize fully substituted pyrimidines in a convenient one-pot process, β-alkylation reactions are applied to alkylate secondary alcohols multiple times with two distinct primary alcohols.

Further, Milstein and co-workers disclosed the dehydrogenative self-coupling of 2-amino alcohols, catalyzed by an acridine-based Mn(I)-pincer complex (Scheme 3) [52]. This catalytic process selectively yielded functionalized 2,5-substituted pyrazine. Interestingly, 2-substituted quinoxaline derivatives were also achieved by the ADC of 1,2-diaminobenzene and 1,2-diols. Several control experiments and kinetic investigations were conducted to gain insight into the plausible mechanism. Based on these findings, it was observed that the formation of quinoxalines through the dehydrogenative coupling of 1,2-diaminobenzene with vicinal diols initiates with the dehydrogenation of the terminal alcohol group within the 1,2-diol system.

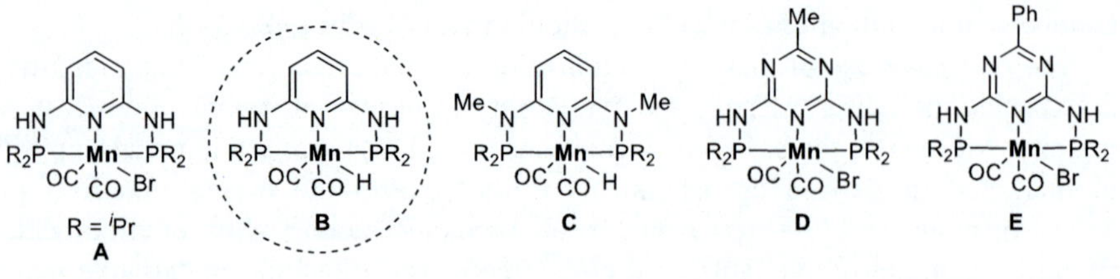

Chart 1 Mn(I)PNP pincer complexes used for quinoline and pyrimidine derivatives [50]

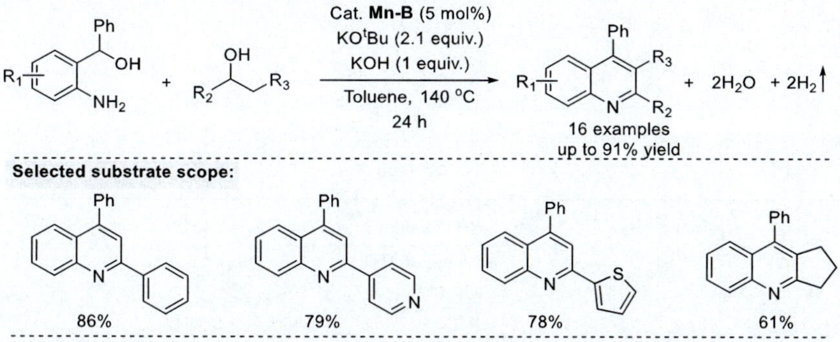

Scheme 1 (**a**) Mn(I)PNP pincer complex catalyzed synthesis of quinolines [50]. (**b**) Mn(I)PNP pincer complex catalyzed synthesis of pyrimidines [50]

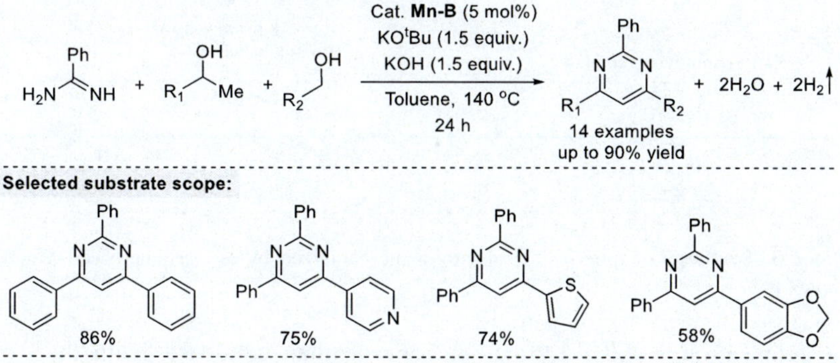

Scheme 1 (continued)

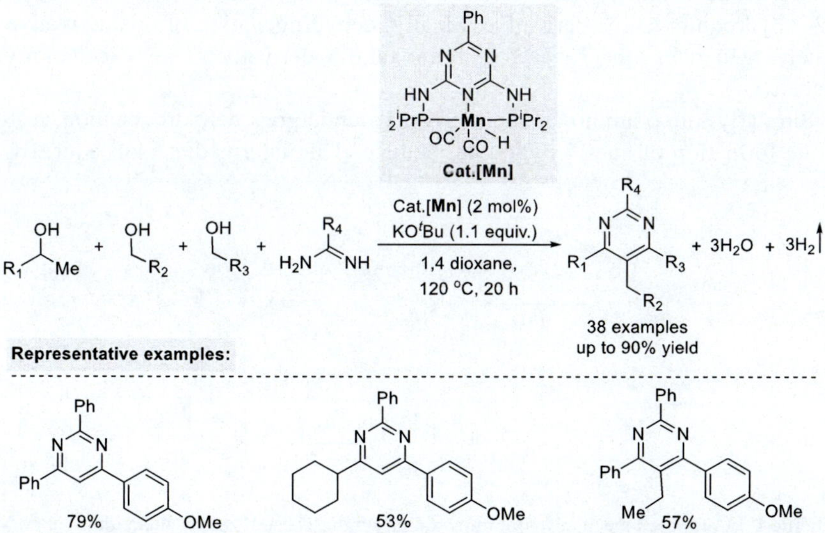

Scheme 2 A PN$_5$P-Mn-pincer complex catalyzed multicomponent synthesis of pyrimidines [51]

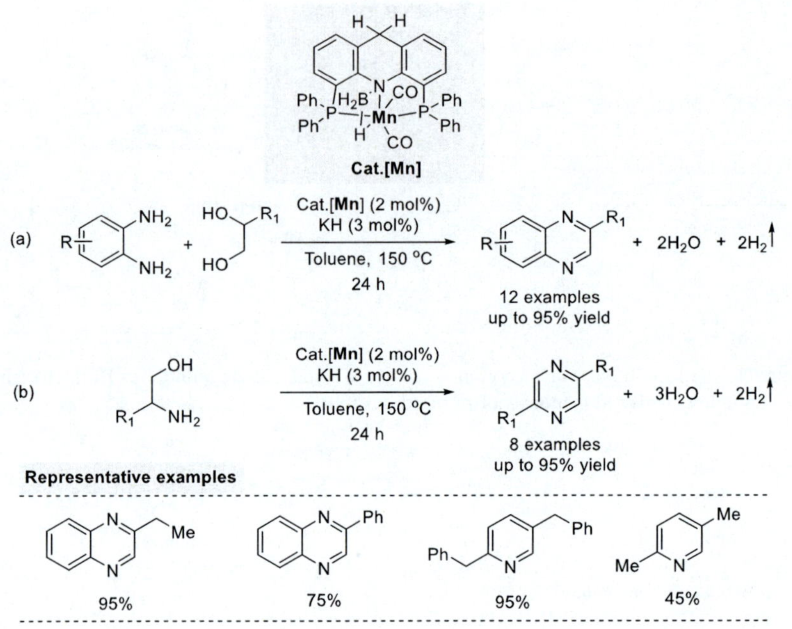

Scheme 3 Synthesis of quinoxaline and pyrazine catalyzed by an acridine-based Mn(I)-pincer complex [52].

The amine group of the 1,2-diaminobenzene then condenses with the carbonyl moiety, forming intermediate **I**. Subsequently, a proton shift results in the formation of intermediate **II**, which undergoes tautomerization to yield intermediate **III**. Further condensation with a second amine group leads to the formation of a 1,2-dihydroquinoxaline derivative. Finally, dehydrogenation of this derivative ultimately generates the expected quinoxaline derivatives in excellent yields (Scheme 4).

Similarly, the β-amino alcohol derivative undergoes dehydrogenation, resulting in the formation of an aldehyde intermediate. This intermediate subsequently self-

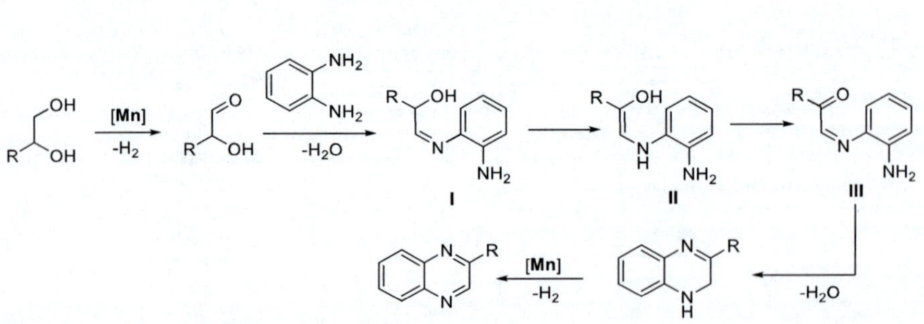

Scheme 4 A proposed mechanism for quinoxaline synthesis catalyzed by an acridine-based Mn(I)-pincer complex [52]

couples with another molecule, producing 2,5-dihydropyrazine derivatives with the elimination of water. The 2,5-dihydropyrazine compound then undergoes dehydrogenation and yields a pyrazine derivative with the liberation of dihydrogen (Scheme 5).

Subsequently, in 2018, Srimani and co-workers reported the synthesis of quinolones, quinoxalines, pyrazines, and benzothiazoles using an NNS-pincer ligand-based Mn(I)-pincer complex (Chart 2 and Scheme 6) [53]. The study

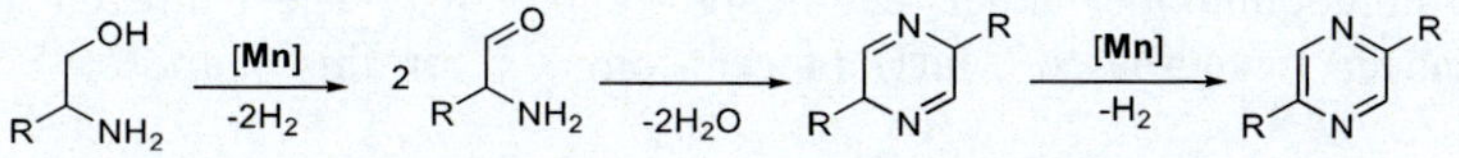

Scheme 5 A proposed mechanism for pyrazine synthesis catalyzed by an acridine-based Mn(I)-pincer complex [52]

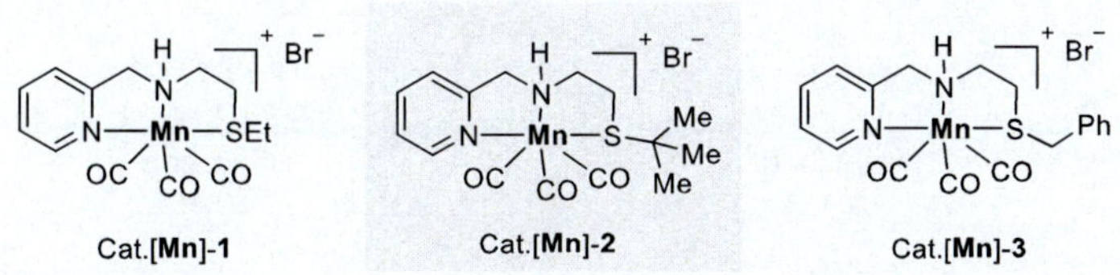

Chart 2 NNS-pincer ligand-based Mn(I)-pincer complexes used for multicomponent synthesis of diverse heterocycles [53]

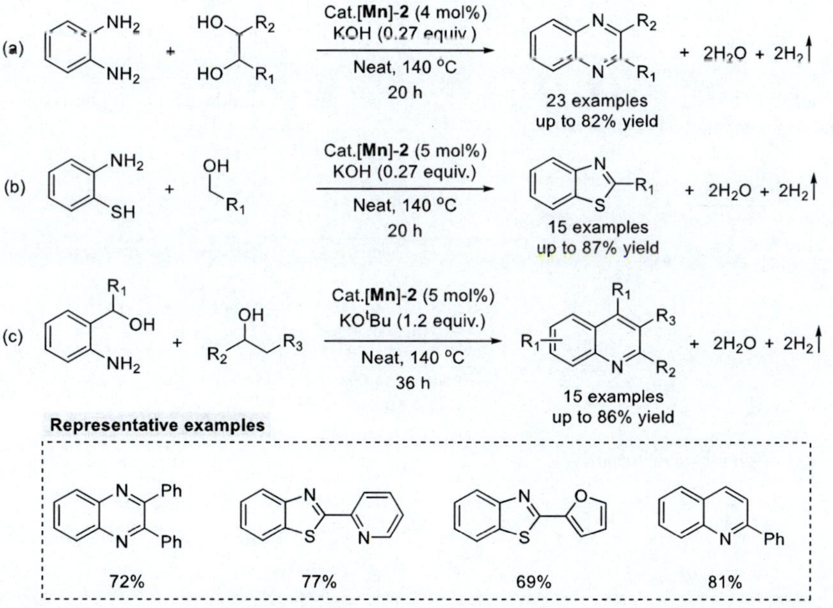

Scheme 6 Synthesis of quinolones, quinoxalines, pyrazines, and benzothiazoles using an NNS-pincer ligand-based Mn(I)-pincer complex [53]

demonstrated solvent-free reaction conditions using alcohols as feedstock starting material. Indeed, desired products were achieved in good to excellent yields with tolerating a diverse array of functional groups under optimized reaction conditions.

Further, the same group reported the acceptorless dehydrogenative coupling of aromatic diamines with primary alcohols catalyzed by the same tridentate NNS-Mn (I) complex. This reaction selectively produced 2-substituted and 1,2-disubstituted benzimidazoles in excellent isolated yields (Scheme 7) [54].

Later, the same group extended their investigation for the synthesis of quinazoline and 2-aminoquinoline under eco-benign conditions. They utilized the same NNS-pincer ligand-based Mn(I) pincer complex for this purpose (Scheme 8)

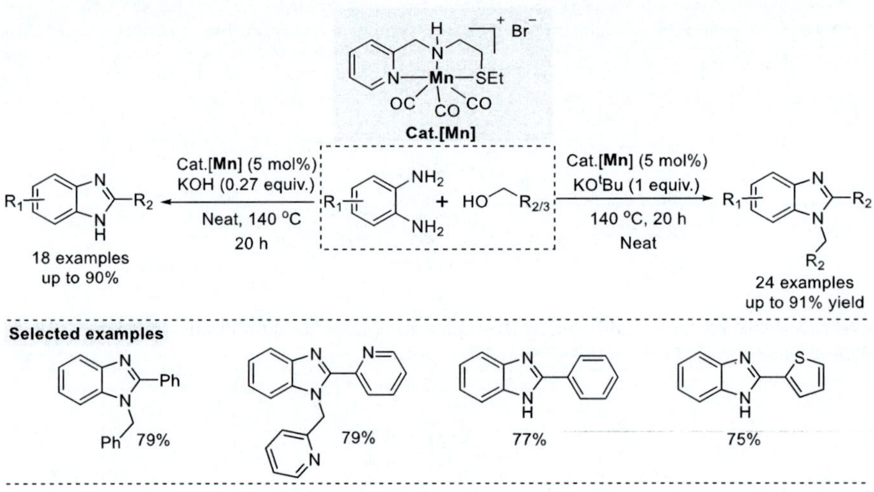

Scheme 7 Synthesis of 2-substituted and 1,2-disubstituted benzimidazoles using an NNS ligand-derived manganese(I) complex [54]

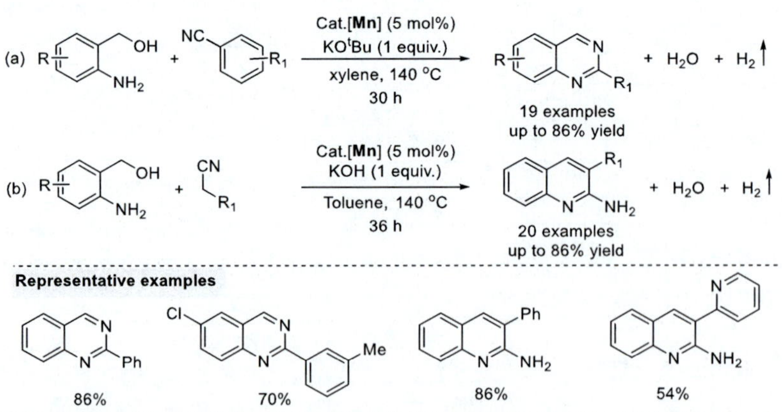

Scheme 8 Synthesis of quinazoline and 2-aminoquinoline using an NNS-pincer ligand-based Mn (I) pincer complex [55]

[55]. The synthesis was accomplished through the dehydrogenative annulation of 2-aminobenzyl alcohol with nitriles.

Subsequently, in 2019, Srimani and co-workers reported an NNS-pincer ligand-based Mn(I)-complex catalyzed dehydrogenative synthesis of structurally important 2,3-dihydro-1*H*-perimidines (Scheme 9) [56]. This catalytic process is very elegant and efficiently used for the benign synthesis of 2,3-dihydro-1*H*-perimidines with high yields using a diverse range of alcohol substrates, including those with electron-donating and electron-withdrawing groups, as well as heteroaromatic compounds.

Recently, in 2020, Balaraman and co-workers developed a manganese(I)-catalyzed direct synthesis of quinoxaline and quinazoline derivatives via ADC of 1,2-diols with 1,2-diaminobenzene and 2-aminobenzyl alcohol with benzamides, respectively (Scheme 10) [57]. This protocol showed remarkable reactivity and enabled the affordable synthesis of numerous biologically important quinoxalines and quinazoline derivatives in excellent yields (Scheme 10a). Noteworthy, a remarkable reactivity was achieved in the presence of $Mn(CO)_5Br$ with a phosphine-free commercially available NNN-Ligand. The synthesis of quinoxalines proceeds by the dehydrogenative annulation of 1,2-diols and 1,2-diaminobenzenes. The reaction begins with the dehydrogenation of the primary alcoholic group of the 1,2-diol system. After that, a condensation reaction of 1,2-diaminobenzene with the carbonyl moiety (**I**) results in the intermediate **II**, which is converted into **III** by acceptorless dehydrogenation. Finally, an intramolecular condensation reaction of **III** delivers quinoxaline (Scheme 10b).

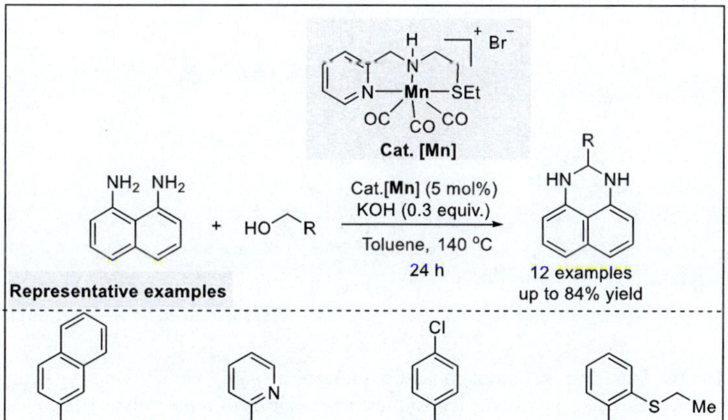

Scheme 9 An NNS-pincer ligand-based Mn(I)-complex catalyzed dehydrogenative synthesis of 2,3-dihydro-1*H*-perimidines [56]

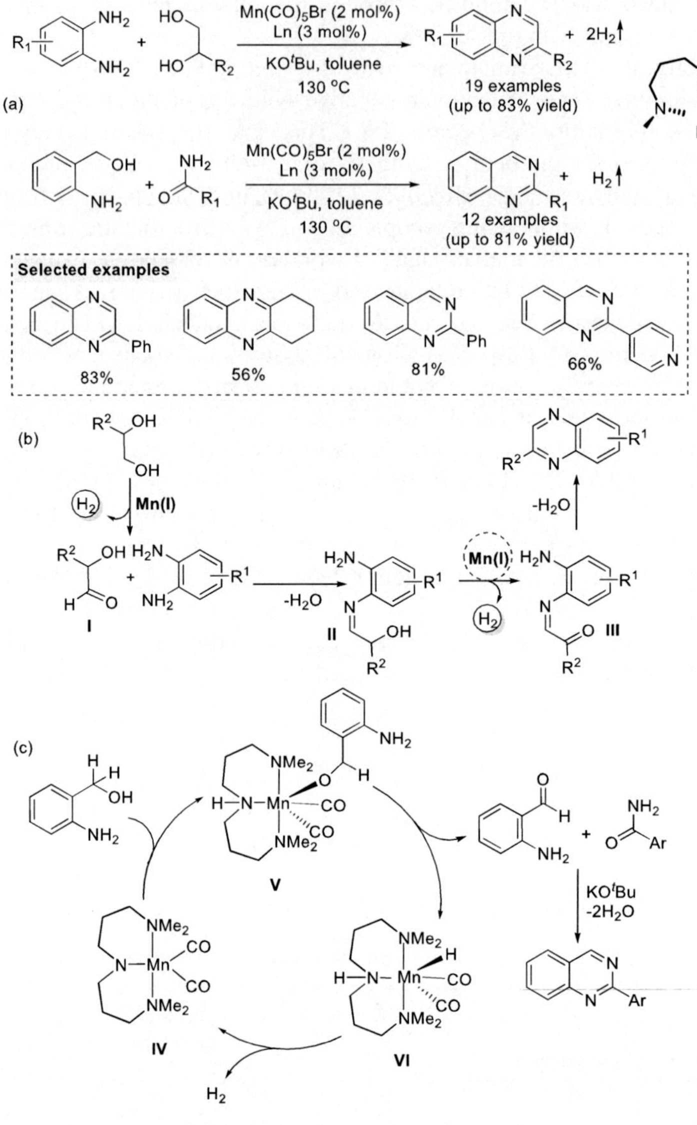

Scheme 10 (**a**) Synthesis of quinoxaline and quinazoline using an NNN-pincer ligand-based Mn (I) pincer complex. (**b**) Proposed mechanism for the synthesis of quinoxaline. (**c**) Catalytic cycle for the ADC of 2-aminobenzyl alcohol with primary amide. Redrawing with permission from Ref. [57]. Copyright 2020 American Chemical Society

The synthesis of quinazoline involves the activation of the O-H bond of 2-aminobenzyl alcohol by the in situ generated active Mn-catalyst (**IV**) through the metal-ligand cooperation (MLC), leading to the intermediate **V**. The intermediate **V** then undergoes β-H elimination and produces 2-aminobenzaldehyde and the intermediate **VI**. Next, 2-aminobenzaldehyde and primary amide couple and eliminate water molecules to form quinazoline. Finally, the liberation of dihydrogen from the intermediate **VI** through the MLC regenerates the active Mn-catalyst **IV** (Scheme 10c).

Further, the same research group successfully demonstrated the synthesis of quinoline and pyridine compounds using a single-molecular PNP-Mn(I)-complex. The reaction operates under mild, solventless conditions with the formation of molecular hydrogen and water as the sole by-products. The current method accommodates a broad range of substrates, including aryl, aliphatic acyclic, and cyclic primary alcohols and amino alcohols, to afford diverse N-heterocyclic compounds in good to excellent yields. Various control and labeling experiments, kinetic, and NMR studies reveal that the reaction follows the acceptorless double dehydrogenative coupling pathway (ADDC), selectively producing desired N-heterocycles (Scheme 11) [58].

In the same year, Srimani and co-workers reported the synthesis of C-3 functionalized indoles from 2-aminophenyl ethanol with primary alcohols using an NNS-pincer ligand-based Mn(I) complex (Scheme 12) [59]. Various substituted 2-aminophenyl ethanol and primary alcohols showed excellent reactivity and afforded the desired products in very good yields (up to 78%).

Subsequently, Kundu and co-workers reported an elaborate synthesis of quinolines, pyrroles, and pyridines using a phosphine-free Mn(II)-complex (Scheme 13) [60]. While screening various ligands, the authors understand the importance of the benzimidazole N–H proton and the amine N–H group in the ligand, which play a

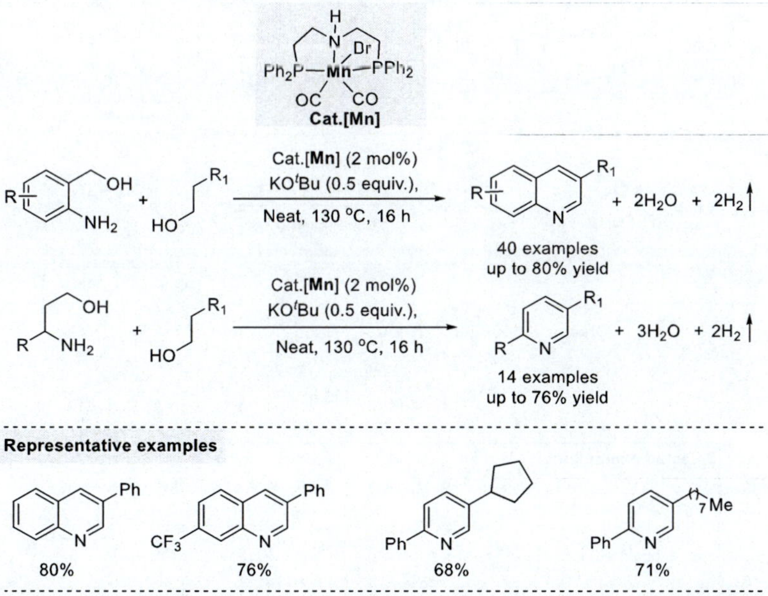

Scheme 11 Synthesis of quinoline and pyridine using a single-molecular Mn(I)-complex [58]

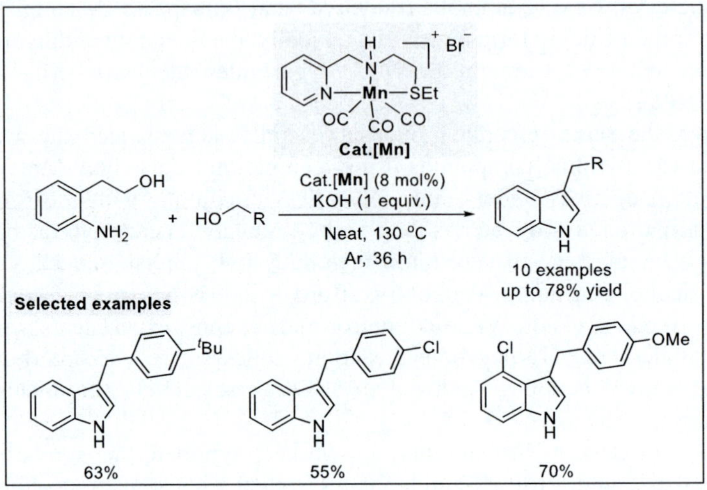

Scheme 12 Synthesis of C-3 functionalized indoles from 2-aminophenyl ethanol with primary alcohols [59]

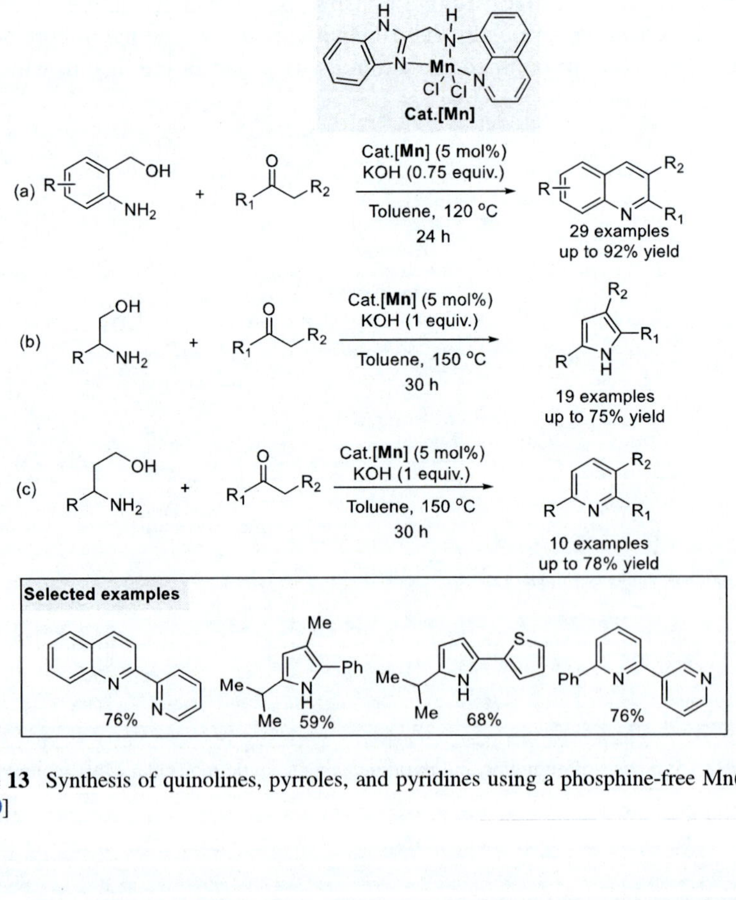

Scheme 13 Synthesis of quinolines, pyrroles, and pyridines using a phosphine-free Mn(II)-complex [60]

critical role in the metal-ligand cooperative mechanism. Several control experiments, kinetic studies, and DFT calculations support the plausible reaction mechanism. The findings from these investigations led to the proposition of a metal-ligand cooperative mechanism (amine to imide). During the course of the reaction 2-aminobenzyl alcohol was activated via the metal-ligand cooperative pathway by the transfer of proton from imido-N in the ligand backbone followed by Mn-alkoxy species formation.

The same research group investigated the use of Mn(I) complexes with benzimidazole and pyridine-containing bidentate ligands for synthesizing pyrimidines and quinolines (Scheme 14) [61]. The authors found that an Mn-complex with a 2-hydroxypyridine-appended benzimidazole ligand, which is sensitive to proton activation, showed excellent efficiency in using alcohols for the synthesis of diverse *N*-heterocycles. Upon the base activation of the precatalyst, a pyridonate ligand fragment is formed, which confirms the formation of the lactam form. They have studied the interconversion of lactam-lactim forms and found that this reaction undergoes an outer sphere mechanism. Remarkably, several substituted pyrimidines and quinolines were successfully synthesized with a significantly reduced catalyst loading (0.05 mol%). Various control experiments and kinetic studies were performed to get mechanistic insights into this catalytic reaction.

Recently, Kempe and co-workers reported a consecutive three-component reaction that allows efficient synthesis of novel N-heterocyclic compounds,

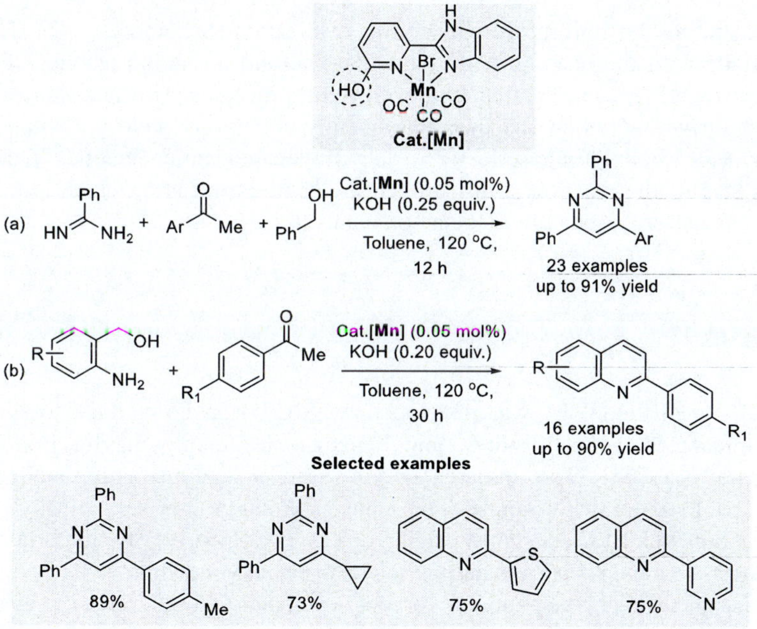

Scheme 14 Benzimidazole and pyridine-containing bidentate ligand-based Mn(I)-complex catalyzed synthesis of pyrimidines and quinolines [61]

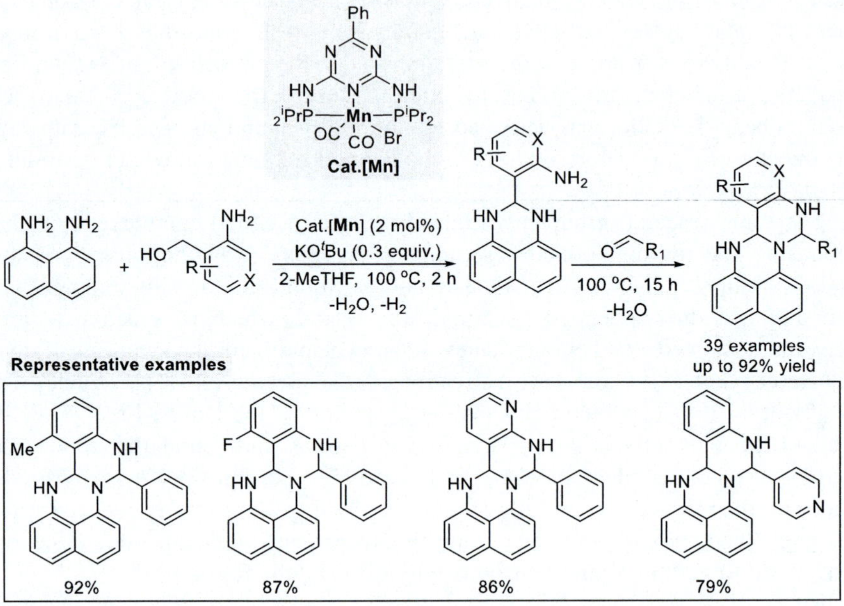

Scheme 15 A consecutive three-component reaction enables easy access to *N*-heterocyclic compounds [62]

demonstrating the concept of regenerative cyclization (Scheme 15) [62]. The reaction starts with the reaction between a diamine and an amino alcohol, involving dehydrogenation, condensation, and cyclization processes. This leads to the formation of a new class of amines that subsequently undergo ring closure with either an aldehyde, carbonyldiimidazole, or a dehydrogenated amino alcohol. During the initial step of the reaction, a dihydrogen molecule is released, which is facilitated by a manganese-based dehydrogenation catalyst.

3 Iron

In 2013, a methodology was developed by Singh and co-workers to synthesize benzimidazoles, benzothiazoles, and benzoxazoles employing an iron catalyst (Scheme 16) [63]. The research group used a commercially available and low-toxic Fe(II)-phthalocyanine as the catalyst. Notably, they successfully executed the reaction with a lower catalyst loading of 1 mol%. However, it is important to note that their substrate scope was narrow to a few primary alcohols.

Subsequently, Nguyen and co-workers developed a highly effective method for the synthesis of benzimidazole and quinoxaline heterocycles through a redox condensation reaction involving *o*-nitroanilines and alcohols (Scheme 17) [64]. The reaction was efficiently facilitated and catalyzed by a combination of sodium sulfide

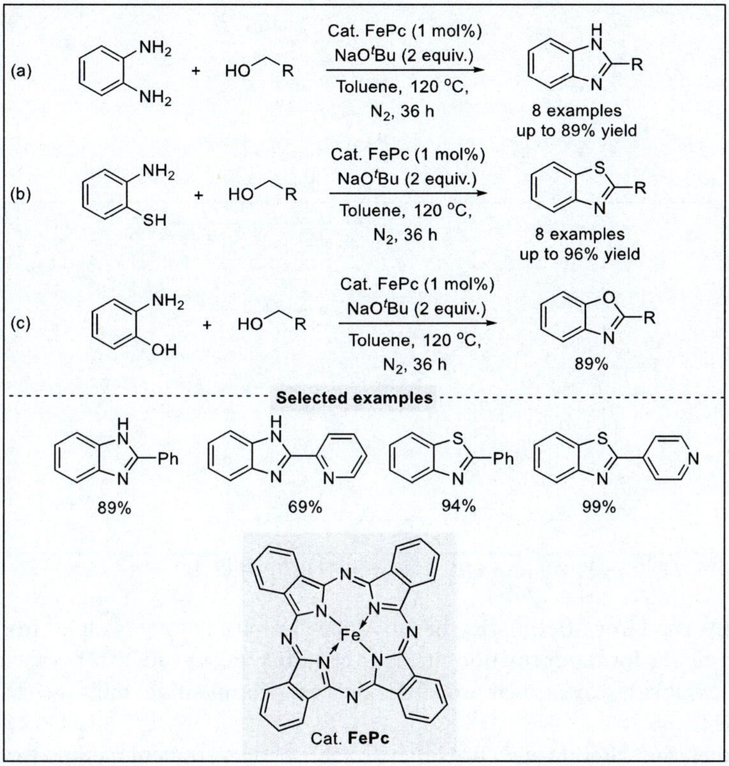

Scheme 16 Fe(II)-phthalocyanine catalyzed synthesis of benzimidazoles, benzothiazoles, and benzoxazole [63]

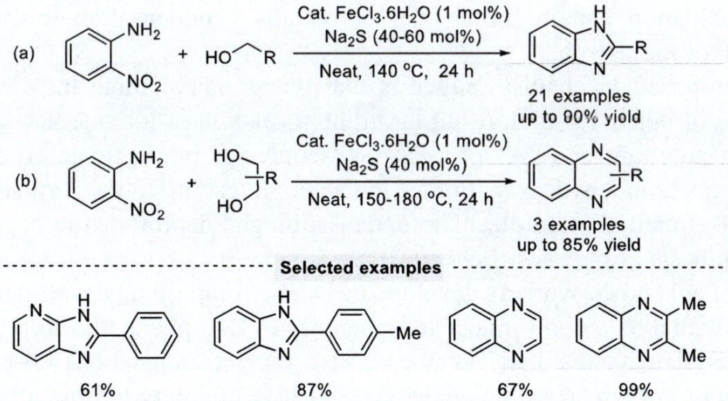

Scheme 17 Fe-catalyzed synthesis of benzimidazole and quinoxaline [64]

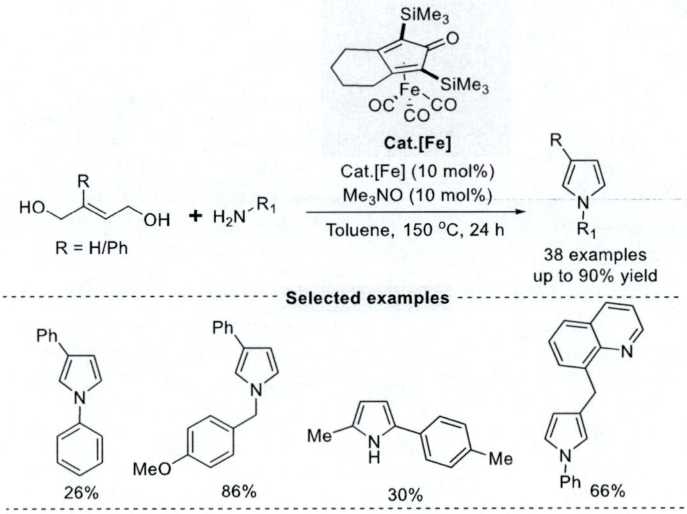

Scheme 18 Fe(0)-catalyzed synthesis of substituted pyrroles [65]

(40 mol%) and iron(III) chloride hexahydrate (1 mol%). Notably, apart from serving as a precursor for the formation of the iron-sulfur (Fe/S) catalyst, hydrated sodium sulfide exhibited exceptional properties as a non-competitive, multi-electron reducing agent.

Further, Sundararaju and co-workers established an efficient method for synthesis of substituted pyrroles using a well-defined, air-stable, molecular iron(0) complex (Scheme 18) [65]. The developed methodology exhibits broad substrate scope and excellent functional group tolerance. Notably, the synthesis of C-2, C-3, and C-2 & C-4 substituted pyrroles achieved in good yields. Based on mechanistic insights, it is proposed that the reaction follows a hydrogen autotransfer process, succeeded by a second oxidation and intramolecular dehydrative condensation, resulting in the formation of the pyrrole.

The proposed mechanism suggests that the primary amine initiates an allylic amination of buten-1,4-diol, resulting in the formation of intermediate **A** through a hydrogen autotransfer (HA) process. Subsequently, intermediate **A** undergoes a second oxidation/dehydrogenation of the alcohol, leading to the formation of intermediate **B**. Finally, intramolecular condensation and deprotonation occur, resulting in the synthesis of pyrroles (Scheme 19).

Later, Paul and co-workers developed a catalyst employing a well-defined Fe(II)-complex with a redox non-innocent 2-phenylazo-(1,10-phenanthroline) ligand. This catalytic system enabled the synthesis of a diverse range of 2,4,6-trisubstituted pyrimidines through an oxidative dehydrogenative coupling of primary and secondary alcohols with amidines (Scheme 20) [66]. The catalytic reaction was conducted under ambient air conditions at a temperature of 100 °C. In order to gain insight into the plausible reaction mechanism, several control experiments were conducted.

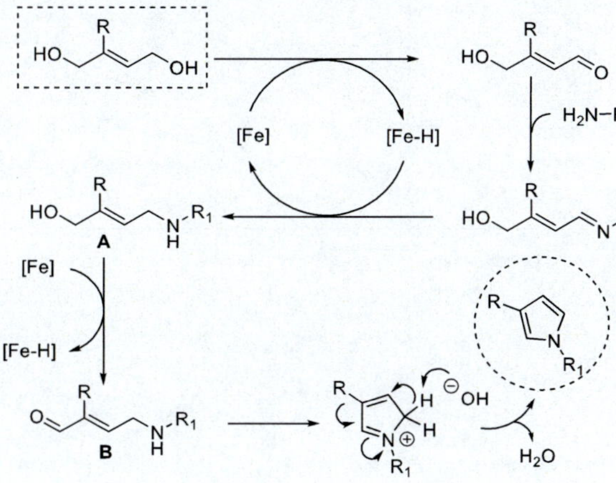

Scheme 19 Proposed mechanism for Fe(0)-catalyzed pyrrole synthesis. Redrawing with permission from Ref. [65]. Copyright 2017 American Chemical Society

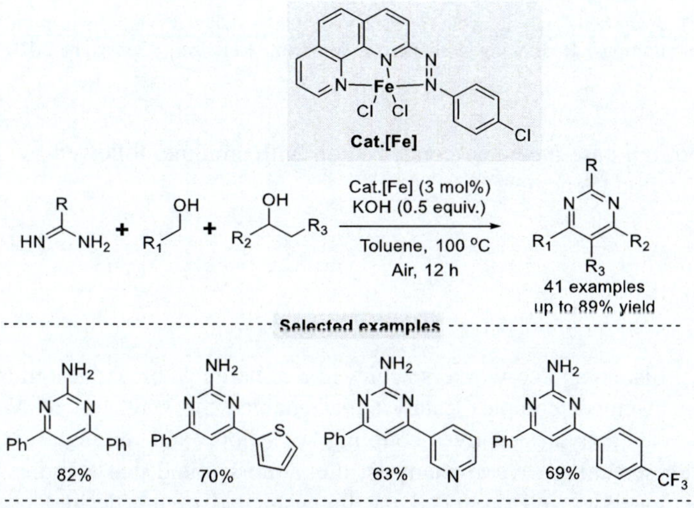

Scheme 20 Fe(II)-complex catalyzed synthesis of pyrimidines under oxidative conditions [66]

In the presence of KO^{t}Bu, the precatalyst undergoes one-electron reduction, leading to the formation of a mono-anionic species containing an azo-anion radical ligand. The iron-stabilized azo-anion radical initiates a hydrogen atom abstraction from the alcohol through a one-electron hydrogen atom transfer (HAT) process, forming a ketyl radical intermediate. The ketyl radical undergoes rapid one-electron oxidation, generating the corresponding carbonyls. The in situ-formed carbonyl compounds undergo base-promoted condensation, leading to the formation of α,-β-unsaturated ketones. Subsequently, the desired pyrimidines are produced in good

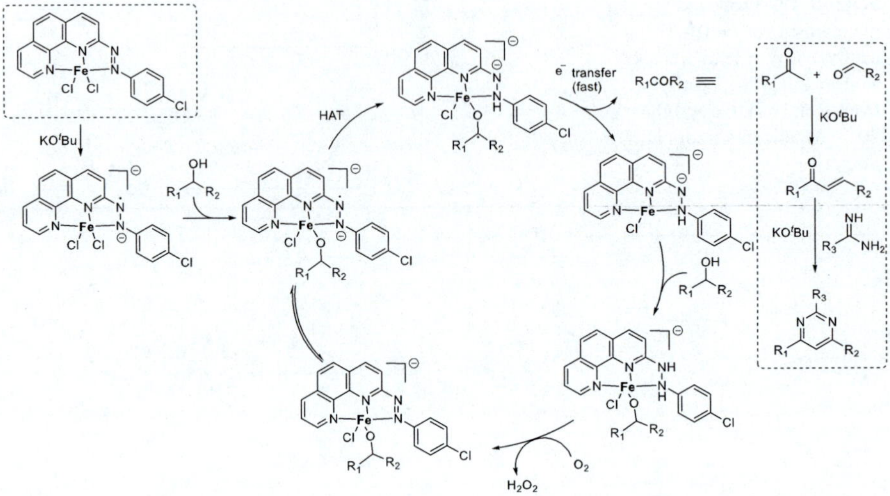

Scheme 21 Proposed catalytic cycle for Fe(II)-complex catalyzed synthesis of pyrimidines under oxidative conditions. Redrawing with permission from Ref. [66]. Copyright 2020 Wiley-VCH Verlag

yields through base-mediated condensation with amidine, followed by intramolecular cyclization (Scheme 21).

4 Cobalt

In 2016, Milstein and co-workers achieved a remarkable breakthrough by developing a cobalt-pincer complex catalyst that enabled the synthesis of *N*-substituted pyrroles via dehydrogenative coupling of diols and amines (Scheme 22) [67]. After screening several catalysts, the authors found that complex **A** showed excellent reactivity and produced the highest yield of the desired product. This innovative reaction produced only water and hydrogen gas as the by-products. The catalytic process, mediated by Co-pincer complexes, exhibited versatility by tolerating a wide range of primary alkyl amines, benzylic amines, aromatic amines, and primary and secondary diols.

The cobalt complex binds with alcohol giving an alkoxy cobalt complex. Dehydrogenation of this complex occurs to form the ketone. The ketone undergoes coupling with the amine and forms the N-substituted pyrrole and water in a Paal-Knorr condensation (Scheme 23).

In 2018, the first phosphine ligand-free, non-precious base-metal catalyst that enables the direct synthesis of *N*-heterocyclic compounds (1*H*-pyrroles, pyridines, quinolines, and pyrazine) from amino alcohols and alcohols as key starting materials with the liberation of hydrogen gas and water as the sole by-products has been

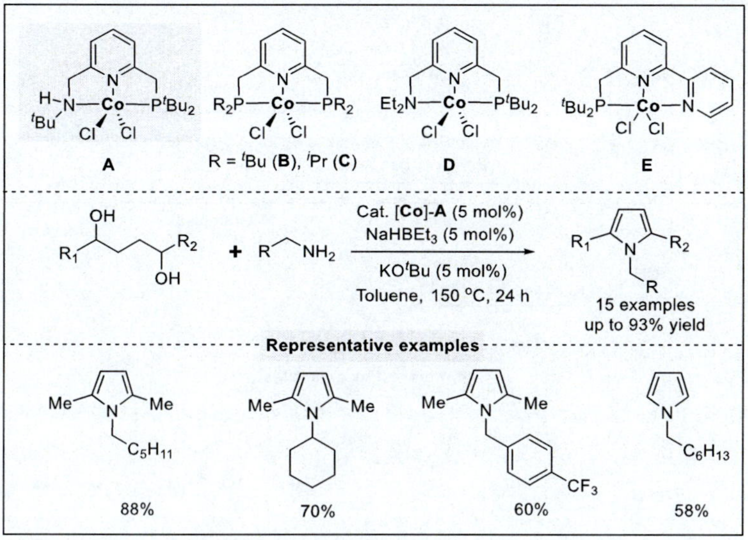

Scheme 22 Cobalt-pinner complex catalyzed synthesis of pyrroles [67]

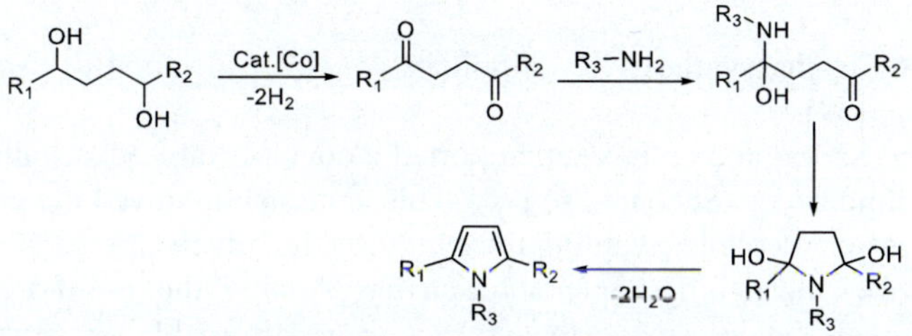

Scheme 23 Cobalt-catalyzed synthesis of pyrroles: A proposed catalytic cycle [67]

described by the research group of Balaraman (Scheme 24) [68]. The air-stable, molecularly defined SNS-cobalt(II) complex catalyzes these reactions. The optimized reaction conditions exhibited excellent reactivity, yielding the corresponding *N*-heterocyclic compounds in good to excellent yields. The authors demonstrated that a sulfur-based ligand is a possible extension to replace the phosphine ligand and worked efficiently in the tandem dehydrogenative annulation reactions.

In the same year, Kundu and co-workers also reported a phosphine-free Co (II) complex for *N*-heterocyclic synthesis. The authors devised a synthesis method for various quinoxalines by dehydrogenative coupling of vicinal diols with o-phenylenediamines or 2-nitroanilines. Moreover, this complex catalyst proved

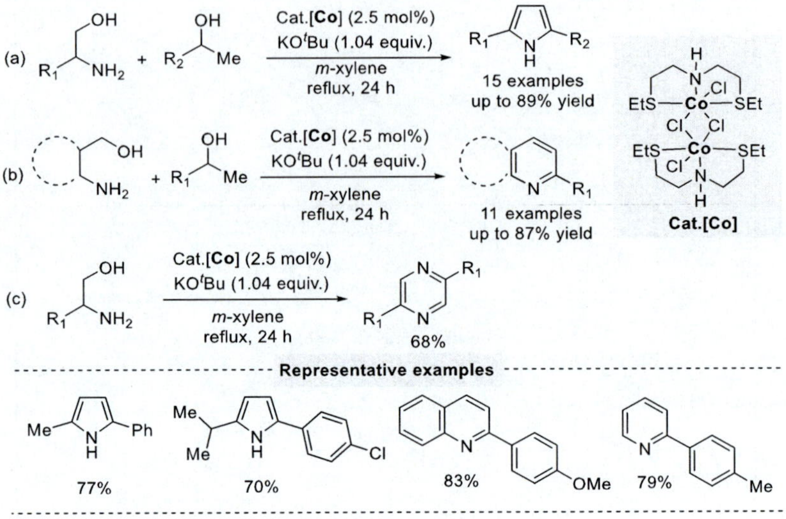

Scheme 24 An SNS-pincer ligand-based Co(II)-complex catalyzed synthesis of *N*-heterocyclic compounds [68]

its efficiency in the synthesis of quinolines as well with good to excellent yields (Scheme 25) [69].

Recently, Sarkar and co-workers reported a cobalt-catalyzed sustainable synthesis of benzimidazoles (Scheme 26) [70]. This approach involved the coupling of *o*-nitroanilines and alcohols, enabling the synthesis of a diverse range of 2-substituted benzimidazoles through hydrogen autotransfer. Notably, the use of a commercially available cobalt catalyst made the process more affordable. A series of control experiments were conducted to gain a comprehensive understanding of the reaction mechanism, providing valuable mechanistic insights into this transformative synthesis strategy. A plausible catalytic cycle for this reaction is as follows, base-assisted oxidation of alcohol leads to the corresponding aldehyde and [Co]-H intermediate, which reduces the nitroaniline to diamine followed by condensation between aldehyde and diamine produces the dihydro-benzimidazole derivative which upon oxidative aromatization delivered the expected product (Scheme 27).

Later, Kundu and co-workers reported on the synthesis of quinoxaline motifs utilizing a nitrogen-doped carbon-supported cobalt catalyst (Co-phen/C-800). This reaction involved transfer hydrogenation and acceptorless dehydrogenative coupling strategies, which enabled the coupling of diamines and diols to form quinoxalines (Scheme 28) [71]. The protocol showed its practical viability through gram-scale synthesis. The catalyst could be recovered and reused with the same catalytic activity for several runs under optimal conditions.

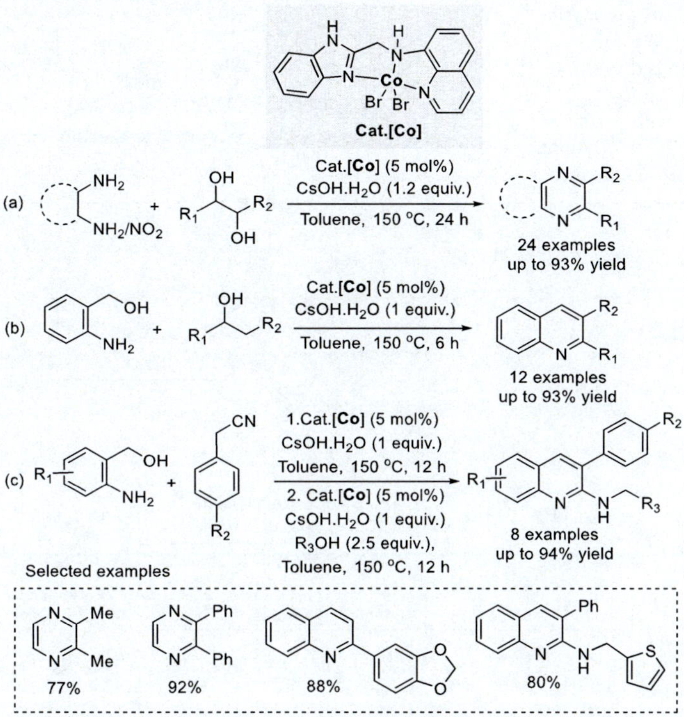

Scheme 25 An NNN-Pinner ligand-based Co(II)-complex catalyzed synthesis of quinoxalines and quinolines [69]

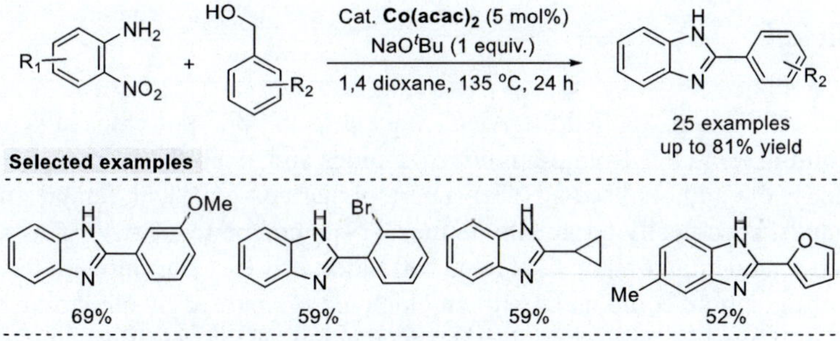

Scheme 26 Cobalt-catalyzed synthesis of benzimidazoles [70]

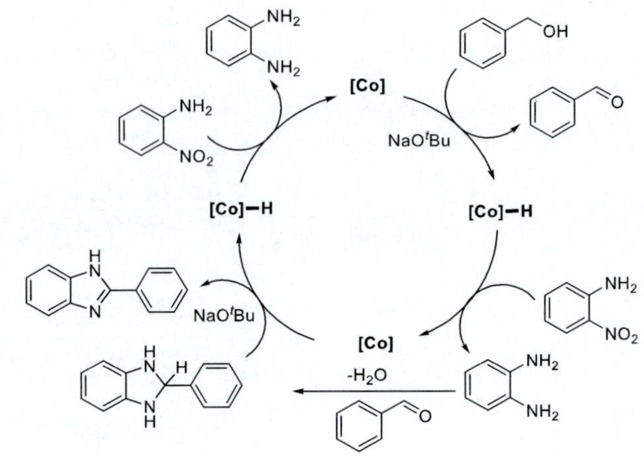

Scheme 27 A proposed catalytic cycle for the cobalt-catalyzed synthesis of benzimidazoles. Redrawing with permission from Ref. [70]. Copyright 2019 American Chemical Society

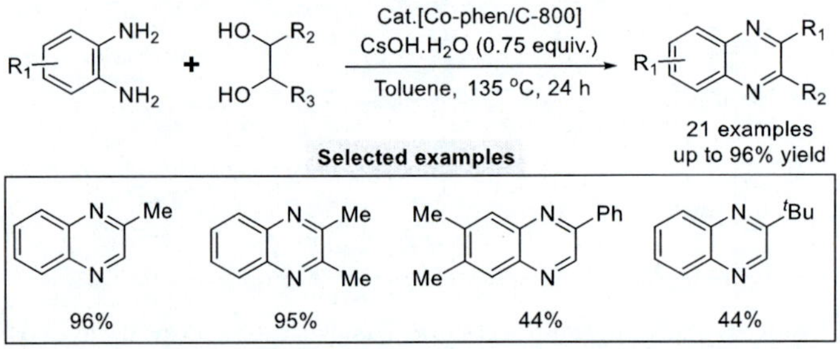

Scheme 28 Co-phen/C-800 catalyzed synthesis of quinoxalines [71]

5 Nickel

Paul and co-workers applied the ADC concept to the one-pot cascade synthesis of quinazolin-4(3*H*)-ones from o-aminobenzamides and alcohols (Scheme 29) [72]. In their study, the authors used a Ni(II) square planar complex with tetraaza macrocyclic ligands, specifically tetramethyltetraaza[14]annulene (MeTAA). This approach involved a sequential double C-N bond formation process, starting with the condensation of the amine component with an aldehyde (generated by alcohol dehydrogenation), followed by an intramolecular cyclization of the resulting amide with an imine.

Further, synthesis of quinolines and pyridines utilizing γ-amino benzyl alcohols and substituted ketones using Ni-complex was developed by Banerjee and co-workers (Scheme 30) [73]. This nickel-catalyzed reaction exhibited a limited substrate scope; however, it yielded moderate to good yields. The reaction conditions involved the use of nickel(II) bromide/1,10-phenanthroline (at a ratio of 5: 6 mol%) as the catalyst, with KO^{t}Bu (1 equivalent) serving as the base.

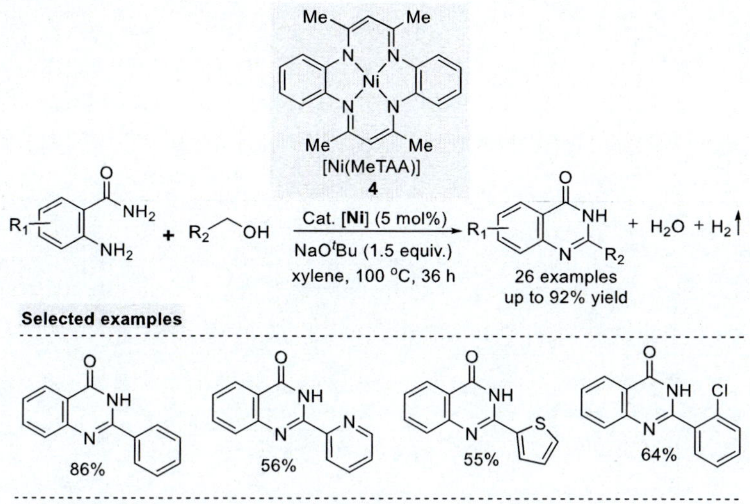

Scheme 29 Ni-catalyzed cyclization of 2-aminobenzamides of alcohols [72]

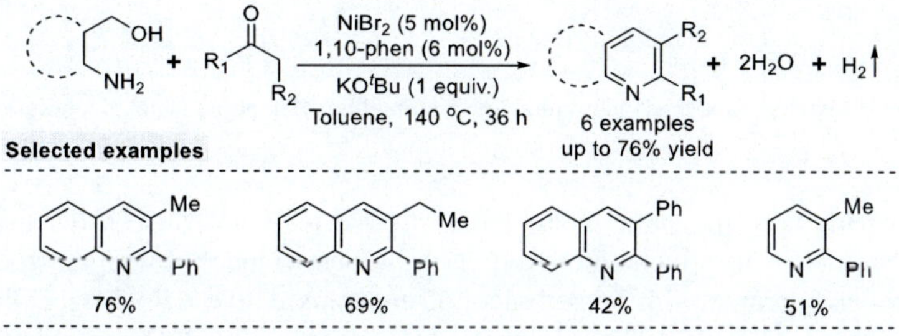

Scheme 30 Synthesis of quinolines and pyridine using ketone and alcohols [73]

Paul and co-workers reported a nickel-catalyzed synthesis of quinolines through mono- and double dehydrogenative cyclization reactions involving o-aminobenzyl alcohols and either ketones or secondary alcohols (Scheme 31a) [74]. They discovered that the application of a Ni catalyst proved highly effective for both ketones and secondary alcohols as coupling partners with o-aminobenzyl alcohols, resulting in the production of a diverse array of substituted quinolines with yields ranging from 22% to 89%. Further, Das and co-workers utilized the same catalyst and conditions to synthesize polysubstituted quinolines from α-2-aminoaryl alcohols (Scheme 31b) [75]. This method offers several advantages, including a low catalyst loading and a broad range of substrates, resulting in isolated yields of up to 93%. A notable aspect of this approach is the utilization of secondary α-2-aminoaryl alcohols, enabling the sequential dehydrogenation and condensation process that leads to the synthesis of diverse 4-substituted quinolines.

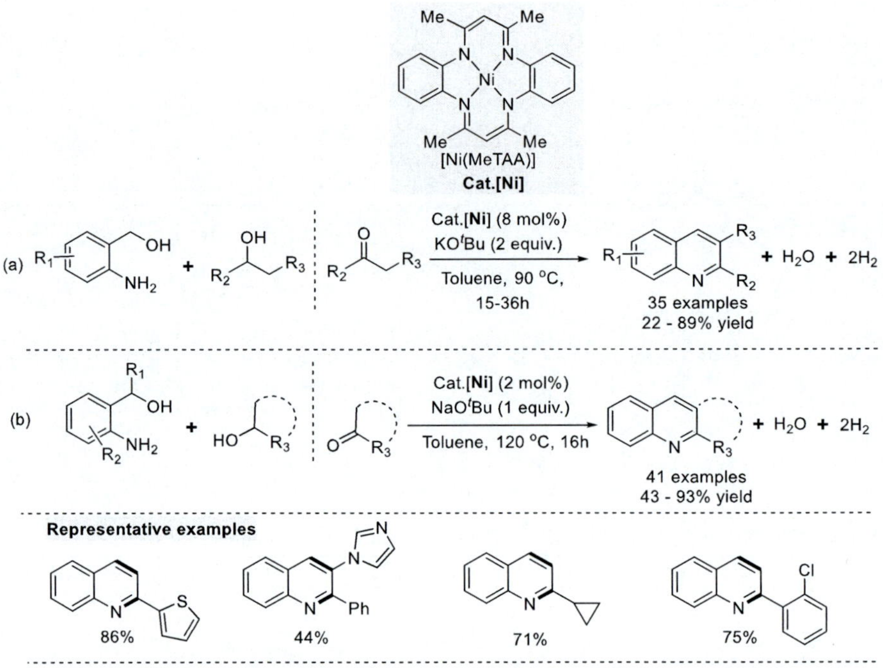

Scheme 31 Nickel-catalyzed annulation of 2-aminobenzylalcohols with alcohols and ketones [74, 75]

Following that, the same nickel catalyst [(NiMeTAA)] was further investigated for the synthesis of quinazolines from 2-aminobenzyl alcohol with benzonitrile and 2-aminobenzylamine with benzyl alcohol under oxidative dehydrogenation conditions (Scheme 32) [76]. The synthesis of various substituted quinazolines was accomplished in moderate to good yields using easily available starting materials, showing a simple approach.

Later, Paul and co-workers extended the annulation of 2-aminobenzyl alcohols with other nucleophiles. They used various nucleophilic partners, such as ketones and nitriles, in a dehydrogenative coupling reaction with 2-aminobenzyl alcohols. This reaction employed a singlet diradical Ni(II) catalyst, leading to successful annulation of the substrates (Scheme 33) [77]. The synthesis of various polysubstituted quinolines, 2-aminoquinolines, and quinazolines was achieved using well-defined singlet diradical Ni(II)-complexes with simple diamine ligands. This methodology involved the dehydrogenative condensation/coupling of 2-aminobenzyl alcohols with ketones, 2-phenylacetonitrile, and nitriles, respectively. Through mechanistic studies, it was revealed that both the nickel catalyst and the coordinated redox-active ligand worked synergistically in the synthesis of these different quinolines, 2-aminoquinolines, and quinazolines.

In 2018, Banerjee and co-workers reported the synthesis of pyrrole, pyridine, and quinoline derivatives through the ADC reaction of β- and γ-amino alcohols with

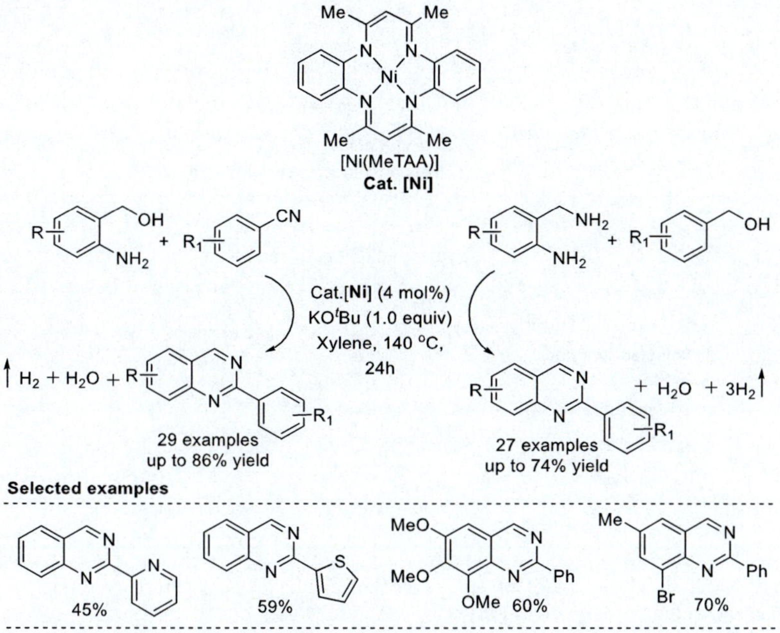

Scheme 32 Dehydrogenative coupling of 2-aminobenzyl alcohols with benzonitriles and 2-aminobenzylamines with benzyl alcohols [76]

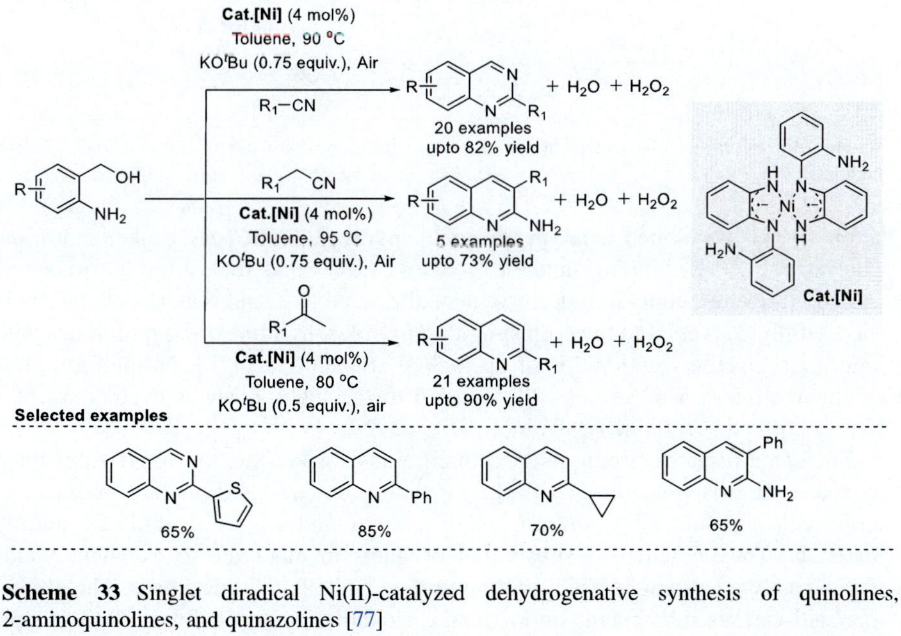

Scheme 33 Singlet diradical Ni(II)-catalyzed dehydrogenative synthesis of quinolines, 2-aminoquinolines, and quinazolines [77]

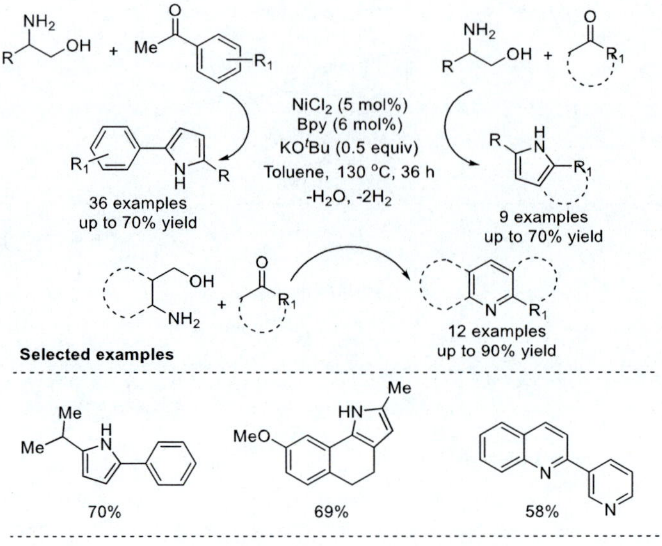

Scheme 34 Ni-catalyzed ADC of amino alcohols with ketones: Direct synthesis of pyrrole, pyridine and quinoline derivatives [78]

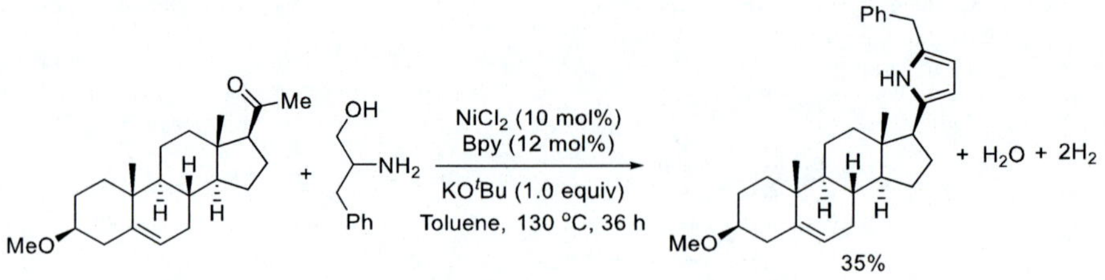

Scheme 35 Intermolecular cyclization of a steroid hormone with phenylalaninol under Ni(II)-catalysis [78]

ketones. The developed catalytic protocol involved a $NiCl_2$/1,10-phenanthroline catalyst system, facilitating tandem C–N and C–C bond formation. Various aryl and alkyl ketones, containing free amine, halides, alkoxy, and N-heterocycles, were successfully converted to the corresponding pyrroles, pyridines, and quinolines with high regioselectivity and yields of up to 90% (Scheme 34) [78]. Furthermore, this synthetic strategy was successfully applied to the intermolecular cyclization of a steroid hormone with phenylalaninol (Scheme 35).

The same research group also explored a sustainable method for *N*-substituted pyrrole synthesis (Scheme 36) [79] by reacting butene-1,4-diols and butyne-1,4-diols with aryl and alkyl amines, using ketones and amino alcohols as starting materials. They obtained *N*-substituted products in moderate to excellent yields with a catalytic system of $NiCl_2$ and bipyridine, and Na_2CO_3 as a base. The system tolerated various substituents on the amino alcohols, such as halides, alkyl groups,

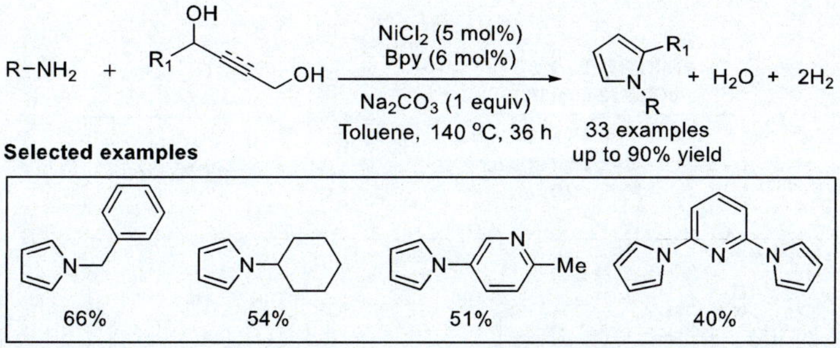

Scheme 36 Synthesis of *N*-substituted pyrroles via the ADC of 1,4-diols with amines [79]

alkoxy groups, heterocycles, and aromatic groups, giving *N*-substituted pyrroles with up to 90% yield.

In 2019, the same group reported the synthesis of substituted pyrroles using a nickel-catalyzed double dehydrogenative coupling method involving secondary and β-amino alcohols (Scheme 37) [80]. Through this process, they successfully obtained a diverse array of 2,5- and 2,3,5-substituted pyrroles with satisfactory to good yields.

The research groups of Tang and Zhou reported the use of $Ni(OTf)_2$ (2 mol%) and 1,3-bis(dicyclohexylphosphine)ethane (dcpe) (2.5 mol%) to enable a divergent cyclization process. This approach involves the reaction of diamine with diols to produce tetrahydroquinoxalines and quinoxaline derivatives using the HA strategy (Scheme 38) [81].

Recently, Banerjee and co-workers reported selective synthesis of substituted benzimidazole and quinoxaline derivatives through the nickel-catalyzed dehydrogenative coupling of aromatic diamines with alcohols (Scheme 39) [82].

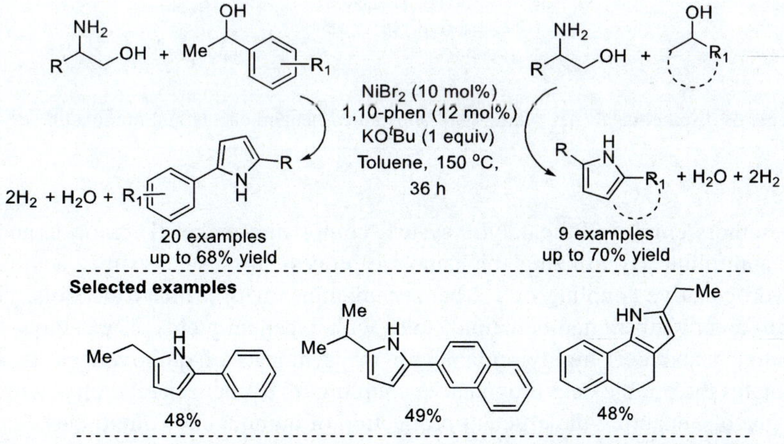

Scheme 37 Synthesis of substituted pyrroles using secondary and β-amino alcohols [80]

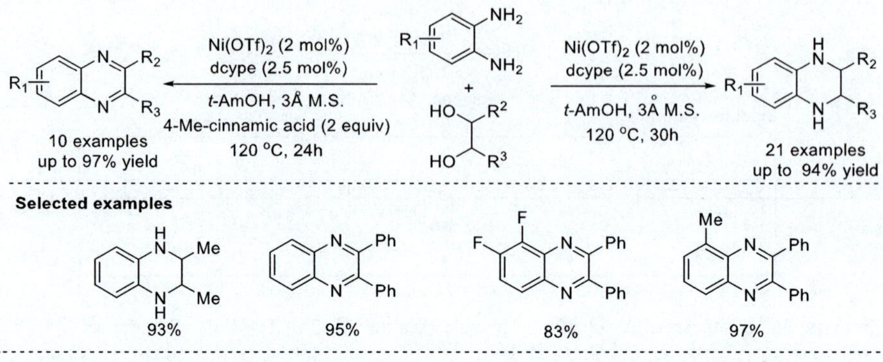

Scheme 38 Cyclization of diamines with diols under nickel-catalysis [81]

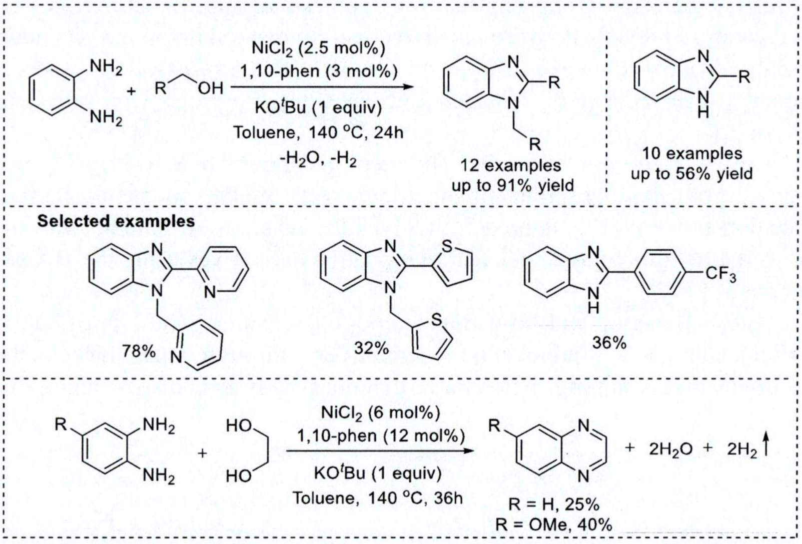

Scheme 39 Nickel-catalyzed synthesis of substituted benzimidazole and quinoxaline derivatives [82]

The authors employed a catalytic system comprising nickel(II) chloride and 1,10-phenanthroline. By utilizing this catalytic process, they successfully achieved the dehydrogenative coupling of 1,2-benzenediamines with primary alcohols, yielding mono- and di-substituted benzimidazoles with excellent yields. The catalyst system mentioned above is highly suitable for the synthesis of quinoxalines as well. It facilitates the double dehydrogenative coupling of 1,2-benzenediamines with ethylene glycols, enabling the efficient production of quinoxaline compounds.

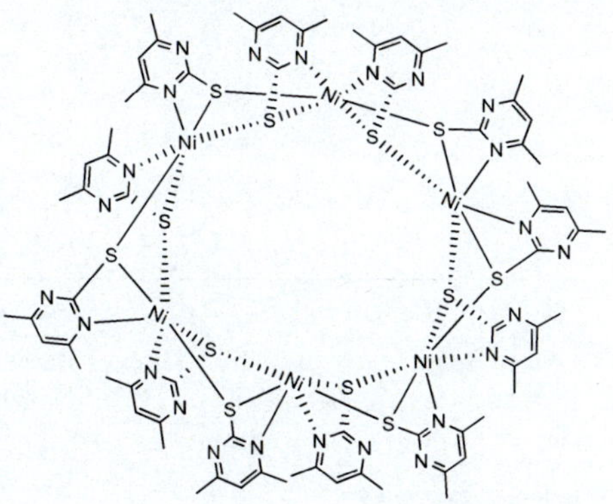

Fig. 2 Structure of Ni(II)/thiolate cluster [Ni(dmpymt)$_2$]$_6$. Redrawing with permission from Ref. [83]. Copyright 2019 Royal Society of Chemistry

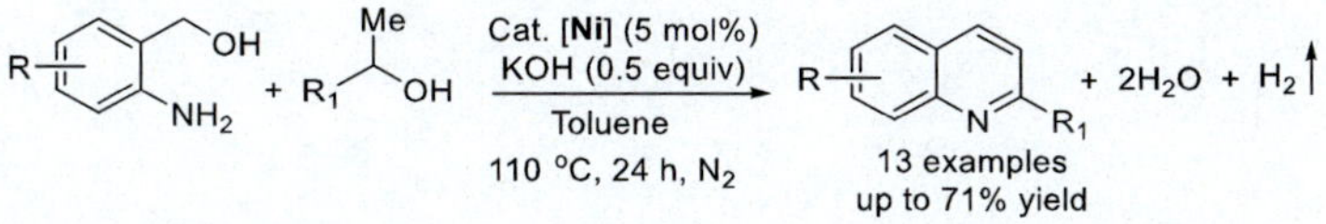

Scheme 40 [Ni(dmpymt)$_2$]$_6$ catalyzed dehydrogenative annulation of 2-aminobenzyl alcohols with secondary alcohols [83]

The research groups of Li and Lang developed a Ni(II)/thiolate cluster catalyzed (Fig. 2) cross-coupling annulation of 2-aminobenzyl alcohols with secondary alcohols. This innovative method enables the synthesis of quinolines under mild reaction conditions (Scheme 40) [83].

Further, Srivastava and co-workers used a pincer nickel complex for the synthesis of benzimidazoles by the dehydrogenative coupling of amines with alcohol (Scheme 41) [84]. They performed the reaction in open vessel at high temperature, using low catalyst loading, under solvent-free condition and got the products in moderate yields with high turnover numbers. The nickel pincer complex synthesized by them also used for the N-alkylation of anilines with benzyl alcohols. From the DFT studies they found that β-hydride elimination is the RDS for alcohol dehydrogenation. They have performed the labeling experiment and found the involvement of benzylic C–H bond in the RDS with k_{CHH}/k_{CHD} value of 2.5.

Recently, Balaraman and co-workers reported the synthesis of *N*-substituted indoles from amino alcohols and alcohols, utilizing a one-pot cascade approach under additive and base-free conditions (Scheme 42) [85]. This process generates water as the sole stoichiometric by-product. They demonstrate the remarkable efficacy of commercially available, stable Ni(OTf)$_2$ salt with 1,2-bis (dicyclohexylphosphino)ethane (dcype) as a catalyst for this transformation. The reaction conditions accommodate a wide range of substrates, including aromatic and aliphatic primary alcohols, cyclic and acyclic secondary alcohols, as well as various substituted 2-aminophenyl ethyl alcohols. As a result, various *N*-alkylated indoles were produced under benign conditions.

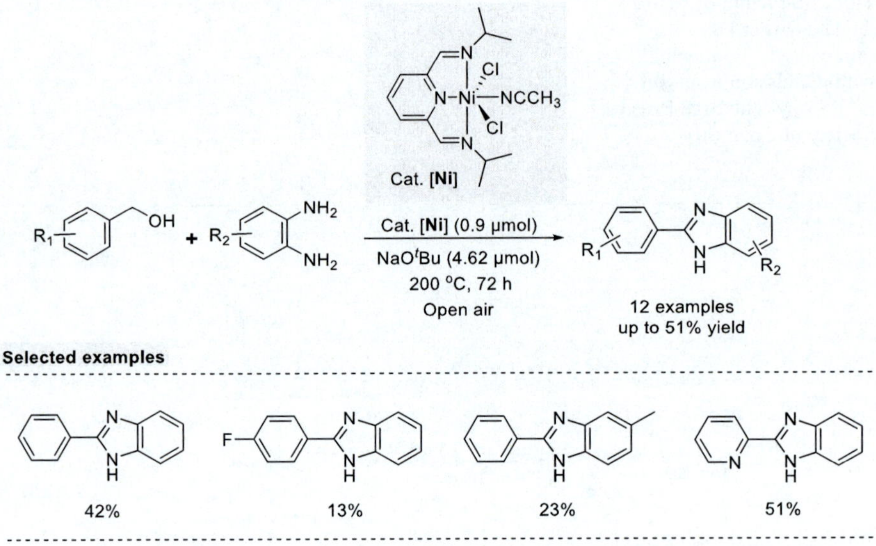

Scheme 41 Ni catalyzed synthesis of benzimidazoles [84]

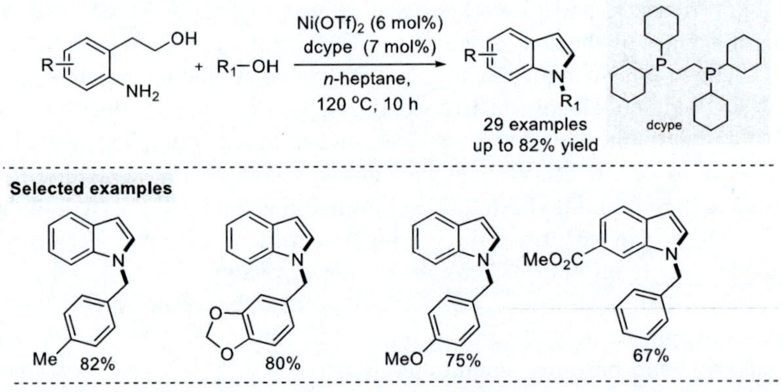

Scheme 42 Nickel-catalyzed direct synthesis of *N*-substituted indoles from amino alcohols and alcohols [85]

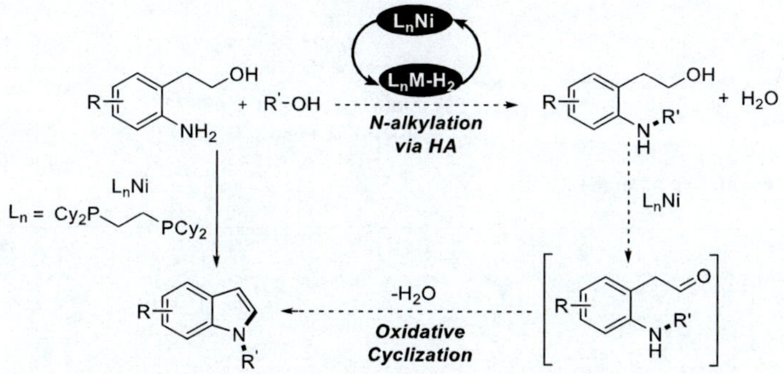

Scheme 43 A proposed catalytic cycle for the one-pot nickel-catalyzed synthesis of *N*-alkylated indoles [85]

The proposed mechanism for this one-pot nickel-catalyzed synthesis of *N*-alkylated indoles suggests that a dehydrogenation/hydrogenation cascade (hydrogen autotransfer) occurs, leading to the generation of an *N*-alkylated alcohol intermediate. Subsequently, the intermediate alcohol undergoes oxidative cyclization, resulting in the formation of the desired *N*-alkylated indole product (Scheme 43).

6 Copper

Chao and co-workers reported the synthesis of quinolines by the coupling of 2-Aminobenzyl alcohol with secondary alcohols in the presence of a copper catalyst combined with MS 4 Å or PEG-2000 (Scheme 44) [85]. They could reuse the catalytic system up to 10 times without the loss of reactivity. Few quinoline derivatives were synthesized in good to excellent yields.

Wu and co-workers reported the synthesis of 2-substituted quinazolines under copper catalyzed cascade reaction of (2-aminophenyl)-methanols, aldehydes, and ammonium chloride (Scheme 45) [87]. A wide variety of substituted quinazoline

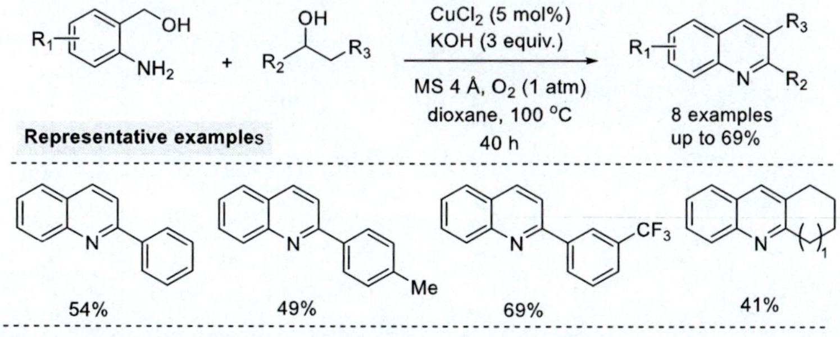

Scheme 44 Cu-catalyzed synthesis of quinolines [86]

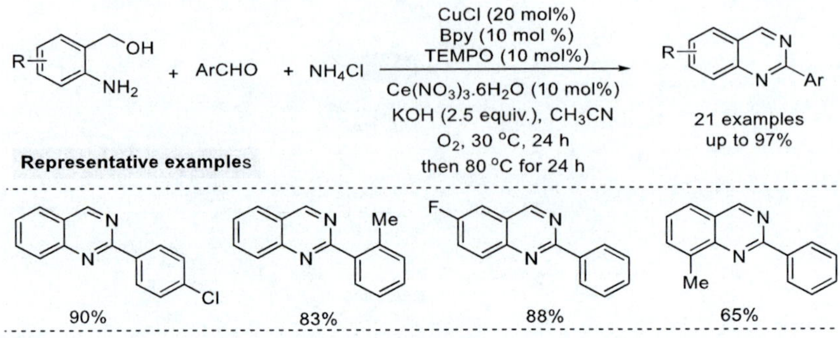

Scheme 45 Cu-catalyzed synthesis of quinazolines [87]

derivatives were synthesized and several control experiments were also performed to get insights into the reaction mechanism.

Based on experimental results, two possible reaction pathways have been proposed for the formation of quinazolines. First, alcohol oxidation occurs, followed by condensation reaction with in situ generated ammonia (from NH_4Cl), which leads to imine formation. Which further undergoes condensation with another carbonyl compounds as a coupling partner then cyclization and aromatization lead to quinazolines in the case of path b, but in the case of path a, it is vice versa (Scheme 46) [87].

The research group of Zhang reported a simple method for the preparation of 1,3,5-triazines by the oxidative coupling between benzamidine and alcohols using $Cu(OAc)_2$ as the catalyst (Scheme 47) [88]. Both aromatic and aliphatic alcohols converted to triazine products in good yields. As the reaction can be performed in atmospheric air and doesn't require any ligands, thus, it makes this approach highly suitable for large-scale synthesis.

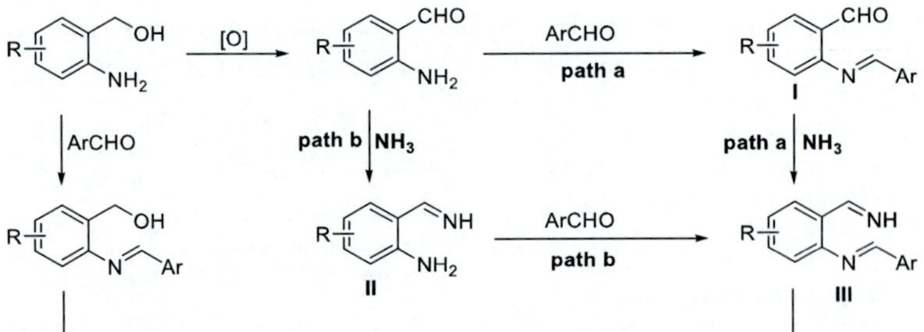

Scheme 46 Proposed mechanism for Cu-catalyzed quinazolines synthesis. Reproduced with permission from Ref. [87]. Copyright 2013 American Chemical Society

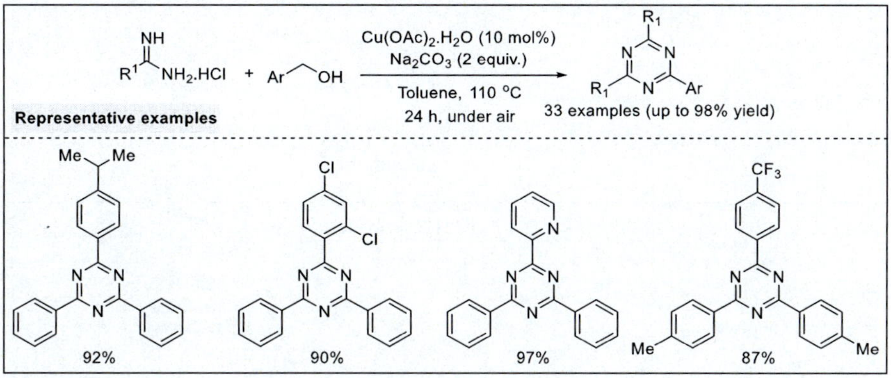

Scheme 47 Cu-catalyzed synthesis of substituted triazines [88]

Scheme 48 Proposed mechanism for Cu-catalyzed triazines synthesis. Redrawing with permission from Ref. [88]. Copyright 2015 Royal Society of Chemistry

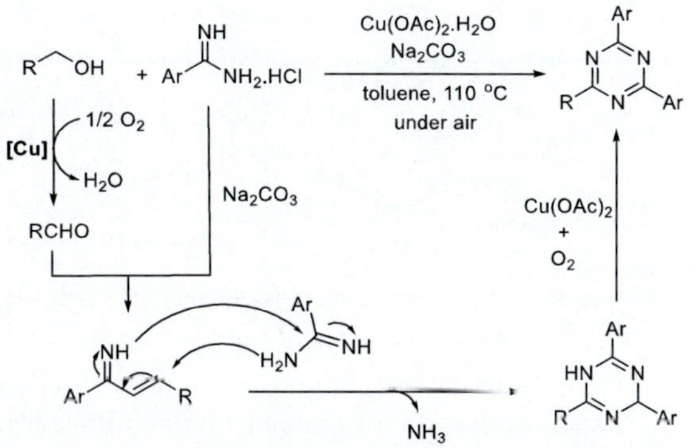

$Cu(OAc)_2$ catalyzed alcohol oxidation in air to form the aldehyde, which then reacts with the amidines, promoted by $Cu(OAc)_2$ and air oxidation giving rise to 1,3,5-triazines (Scheme 48) [88].

In 2017, Wang and co-workers reported the synthesis of benzimidazoles using copper catalyst. Copper catalyst used by them exhibited excellent and tunable catalytic activity for both dehydrogenation and borrowing hydrogen reactions and they demonstrated more than 80 examples in good yields (Scheme 49) [89]. Ligand played a crucial role in promoting copper-catalyzed dehydrogenation and borrowing hydrogen reactions. To better understand the mechanism, deuterium labeling experiments were performed which showed that the initial reversible alcohol dehydrogenation step involved a copper hydride intermediate, which confirmed a copper-catalyzed dehydrogenation reaction.

Initially, in the presence of Cu-catalyst alcohol undergoes dehydrogenation and yields aldehyde which then undergoes condensation with amine to give imine. For pathway **I**, bismine formation from condensation, cyclisation, and rearrangement

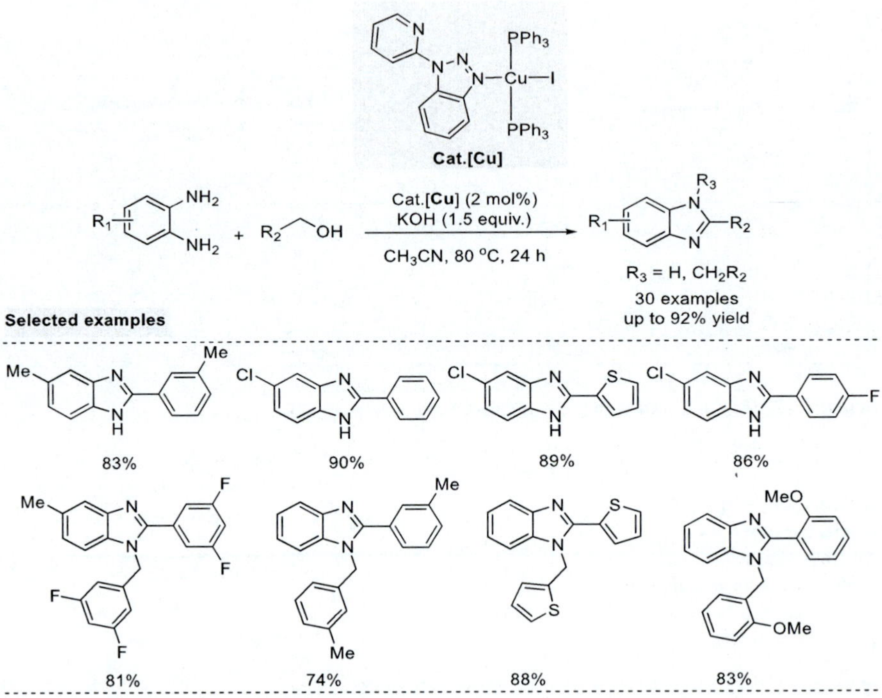

Scheme 49 Cu-catalyzed synthesis of benzimidazoles [89]

occurs to form the benzimidazole. Similarly, a pathway **II** can also be followed, but the exact mechanism of this reaction is not clear (Scheme 50).

The same strategy for the synthesis of benzimidazole derivatives was extended by Wang and co-workers using a heterogeneous hydrotalcite supported BINAP-Cu system (Scheme 51) [90]. The catalyst can be recycled up to the 5th run without any considerable loss in activity and selectivity. Notably, water was used as a solvent for this reaction (Scheme 51).

Subsequently, Zhang and co-workers reported the one-pot three-component synthesis of multi-substituted pyrimidine derivatives by coupling amidine, benzyl alcohol, and 1-phenyl alcohol with the use of an inexpensive copper catalyst (Scheme 52) [91]. Key benefits of this protocol are this approach doesn't require the use of ligands and can be performed under air. This protocol showed excellent functional group tolerance.

Based on the experimental results and literature reports, possible mechanism is proposed. Initially oxidation of both primary and secondary alcohols occurs in presence of copper catalyst and molecular oxygen. Then, the carbonyl compounds undergo aldol type condensation to form α,β-unsaturated ketone followed by reaction with amidine via intermolecular Michael addition and condensation to lead to

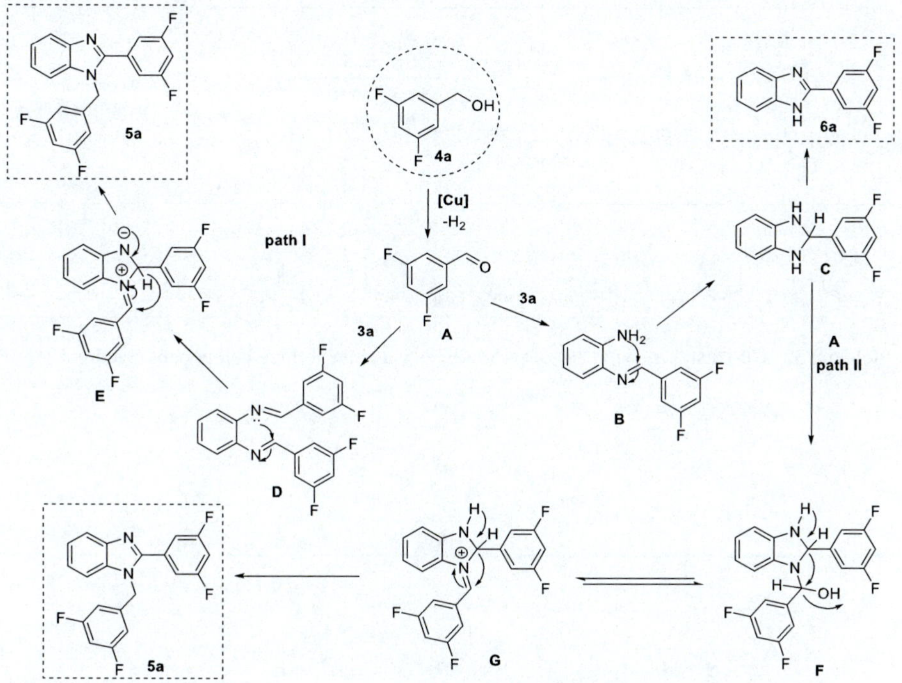

Scheme 50 Plausible pathways for the Cu-catalyzed benzimidazole synthesis. Redrawing with permission from Ref. [89]. Copyright 2017 Wiley-VCH Verlag

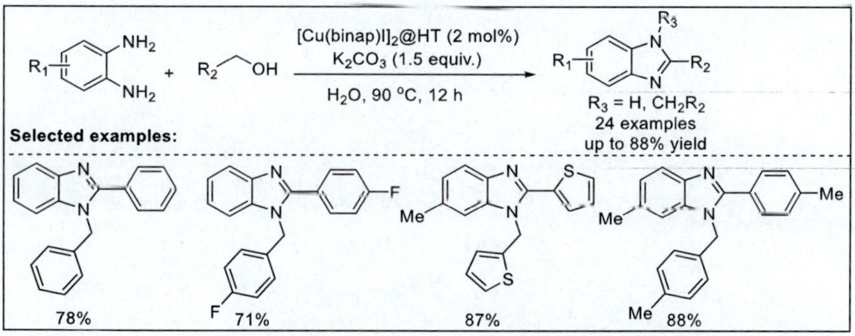

Scheme 51 Heterogeneous hydrotalcite supported Cu-catalyzed synthesis of benzimidazoles [90]

the intermediate **D**. Later, aromatization of **D** in the presence of [Cu]/O_2 forms the desired product pyrimidines (Scheme 53).

Li and co-workers developed a simple and efficient method to synthesize various quinazolinones from 2-aminobenzonitriles and benzyl alcohols using a copper catalyst and air as the sole oxidant (Scheme 54) [92]. They achieved good yields

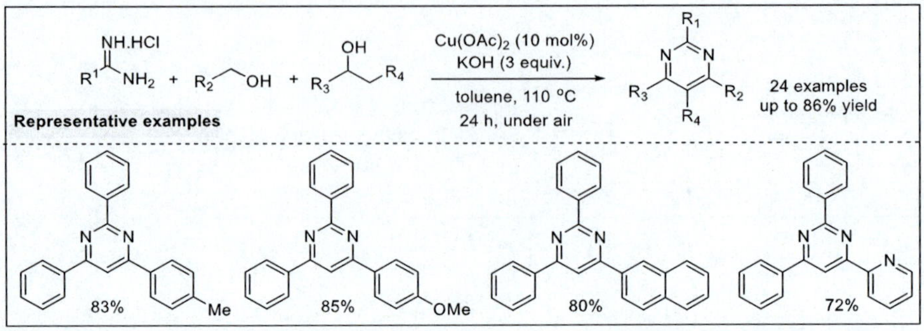

Scheme 52 Cu-catalyzed synthesis of pyrimidines via tandem three-component reaction [91]

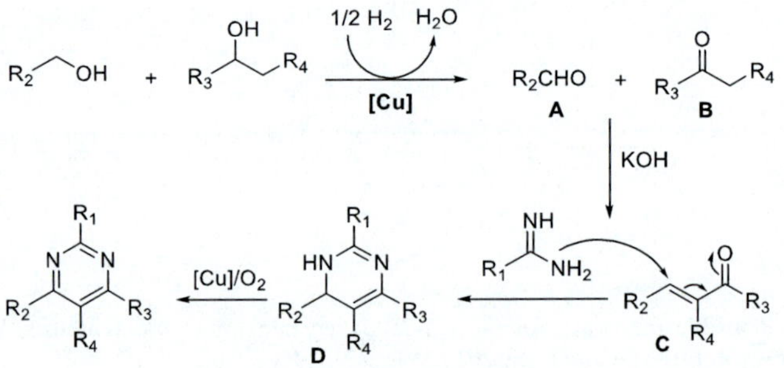

Scheme 53 Proposed mechanism for the Cu-catalyzed synthesis of pyrimidines [91]

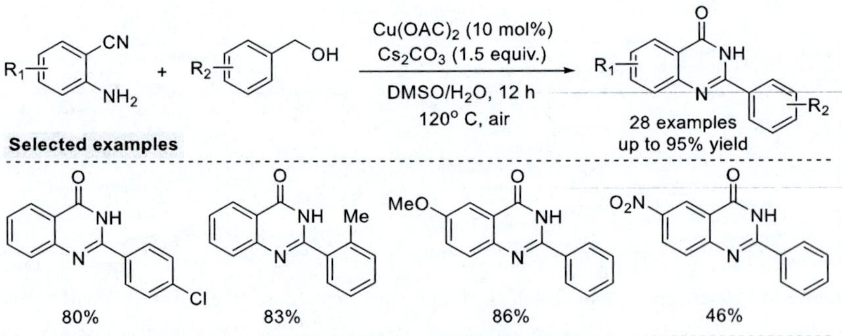

Scheme 54 Cu-catalyzed synthesis of quinazolinones [92]

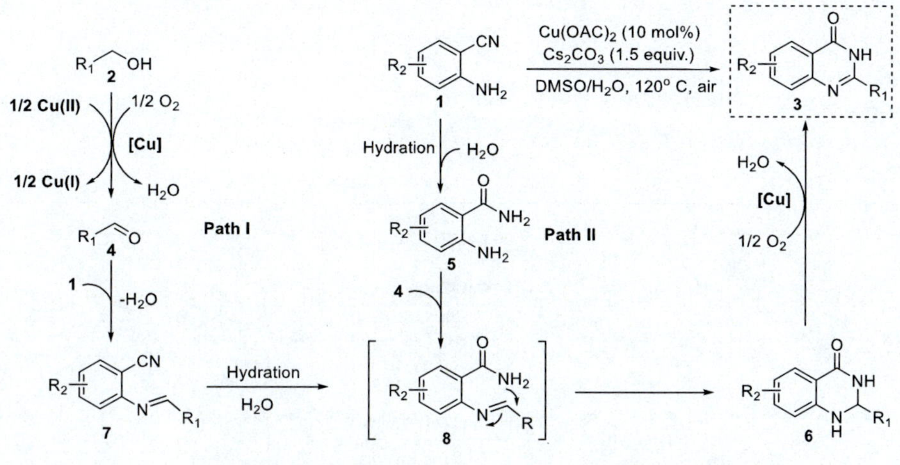

Scheme 55 Proposed mechanism for the Cu-catalyzed quinazolinones synthesis. Redrawing with permission from Ref. [92]. Copyright 2019 Royal Society of Chemistry

for a wide range of substrates. The proposed reaction mechanism is based on the control experiments and the literature.

Two possible mechanistic pathways for the formation of quinazolinones are proposed. First, $Cu(OAc)_2$ catalyzes the aerobic oxidation of primary alcohols. The imine intermediate **7** formed by the condensation reaction of **4** and **5** transforms into **8** by hydration. Then intramolecular cyclization occurs, followed by oxidation to form the quinazolinone (Scheme 55) [92].

7 Zinc

In 2021, Paul and co-workers reported the method to synthesize *N*-heterocycles by employing a Zn(II)-stabilized azo-anion radical complex as the catalyst that involved the dehydrogenative coupling of alcohols under air. This method proved quinazolin-4(3*H*)-ones, quinazolines, and quinolines (Scheme 56) [93]. A mechanistic study showed that the dehydrogenation reactions follow a pathway involving a one-electron hydrogen atom transfer (HAT). In this process, the zinc-stabilized azo-anion radical ligand abstracts the hydrogen atom from the alcohols/amino alcohols. The entire catalytic cycle proceeds exclusively through redox events centered on the ligand, with the zinc serving solely as a template.

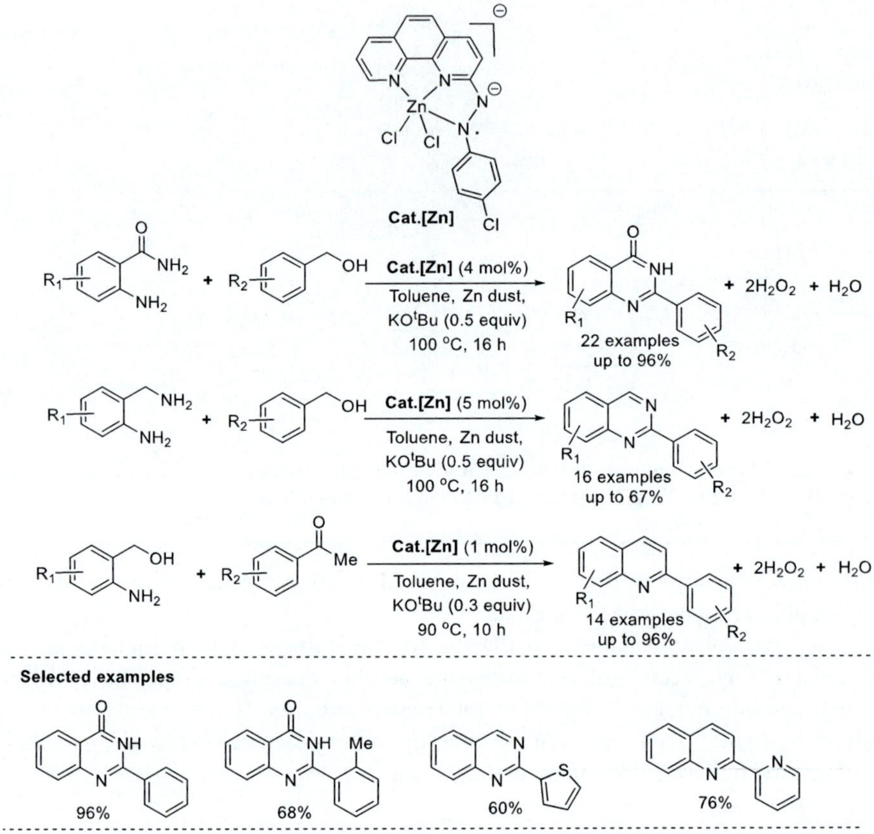

Scheme 56 zinc-catalyzed synthesis of quinazolin-4(3*H*)-ones, quinazolines, and quinolines [93]

8 Summary

This chapter summarizes the recent progress in the 3d-transition metal-catalyzed dehydrogenative coupling of alcohols and amino alcohols to produce various heterocycles. This method enables a wide range of new reactions and transformations using a dehydrogenation and coupling strategy. The use of safe and less harmful reagents such as alcohols and amines offers a new avenue in synthetic chemistry. Importantly, the generation of non-toxic by-products, such as water and/or dihydrogen, makes this protocol highly attractive for the green synthesis of value-added chemicals under atom-efficient and eco-friendly conditions. The use of abundant base-metal catalysts improves reactivity and facilitates the direct synthesis of both simple and complex molecules. The remarkable improvements in catalytic systems have led to different prefix terms for the catalyst, such as acceptorless, solventless, and base-free, making this synthetic approach an alternative to conventional methods for academic and industrial purposes. Moreover, the development of low-cost, air-stable catalysts and their easy handling have gained significant recognition in the field of sustainable catalysis.

Acknowledgments

This work is supported by CSIR (Project No: 01/(3030)/21/EMR-II). EB is a Swarnajayanti Fellow (Award No: SERB/F/5892/20202021) and gratefully acknowledges support from the Alexander-von-Humboldt (AvH) Foundation. G.S. acknowledges IISER-Tirupati and A.K.S. thanks UGC for their fellowships.
Declaration of Competing Interest The authors declare that they have no known competing financial interests.

References

1. Amin A, Qadir T, Sharma PK et al (2022) A review on the medicinal and industrial applications of N-containing heterocycles. Open Med Chem J 16
2. Gomtsyan A (2012) Heterocycles in drugs and drug discovery. Chem Heterocycl Compd 48:7–10
3. Katritzky A, Ramsden C, Scriven E, Taylor R (2008) Comprehensive heterocyclic chemistry III, vol 7, pp 217–308
4. Kerru N, Gummidi L, Maddila S et al (2020) A review on recent advances in nitrogen-containing molecules and their biological applications. Molecules 25:1909
5. Nepali K, Lee H-Y, Liou J-P (2019) Nitro-group-containing drugs. J Med Chem 62:2851–2893
6. Stockman R (2002) Heterocyclic chemistry. Annu Rep Prog Chem Sect B Org Chem 98(103): 107–124
7. Joule JA, Mills K (2010) Heterocyclic chemistry. Wiley
8. Barta K, Ford PC (2014) Catalytic conversion of nonfood Woody biomass solids to organic liquids. Acc Chem Res 47:1503–1512
9. Faisca Phillips AM, Pombeiro A, Kopylovich M (2016) Recent advances in Cascade reactions initiated by alcohol oxidation. ChemCatChem 9:217–246
10. Besson M, Gallezot P, Pinel C (2014) Conversion of biomass into chemicals over metal catalysts. Chem Rev 114:1827–1870
11. Afanasyev OI, Kuchuk E, Usanov DL, Chusov D (2019) Reductive amination in the synthesis of pharmaceuticals. Chem Rev 119:11857–11911
12. Murugesan K, Senthamarai T, Chandrashekhar VG et al (2020) Catalytic reductive aminations using molecular hydrogen for synthesis of different kinds of amines. Chem Soc Rev 49:6273–6328
13. Sheikh Abdul Hamid MH, Slatford P, Williams J (2007) Borrowing hydrogen in the activation of alcohols. Adv Synth Catal 349:1555–1575
14. Dobereiner GE, Crabtree RH (2010) Dehydrogenation as a substrate-activating strategy in homogeneous transition-metal catalysis. Chem Rev 110:681–703
15. Choi J, MacArthur AHR, Brookhart M, Goldman AS (2011) Dehydrogenation and related reactions catalyzed by iridium pincer complexes. Chem Rev 111:1761–1779
16. Gunanathan C, Milstein D (2013) Applications of acceptorless dehydrogenation and related transformations in chemical synthesis. Science 341:1229712
17. Watson A, Williams J (2010) The give and take of alcohol activation. Science 329:635–636
18. Guillena G, J Ramón D, Yus M (2010) Hydrogen autotransfer in the N-alkylation of amines and related compounds using alcohols and amines as electrophiles. Chem Rev 110:1611–1641

19. Bower JF, Krische MJ (2011) Formation of C-C bonds via iridium-catalyzed hydrogenation and transfer hydrogenation. Top Organomet Chem 34:107–138
20. Baehn S, Imm S, Neubert LK et al (2011) The catalytic amination of alcohols. ChemCatChem 3:1853–1864
21. Yang Q, Wang Q, Yu Z (2015) ChemInform abstract: substitution of alcohols by N-nucleophiles via transition metal-catalyzed dehydrogenation. Chem Soc Rev 44:2305–2329
22. Bower JF, Skucas E, Patman RL, Krische MJ (2007) Catalytic C-C coupling via transfer hydrogenation: reverse prenylation, crotylation, and allylation from the alcohol or aldehyde oxidation level. J Am Chem Soc 129(49):15134–15135
23. Guillena G, Ramón D, Yus M (2007) Alcohols as electrophiles in C-C bond-forming reactions: the hydrogen autotransfer process. Angew Chem Int Ed 46:2358–2364
24. Corma A, Navas J, Sabater MJ (2018) Advances in one-pot synthesis through borrowing hydrogen catalysis. Chem Rev 118:1410–1459
25. Irrgang T, Kempe R (2020) Transition-metal-catalyzed reductive amination employing hydrogen. Chem Rev 120:9583–9674
26. Reed-Berendt BG, Latham DE, Dambatta MB, Morrill LC (2021) Borrowing hydrogen for organic synthesis. ACS Cent Sci 7:570–585
27. Podyacheva E, Afanasyev O, Vasilyev D, Chusov D (2022) Borrowing hydrogen amination reactions: a complex analysis of trends and correlations of the various reaction parameters. ACS Catal 12:7142–7198
28. Yamaguchi R, Fujita K, Zhu M (2010) Recent progress of new catalytic synthetic methods for nitrogen heterocycles based on hydrogen transfer reactions. ChemInform 81:1093–1140
29. Michlik S, Kempe R (2013) Regioselektiv funktionalisierte Pyridine aus nachhaltigen Ressourcen. Angew Chemie 52:6326–6329
30. Schranck J, Tlili A, Beller M (2013) More sustainable formation of C-N and C-C bonds for the synthesis of N-heterocycles. Angew Chem Int Ed Engl 52:7642–7644
31. Chelucci G (2017) Metal-catalyzed dehydrogenative synthesis of pyrroles and indoles from alcohols. Coord Chem Rev 331:37–53
32. Daw P, Ben-David Y, Milstein D (2018) Acceptorless dehydrogenative coupling using ammonia: direct synthesis of N-heteroaromatics from diols catalyzed by ruthenium. J Am Chem Soc 140:11931–11934
33. Guo W, Zhao M, Tan W et al (2019) Developments towards synthesis of N-heterocycles from amidines via C–N/C–C bond formation. Org Chem Front 6:2120–2141
34. Maji M, Panja D, Borthakur I, Kundu S (2021) Recent advancement on sustainable synthesis of N-heterocycles following acceptorless dehydrogenative coupling protocol using alcohols. Org Chem Front 8:2673–2709
35. Nandakumar A, Midya SP, Landge VG, Balaraman E (2015) Transition-metal-catalyzed hydrogen-transfer annulations: access to heterocyclic scaffolds. Angew Chem Int Ed Engl 54: 11022–11034
36. Werkmeister S, Neumann J, Junge K, Beller M (2015) Pincer-type complexes for catalytic (De)hydrogenation and transfer (De)hydrogenation reactions: recent progress. Chemistry 21: 12226–12250
37. Berlet, Christopher JKS (2023) MineralPrices.com. http://mineralprices.com/default.aspx#rar, http://www.infomine.com/investment/metalprices/. Accessed 9 Aug 2023
38. Medici S, Peana M, Zoroddu M (2018) Noble metals in pharmaceuticals: applications and limitations-biomedical applications of metals. Springer, Cham, pp 3–48
39. Balaraman E, Nandakumar A, Jaiswal G, Sahoo MK (2017) Iron-catalyzed dehydrogenation reactions and their applications in sustainable energy and catalysis. Cat Sci Technol 7:3177–3195
40. Filonenko GA, van Putten R, Hensen EJM, Pidko EA (2018) Catalytic (de)hydrogenation promoted by non-precious metals – Co, Fe and Mn: recent advances in an emerging field. Chem Soc Rev 47:1459–1483

41. Kallmeier F, Kempe R (2018) Manganese complexes for (De)hydrogenation catalysis: a comparison to cobalt and iron catalysts. Angew Chemie 57(1):46–60
42. Reed-Berendt BG, Polidano K, Morrill LC (2019) Recent advances in homogeneous borrowing hydrogen catalysis using earth-abundant first row transition metals. Org Biomol Chem 17: 1595–1607
43. Subaramanian M, Sivakumar G, Balaraman E (2021b) Recent advances in nickel-catalyzed C–C and C–N bond formation via HA and ADC reactions. Org Biomol Chem 19:4213–4227
44. Subaramanian M, Sivakumar G, Balaraman E (2021a) First-row transition-metal catalyzed acceptorless dehydrogenation and related reactions: a personal account. Chem Rec 21:3839–3871
45. Bullock RM (2011) Catalysis without precious metals. Wiley
46. Zell T, Langer R (2018) From ruthenium to iron and manganese – a mechanistic view on challenges and design principles of base-metal hydrogenation catalysts. ChemCatChem 10: 1930–1940
47. Cai Y, Li F, Li Y-Q et al (2018) Base metal-catalyzed alcohol C-C couplings under hydrogen transfer conditions. Tetrahedron Lett 59:1073–1079
48. Mukherjee A, Milstein D (2018) Homogeneous catalysis by cobalt and manganese pincer complexes. ACS Catal 8:11435–11469
49. Irrgang T, Kempe R (2019) 3d-metal catalyzed N- and C-alkylation reactions via borrowing hydrogen or hydrogen autotransfer. Chem Rev 119:2524–2549
50. Mastalir M, Glatz M, Pittenauer E et al (2016) Sustainable synthesis of quinolines and pyrimidines catalyzed by manganese PNP pincer complexes. J Am Chem Soc 138:15543–15546
51. Deibl N, Kempe R (2017) Manganese-catalyzed multicomponent synthesis of pyrimidines from alcohols and amidines. Angew Chem Int Ed 56:1663–1666
52. Daw P, Kumar A, Espinosa-Jalapa NA et al (2018) Synthesis of pyrazines and quinoxalines via acceptorless dehydrogenative coupling routes catalyzed by manganese pincer complexes. ACS Catal 8:7734–7741
53. Das K, Mondal A, Srimani D (2018) Phosphine free Mn-complex catalysed dehydrogenative C–C and C–heteroatom bond formation: a sustainable approach to synthesize quinoxaline, pyrazine, benzothiazole and quinoline derivatives. Chem Commun 54:10582–10585
54. Das K, Mondal A, Srimani D (2018) Selective synthesis of 2-substituted and 1,2-disubstituted benzimidazoles directly from aromatic diamines and alcohols catalyzed by molecularly defined nonphosphine manganese(I) complex. J Org Chem 83:9553–9560
55. Das K, Mondal A, Pal D, Srimani D (2019) Sustainable synthesis of quinazoline and 2-aminoquinoline via dehydrogenative coupling of 2-aminobenzyl alcohol and nitrile catalyzed by phosphine-free manganese pincer complex. Org Lett 21:3223–3227
56. Das K, Mondal A, Pal D et al (2019) Phosphine-free well-defined Mn(I) complex-catalyzed synthesis of amine, imine, and 2,3-dihydro-1H-perimidine via hydrogen autotransfer or acceptorless dehydrogenative coupling of amine and alcohol. Organometallics 38:1815–1825
57. Mondal A, Sahoo MK, Subaramanian M, Balaraman E (2020) Manganese(I)-catalyzed sustainable synthesis of quinoxaline and quinazoline derivatives with the liberation of dihydrogen. J Org Chem 85:7181–7191
58. Balaraman GSA-GSA-MSA-E (2022) Single-molecular Mn(I)-complex-catalyzed tandem double dehydrogenation cross-coupling of (amino)alcohols under solventless conditions with the liberation of H_2 and H_2O. ACS Sustain Chem Eng 10:7362–7373
59. Mondal A, Sharma R, Dutta B et al (2022) Well-defined NNS-Mn complex catalyzed selective synthesis of C-3 alkylated indoles and bisindolylmethanes using alcohols. J Org Chem 87: 3989–4000
60. Maji A, Gupta S, Maji M, Kundu S (2022) Well-defined phosphine-free manganese(II)-complex-catalyzed synthesis of quinolines, pyrroles, and pyridines. J Org Chem 87:8351–8367

61. Nandi S, Borthakur I, Ganguli K, Kundu S (2023) 2-(2-Benzimidazolyl)pyridine Mn (I) complexes: synthesis and exploration of catalytic activity toward synthesis of pyrimidine and quinoline. Organometallics 14:1793–1802
62. Fertig R, Leowsky-Künstler F, Irrgang T, Kempe R (2023) Rational design of N-heterocyclic compound classes via regenerative cyclization of diamines. Nat Commun 14:595
63. Bala M, Verma PK, Sharma U et al (2013) Iron phthalocyanine as an efficient and versatile catalyst for N-alkylation of heterocyclic amines with alcohols: one-pot synthesis of 2-substituted benzimidazoles, benzothiazoles and benzoxazoles. Green Chem 15:1687–1693
64. Nguyen B, Ermolenko L, Al-mourabit A (2015) ChemInform abstract: sodium sulfide: a sustainable solution for unbalanced redox condensation reaction between o-nitroanilines and alcohols catalyzed by an iron-sulfur system. Synthesis (Stuttg) 47:1741–1748
65. Emayavaramban B, Sen M, Sundararaju B (2017) Iron-catalyzed sustainable synthesis of pyrrole. Org Lett 19:6–9
66. Mondal R, Sinha S, Das S et al (2020) Iron catalyzed synthesis of pyrimidines under air. Adv Synth Catal 362:594–600
67. Daw P, Chakraborty S, Garg JA et al (2016) Direct synthesis of pyrroles by dehydrogenative coupling of diols and amines catalyzed by cobalt pincer complexes. Angew Chem 55(46): 14373–14377
68. Midya SP, Landge VG, Sahoo MK et al (2018) Cobalt-catalyzed acceptorless dehydrogenative coupling of aminoalcohols with alcohols: direct access to pyrrole, pyridine and pyrazine derivatives. Chem Commun 54:90–93
69. Shee S, Ganguli K, Jana K, Kundu S (2018) Cobalt complex catalyzed atom-economical synthesis of quinoxaline, quinoline and 2-alkylaminoquinoline derivatives. Chem Commun (Camb) 54:6883–6886
70. Das S, Mallick S, De Sarkar S (2019) Cobalt-catalyzed sustainable synthesis of benzimidazoles by redox-economical coupling of o-nitroanilines and alcohols. J Org Chem 84:12111–12119
71. Panja D, Paul B, Balasubramaniam B et al (2020) Application of a reusable Co-based nanocatalyst in alcohol dehydrogenative coupling strategy: synthesis of quinoxaline and imine scaffolds. Catal Commun 137:105927
72. Parua S, Das S, Sikari R et al (2017) One-pot cascade synthesis of quinazolin-4(3H)-ones via nickel-catalyzed dehydrogenative coupling of o-aminobenzamides with alcohols. J Org Chem 82:7165–7175
73. Das J, Singh K, Vellakkaran M, Banerjee D (2018) Nickel-catalyzed hydrogen-borrowing strategy for α-alkylation of ketones with alcohols: a new route to branched gem-bis(alkyl) ketones. Org Lett 20:5587–5591
74. Das S, Maiti D, De Sarkar S (2018) Synthesis of polysubstituted quinolines from α-2-aminoaryl alcohols via nickel-catalyzed dehydrogenative coupling. J Org Chem 83:2309–2316
75. Parua S, Sikari R, Sinha S et al (2018) A nickel catalyzed acceptorless dehydrogenative approach to quinolines. Org Biomol Chem 16:274–284
76. Parua S, Sikari R, Sinha S et al (2018) Accessing polysubstituted quinazolines via nickel catalyzed acceptorless dehydrogenative coupling. J Org Chem 83:11154–11166
77. Chakraborty G, Sikari R, Das S et al (2019) Dehydrogenative synthesis of quinolines, 2-aminoquinolines, and quinazolines using singlet diradical Ni(II)-catalysts. J Org Chem 84: 2626–2641
78. Singh K, Vellakkaran M, Banerjee D (2018) A nitrogen-ligated nickel-catalyst enables selective intermolecular cyclisation of β- and γ-amino alcohols with ketones: access to five and six-membered N-heterocycles. Green Chem 20:2250–2256
79. Singh K, Kabadwal LM, Bera S et al (2018) Nickel-catalyzed synthesis of N-substituted pyrroles using diols with aryl- and alkylamines. J Org Chem 83:15406–15414
80. Alanthadka A, Bera S, Vellakkaran M, Banerjee D (2019) Nickel-catalyzed double dehydrogenative coupling of secondary alcohols and β-amino alcohols to access substituted pyrroles. J Org Chem 84:13557–13564

81. Yang P, Zhang C, Gao W-C et al (2019) Nickel-catalyzed borrowing hydrogen annulations: access to diversified N-heterocycles. Chem Commun 55:7844–7847
82. Bera A, Sk M, Singh K, Banerjee D (2019) Nickel-catalysed dehydrogenative coupling of aromatic diamines with alcohols: selective synthesis of substituted benzimidazoles and quinoxalines. Chem Commun 55:5958–5961
83. Zhang M, Li H, Young DJ et al (2019) Reaction condition controlled nickel(ii)-catalyzed C-C cross-coupling of alcohols. Org Biomol Chem 17(14):3567–3574
84. Arora V, Dutta M, Das K et al (2020) Solvent-free N-alkylation and dehydrogenative coupling catalyzed by a highly active pincer-nickel complex. Organometallics 39:2162–2176
85. Yadav V, Jagtap SG, Balaraman E, Mhaske SB (2022) Nickel-catalyzed direct synthesis of N-substituted indoles from amino alcohols and alcohols. Org Lett 24:9054–9059
86. Cho CS, Ren WX, Yoon N-S (2009) A recyclable copper catalysis in modified Friedländer quinoline synthesis. J Mol Catal A Chem 299:117–120
87. Chen Z, Chen J, Liu M et al (2013) Unexpected copper-catalyzed cascade synthesis of quinazoline derivatives. J Org Chem 78:11342–11348
88. You Q, Wang F, Wu C et al (2015) Synthesis of 1,3,5-triazines via Cu(OAc)2-catalyzed aerobic oxidative coupling of alcohols and amidine hydrochlorides. Org Biomol Chem 13:6723–6727
89. Xu Z, Wang D-S, Yu X et al (2017) Tunable triazole-phosphine-copper catalysts for the synthesis of 2-aryl-1H-benzo[d]imidazoles from benzyl alcohols and diamines by acceptorless dehydrogenation and borrowing hydrogen reactions. Adv Synth Catal 359:3332–3340
90. Xu Z, Yu X, Sang X, Wang D (2018) BINAP-copper supported by hydrotalcite as an efficient catalyst for the borrowing hydrogen reaction and dehydrogenation cyclization under water or solvent-free conditions. Green Chem 20:2571–2577
91. Shi T, Qin F, Li Q, Zhang W (2018) Copper-catalyzed three-component synthesis of pyrimidines from amidines and alcohols. Org Biomol Chem 16:9487–9491
92. Hu Y, Li S, Li H et al (2019) Copper-catalyzed tandem oxidative synthesis of quinazolinones from 2-aminobenzonitriles and benzyl alcohols. Org Chem Front 6:2744–2748
93. Das S, Mondal R, Chakraborty G et al (2021) Zinc stabilized azo-anion radical in dehydrogenative synthesis of N-heterocycles. An exclusively ligand centered redox controlled approach. ACS Catal 11:7498–7512

Top Organomet Chem (2024) 73: 173–198
https://doi.org/10.1007/3418_2023_109

Published online: 21 November 2023

Catalytic Methylation Using Methanol as C1 Source

Mohamed Elghobashy and Osama El-Sepelgy

Contents

Abstract The development of 3d metal-catalyzed molecular transformations has been a key focus of research in recent decades. One significant advancement is the discovery of the homogenous iron, cobalt, and manganese-catalyzed (de)-hydrogenation processes. Among these redox transformations, the "Borrowing Hydrogen" (BH) principle, also known as hydrogen auto-transfer, stands out as an

M. Elghobashy and O. El-Sepelgy (✉)
Leibniz Institute for Catalysis e.V., Rostock, Germany
e-mail: Osama.Elsepelgy@Catalysis.de

elegant and eco-friendly method that facilitates the self-transfer of hydrogen between reaction molecules and intermediates, eliminating the need for external hydrogen donors or acceptors. This concept allows for the eco-friendly use of alcohols, such as methanol, as environmentally benign C1 synthons for the alkylation of organic molecules, including pharmaceutically relevant candidates. In this context, the methyl group represents one of the most prevalent carbon fragments in small-molecule drugs. In this book chapter, we summarize the discovery and recent advancements in the use of 3d metal complexes for (multi)methylation of organic compounds using methanol via the hydrogen borrowing methodology. Additionally, we discuss current limitations, challenges, and the future prospects of this field.

Keywords methanol · hydrogen borrowing · catalysis · 3d-metals · C1 synthons

1 Introduction

The "magic methyl" group represents one of the most prevalent carbon fragments in small-molecule drugs. This basic alkyl segment is present in over 67% of top-selling drugs in 2011. It holds significant importance in shaping the pharmaceutical characteristics of diverse bioactive compounds by influencing binding affinities, binding selectivity, solubility, half-life, and various metabolic as well as pharmacokinetic/pharmacodynamic properties [1, 2].

Classical methods for introduction of methyl groups involve the use of toxic and hazardous and waste producing methylating agents such as methyl iodide, diazomethane, methyl triflate, trimethyl oxonium tetrafluoroborate, and dimethyl sulfate [3]. An industrially applied method for the methylation and dimethylation of amines involves the application of Eschweiler-Clarke reaction [4]. The reaction involves the reductive amination with the carcinogenic formaldehyde followed by hydrogenation. For sustainability reasons, the use of greener C1 alternatives such as methanol is of particular interest (*see* Fig. 1a, b). However, a major challenge lies in the energetically demanding dehydrogenation of methanol ($\Delta H = +84$ kJ mol–1) [5].

The key to the success of this chemistry is the use of appropriate transition metal catalysis that can enable the transformation using the concept of hydrogen borrowing or hydrogen autotransfer (*see* Fig. 1a, b) [6]. The initial step involves the transition-metal-catalyzed dehydrogenation of methanol, leading to the in-situ production of formaldehyde while storing the hydrogen on the transition metal catalyst. Then, base-catalyzed condensation between formaldehyde and the C- or N- nucleophiles, such as amines or active methylene or methyl compound, leads to the formation of imines or electron-deficient olefins. These intermediates can undergo hydrogenation with the borrowed hydrogen on the transition metal catalyst, leading to the formation of the C- or N-methylated products without the need for the use of and external oxidant or reductant (*see* Fig. 1c, d).

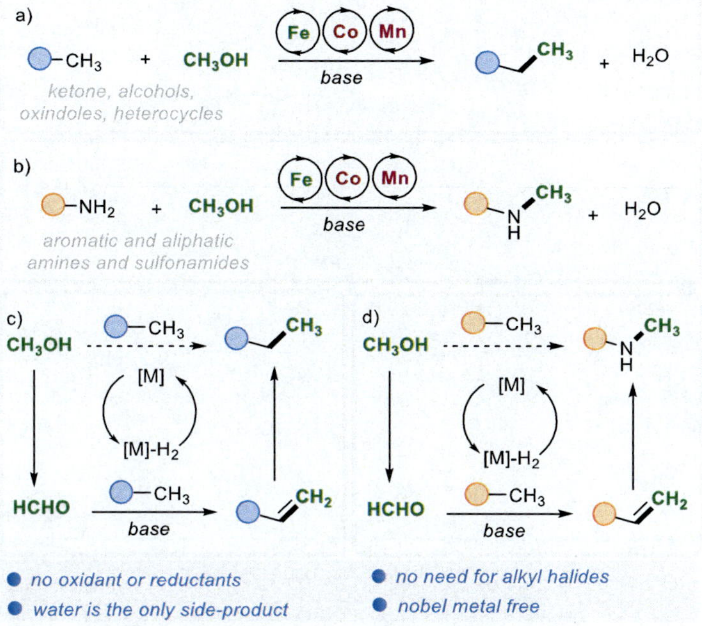

Fig. 1 Methylation using methanol via hydrogen borrowing concept

1.1 Catalyst Development

The use of alcohols, including methanol, as alkylating reagents was initially developed using noble-metal catalysis such as ruthenium and iridium catalysis. Yet, in recent times, a notable headway has been achieved in substituting these precious metals with base-metal alternatives, with a focus on iron, cobalt, and manganese [7]. In this context, we provide a succinct overview of the most important 3d transition metal catalytic systems that have been harnessed for C- and N-methylation reactions utilizing methanol through the concept of hydrogen borrowing (depicted in Fig. 2).

Iron catalysis: In 2007, Caesy [8] initially discovered the distinct catalytic capability of the iron hydride hydroxycyclopentadienyl complex **Fe-1a** [9] for heterolytic hydrogen activation. Subsequently, multiple research teams elucidated the in-situ formation of bifunctional iron (II) hydride complexes. This is achieved through the utilization of air- and moisture-stable iron (0) cyclopentadienone complexes, such as **Fe-1** and **Fe-2**. These pre-catalysts can be in situ activated through the removal of one carbonyl ligand using trimethylamine *N*-oxide [10], a suitable inorganic base [11], or light exposure [12] (*see* Fig. 3a). Notably, these phosphine-free complexes can be readily synthesized from economical materials, suggesting a promising potential for extensive industrial application [13].

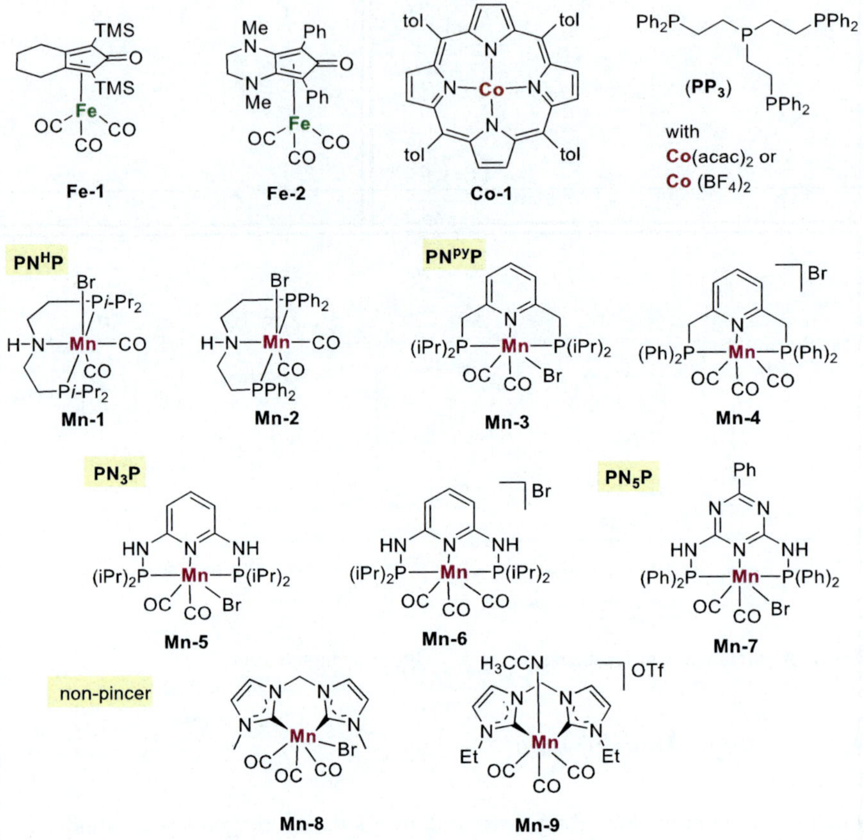

Fig. 2 3d transition metal catalysis used for methylation with methanol

Cobalt catalysis: Compared to iron and manganese, cooperative metal-ligand complexes based on cobalt have received relatively less attention. Nonetheless, a simple mixture of cobalt salts like $Co(acac)_2$ or $Co(BF_4)_2$ with tri- or tetra-dentate phosphines, such as 1,1,1-Tris(diphenylphosphinomethyl)ethane (Triphos) [14] or bis(diphenylphosphinoethyl)phenylphosphine (PP_3) [15], has demonstrated significant utility in (de)hydrogenation reactions, including the innovative concept of hydrogen borrowing methylation. Mechanistically, it involves the hydrogen activation through the classical monohydride pathway.

Manganese catalysis: In the spring of 2016, the breakthrough discovery of manganese hydrogenation catalysis was made by the research teams led by Beller [16] and Milstein [17]. Subsequent to this pioneering work, there has been a remarkable acceleration in advancements, propelling manganese to the forefront of 3d transition metals utilized in (de)hydrogenation transformations in under 3 years [18–20].

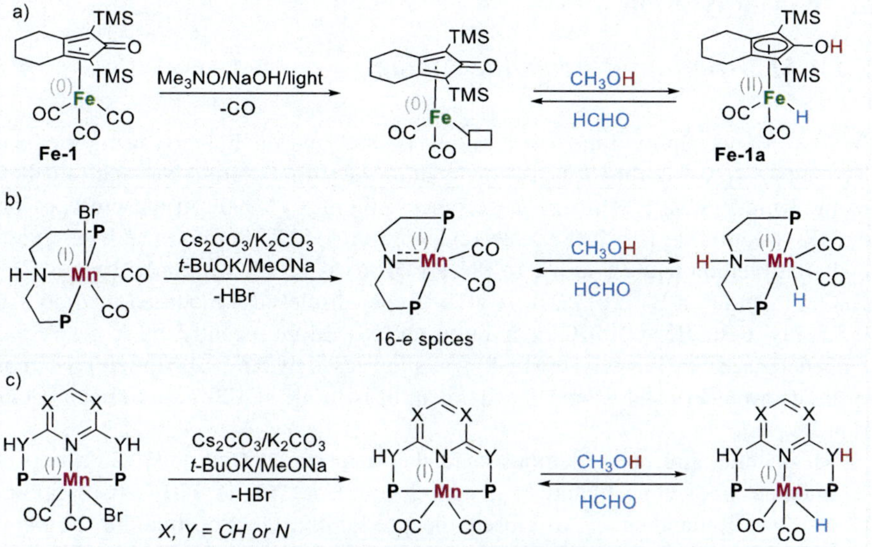

Fig. 3 Methanol dehydrogenation via metal ligand cooperation (MLC)

In terms of the mechanism, manganese catalysis employing the aliphatic PNHP "aminopincer" motif (such as **Mn-1 to Mn-2**), sharing structural similarities with Noyori's original catalysts, can undergo in situ activation through a suitable base. This leads to the formation of a catalytically active 16-electron species, as illustrated in Fig. 3b. Furthermore, the pyridine-based pincer ligands $PN^{Py}P$ and PN_3P, along with the trizine ligand PN_5P, have also seen extensive application in manganese (de)hydrogenation catalysis (**Mn-3** to **Mn-7**). Upon the introduction of a base, deprotonation of a CH_2 or NH group within the backbone results in the generation of an anionic, "dearomatized" pincer ligand with amide-like characteristics. Such dearomatized complexes can engage in stoichiometric, cooperative hydrogen activation through 1,3 addition across the metal-ligand framework, facilitated by the presence of a base (*see* Fig. 3c).

More recently, non-metal-ligand cooperative (MLC) pathways have also gained prominence in manganese (de)hydrogenation catalysis, exemplified by cases like **Mn-6** and **Mn-7**.

This book chapter provides a comprehensive overview of endeavors focused on the direct incorporation of methyl groups onto C–H and N–H bonds, utilizing methanol as a valuable C1 source. This advancement is made possible through the application of homogeneous catalysis anchored in 3d metals. Distinguishing itself from recent reviews [21–24], this chapter offers a critical synthesis of advancements in this domain, dissecting the progress achieved in each transformation. Notably, this chapter undertakes a systematic comparison of diverse catalytic systems, shedding light on their respective merits and limitations within this field of research.

2 N-Methylation

2.1 *Methylation of Aromatic Amines*

In 2016, Beller reported the first base-metal-catalyzed methylation using methanol. The authors have outlined the N-methylation of aniline derivatives with methanol using 3 mol% of a PNP manganese pincer complex (**Mn-1**) in the presence of *t*-BuOK (1 equiv.) at 100°C (Scheme 1, Condition A) [25]. Soratis et al. later reported a similar method using 5 mol% of PN_3P (**Mn-6**) at 120°C with only 20 mol% of *t*-BuOK (Condition B) [26]. Shortly after, Beller disclosed a modified method using 0.5 equiv. t-BuOK at 100°C (Condition C) [27]. More recently, an *N*-heterocyclic carbene-based Mn(I) catalyst (**Mn-8**) was showcased by Ke et al., utilizing 1.5 mol% of the non-pincer catalyst and 1 equivalent of *t*-BuOK at 120°C in toluene (Condition D) [28].

Meanwhile, Liu et al. demonstrated that a mixture of 5 mol% $Co(acac)_2$ and tetradentate phosphine ligand PP_3, with 1 equiv. K_3PO_4 at 140°C (Condition E) [29]. Also, Renaud et al. disclosed the use of the electron-rich Fe tricarbonyl complex (**Fe-2**) and only 10 mol% of CsOH at 110°C leading to similar results (Condition F) [30]. All the previously mentioned studies showed a great tolerance for a wide range of substitutions with electron-donating and electron withdrawing groups in ortho, meta and para positions. Interestingly, several hydrogen sensitive functional groups such as nitro, ketone, ester, amide, and olefins were tolerated under these conditions. These methodologies were also found to tolerate different heteroaromatic amines such as pyridines.

2.2 *Reductive Methylation of Nitrobenzene Derivatives*

In 2020, Morrill et al. demonstrated the direct conversion of nitrobenzene derivatives to *N*-methylarylamines using 5 mol% of Mn-PN_3P pincer complex (**Mn-6**) with KOH (2 equiv.) and 4 Å MS at 110°C [31]. The process involves tandem transfer hydrogenation followed by methylation, with methanol serving as both a hydrogen donor and a methylation reagent (Scheme 2).

The optimized method exhibited compatibility with meta- and para-electron-donating and electron-withdrawing substituents, but not with ortho-substitution due to hindered hydrogenation. The protocol also tolerated hydrogen-sensitive functional groups such as alcohols and alkenes, yielding moderate results.

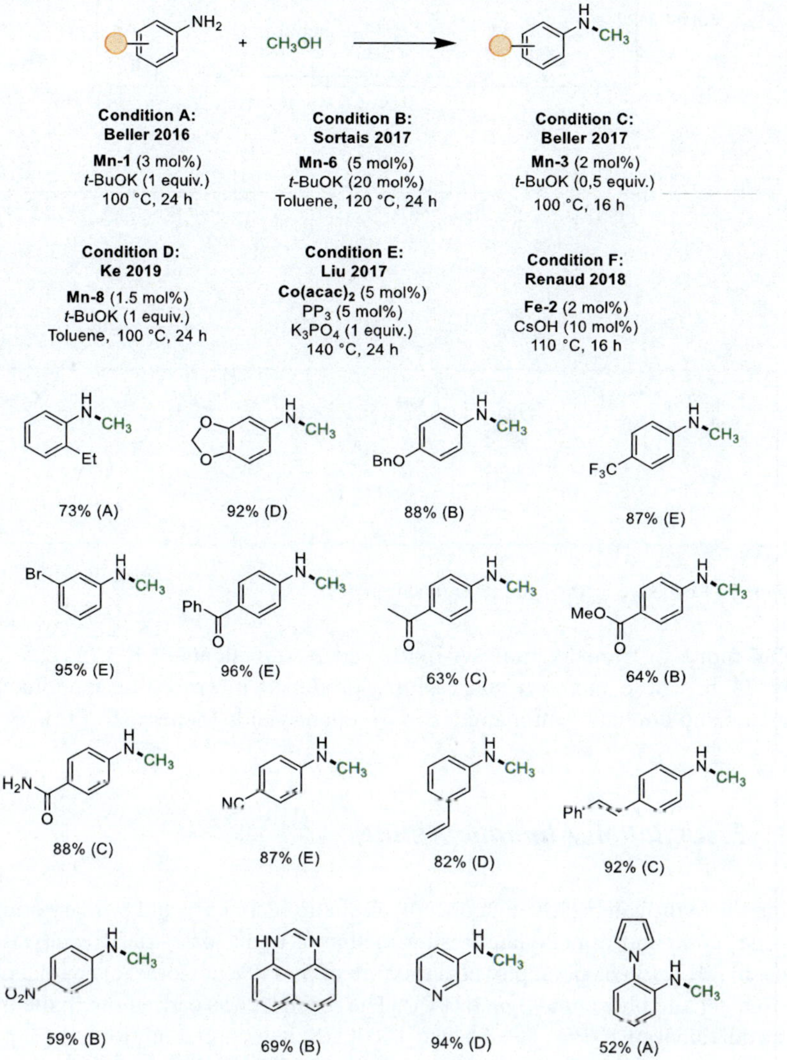

Scheme 1 N-methylation of arylamines

2.3 *Methylation of Sulfonamides*

Regarding sulfonamides methylation, Soratis et al. employed the PN_3P manganese catalyst **Mn-6** (Condition A). The procedure utilized 5 mol% of the Mn-pincer complex and 1.2 equivalents of *t*-BuOK as the base. The reaction occurred at 120° C in toluene as a cosolvent for 60 h [31]. Soon after, Morrill et al. introduced iron catalysts for the same process (Condition B). This involved 8 mol% of **Fe-2**, along

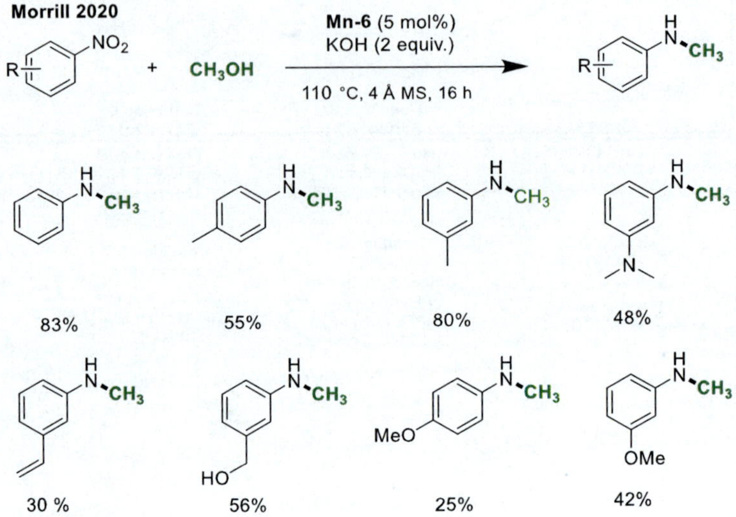

Scheme 2 Reductive methylation of nitrobenzene

with 4 mol% of trimethylamine *N*-oxide and 2 equivalents of K_2CO_3 at 100°C for only 24 h. These methods successfully produced a variety of monomethylated aromatic and aliphatic sulfonamides in excellent yields (Scheme 3) [32].

2.4 *Methylation Aliphatic Amines*

Using the hydrogen borrowing concept, aliphatic secondary and primary amines can also be mono-and dimethylated using methanol under base-metal catalysis conditions. In 2017, Liu et al. employed a mixture of 5 mol% of $Co(acac)_2$ with 5 mol% of the tetradentate phosphine ligand (PP_3). The reaction was carried out in the presence of 1 equivalent of K_3PO_4 as a base at 140°C (Scheme 4, Condition A) [29]. A year later, the groups of Renaud [30] and Morril [32] independently reported two methodologies based on the use of iron carbonyl complexes. Renaud et al. utilized 2 mol% **Fe-2** and 10 mol% CsOH at 110°C (Condition B). However, to ensure good yields, the reaction was conducted under 10 atm of H_2. On the other hand, the use of **Fe-1** (4 mol%) and Me_3NO (8 mol%) and 2 equivalents of K_2CO_3 led to similar results without the need of additional hydrogen gas (Condition C).

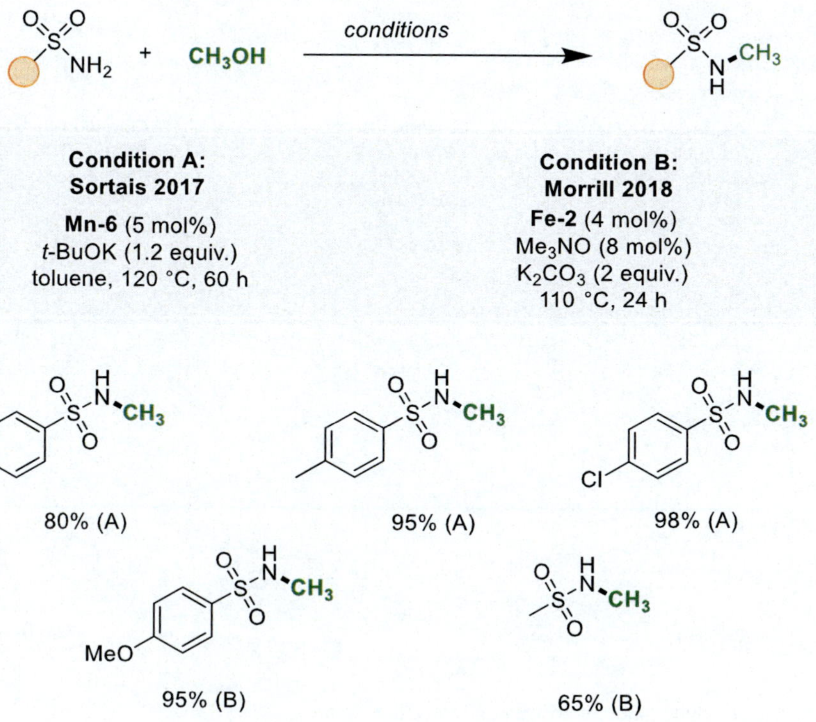

Scheme 3 Methylation of sulfonamides

3 C-Methylation

3.1 α-Methylation of Ketones

In 2019, El-Sepelgy/Rueping demonstrated a mild manganese-catalyzed approach for α-methylation and dimethylation of aromatic and aliphatic ketones. 2.5 mol% of the PNP manganese complex **Mn-4** was used, along with 2 equivalents of Cs_2CO_3 at 85°C (Scheme 5, Condition A) [33]. Meanwhile, Soratis et al. presented a related manganese-catalyzed α-methylation of ketones using 3 mol% of PN_3P complex (**Mn-6**) and 50 mol% of *t*-BuONa as a base at 120°C (Condition B) [34].

The iron version of this transformation was first reported by Morill et al. However, the scope of the iron-catalyzed transformation was limited to the relatively more active aromatic ketones. The iron-catalyzed methylation of ketones involves the use of 2 mol% of **Fe-1**, 4 mol% of Me_3NO and 2 equivalents of K_2CO_3 at 80 (Condition C) [32]. Very recently, Sundararaju et al. reported that 4 mol% of **Fe-1** in the presence of 2 equivalents of *t*-BuOK for 24 h at 42°C under visible light

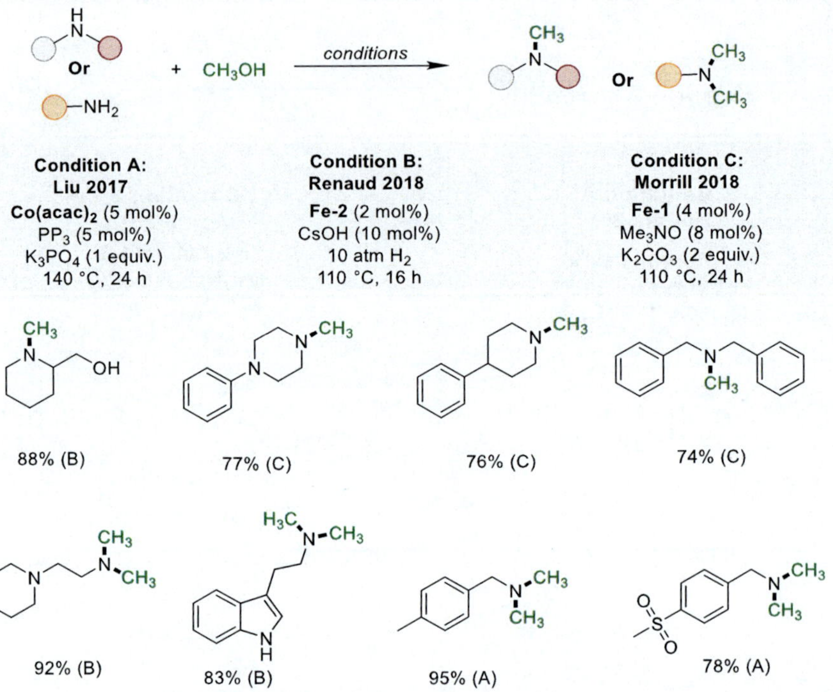

Scheme 4 Methylation of primary and secondary amines

irradiation (Condition D) [35]. Control experiments revealed that no reaction was observed in the absence of light at 40°C.

Furthermore, Liu et al. also disclosed cobalt catalytic system for the α-methylation of ketones with methanol. The authors demonstrated the need of 1 mol% $Co(BF_4)_2{\cdot}6H_2O$ and 1 mol% of a tetradentate phosphine ligand, $P(CH_2CH_2PPh_2)_3$ in the presence of 1 equivalent of K_3PO_4 at 100°C (Condition E) [36]. Similar to the iron catalytic system the reaction scope was limited to aromatic ketones.

Importantly, El-Sepelgy and Rueping have demonstrated the usefulness of the metal-ligand cooperation concept for the methylation of ketones with highly selective isotope labelling. An example is shown in Scheme 6, the methylation of propiophenone with four different labeled methanol, led to the formation of 4 different isotope-labeled isobutyrophenone derivatives.

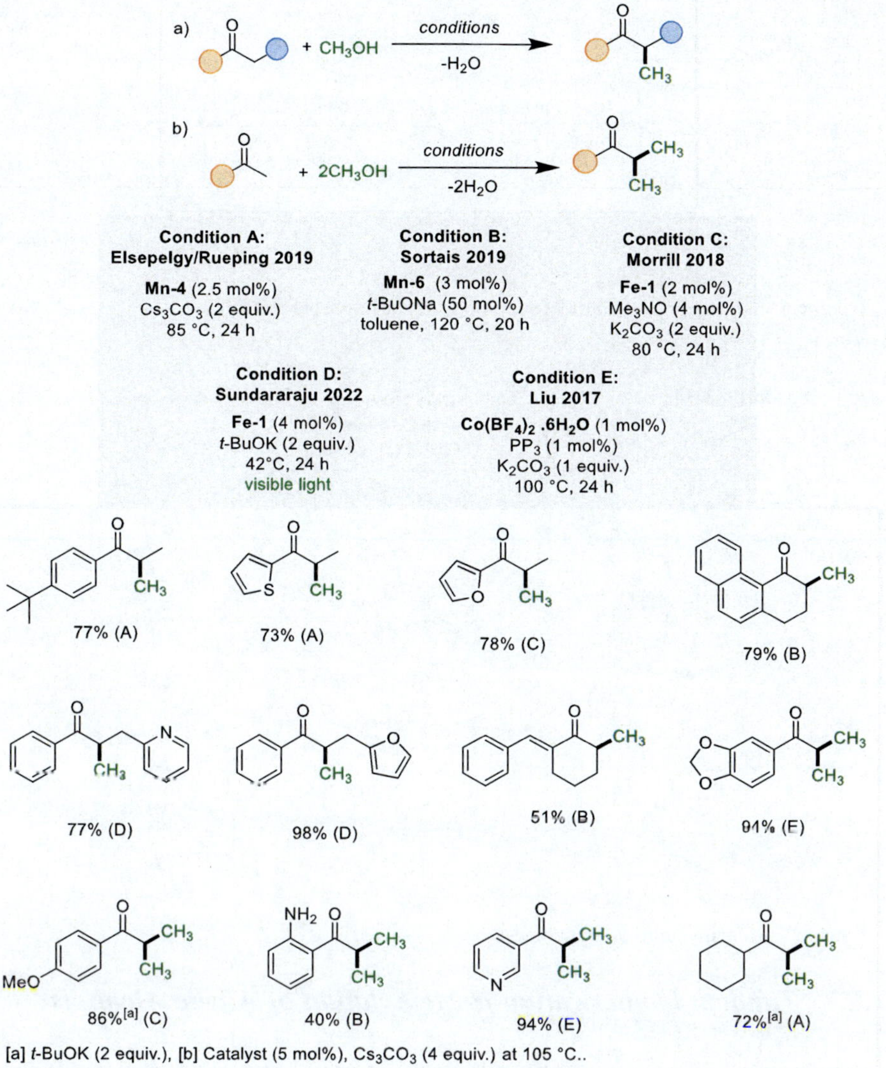

Scheme 5 α-methylation and double methylation of ketones

3.2 α-Methylation of Oxindoles

The α-methylation of ketones was further extended to activated amides (oxindole). In this regard, Morrill demonstrated that 2 mol% of **Fe-1** activated by 4 mol% of Me_3NO could be used for the methylation of oxindoles in the presence of 2 equivalents of K_2CO_3 at 80 to 110°C (Scheme 7). Interestingly, unprotected oxindoles bearing free NH was also amendable to the iron catalytic system [32].

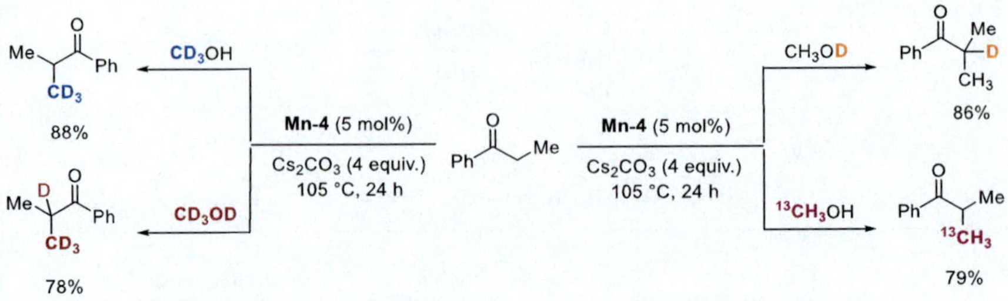

Scheme 6 Selective methylation of propiophenone using label methanol

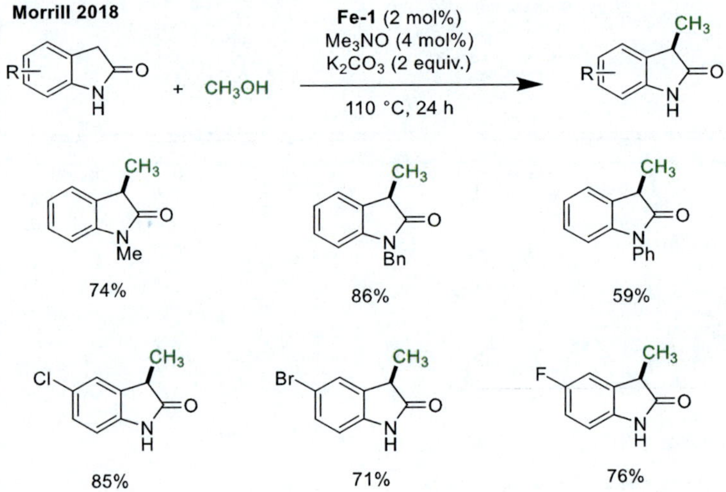

Scheme 7 α-methylation of oxindoles

3.3 *Tandem Isomerization α-Methylation of Allylic Alcohols*

In 2009, Morrill et al. demonstrated that α-methylated ketones can be synthetized upon the use of allylic alcohols instead of ketones (Scheme 8). The authors utilized 2 mol% of **Fe-2** catalyst in the presence of 2 equivalents of K_2CO_3 and 4 mol% of Me_3NO at 130°C for 24 h. This process involves tandem iron-catalyzed isomerization of the allylic alcohol to the corresponding ketone followed by iron-catalyzed methylation with methanol. The developed protocol was found to tolerate a wide range of allylic alcohols including aliphatic alcohols and heterocylic-containing substrates [37].

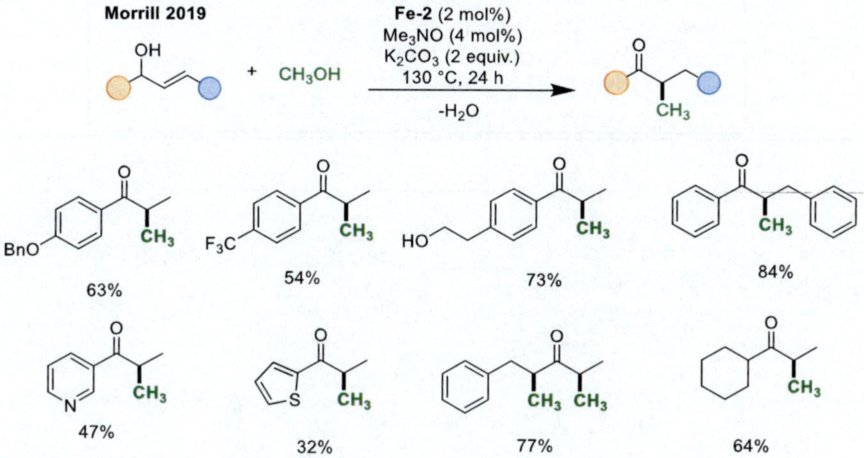

Scheme 8 Isomerization α-methylation of allylic alcohols

3.4 *Tandem Hydrogenation α-Methylation of α,β-Unsaturated Ketones*

Similarly, to the previous protocol, Kundu et al. decided to synthetize α-methylated ketones starting from α,β-unsaturated ketones. The authors have used a phosphine free bis-NHC manganese complex **Mn-9** combined with 0.5 equivalents of the strong base NaOH. The reaction was conducted at 120°C for 5 h [38]. The tandem transformation starts with manganese-catalyzed transfer hydrogenation to form ketones using methanol as a hydrogen donor followed by the typical manganese-catalyzed methylation. The reaction also showed a good tolerance to different functional groups including hydrogenation sensitive moieties (Scheme 9).

3.5 *β-Methylation of 2-Arylethanol*

Next to ketones, several groups have also investigated β-Methylation of alcohols with methanol. The reaction involves double hydrogen auto-transfer, where both methanol and the alcohol substrate undergo dehydrogenation followed by condensation. The formed α,β-unsaturated ketone then underwent double hydrogenation to produce the desired β-methylated alcohol.

The first base-metal-catalyzed alcohol methylation protocol was reported by Renaud in 2019. The catalytic system involves the use of the iron carbonyl complex **Fe-2** together with NaOH (10 mol%) and *t*-BuONa at 110°C for 40 h (Scheme 10, Condition A) [39]. However, the iron-catalyzed transformation is limited to the relatively high activation of 2-arylethanol. One year later, Leitner et al. developed

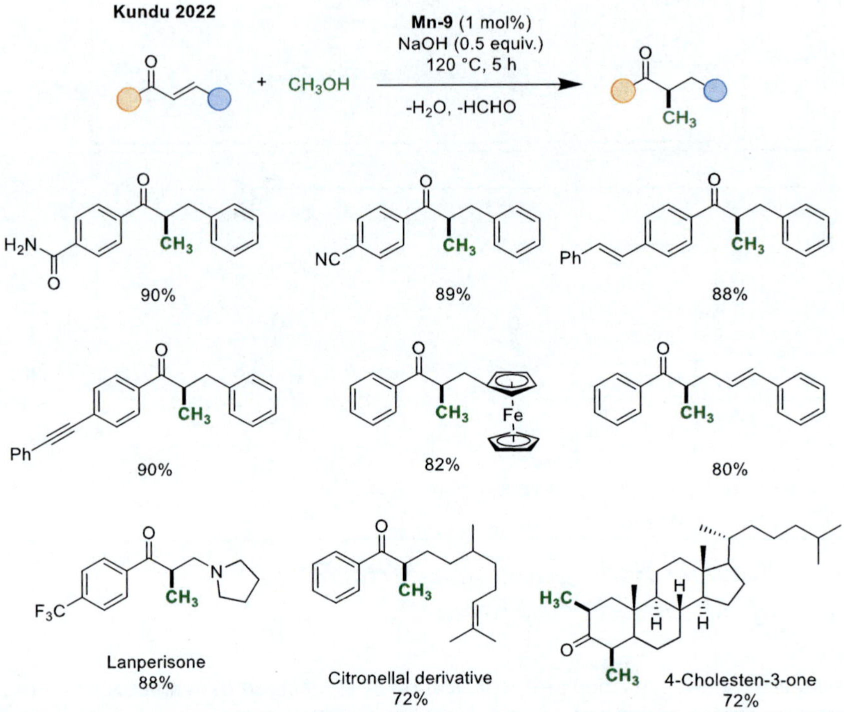

Scheme 9 Tandem hydrogenation α-methylation of α,β-unsaturated ketones

highly selective approach for the methylation of the same substates using 0.5 mol% of **Mn-1** and 2 equivalents NaOMe at 150°C (Scheme 10, Condition B) [40].

3.6 β-Methylation of Alcohols

Interestingly, the Leitner's protocol for the β-Methylation of 2-arylethanol was also found to be effective for the methylation of more challenging alcohols including purely aliphatic examples (Scheme 11, Condition A) [40]. Almost at the same time, Kempe et al. introduced another manganese-catalyzed methylation and multi-methylation of challenging alcohols using only 0.1 mol% of the **Mn-7** (Condition B) [41]. Selected examples from both protocols are shown in Scheme 11. These methodologies tolerate a diverse range of sensitive functional groups to produce the desired methylated and multi-methylated alcohols in high yields. Impressively, the methodologies exhibited good reactivity in the methylation of various aliphatic alcohols, encompassing those derived from biomass, diols, natural products, and drug molecules.

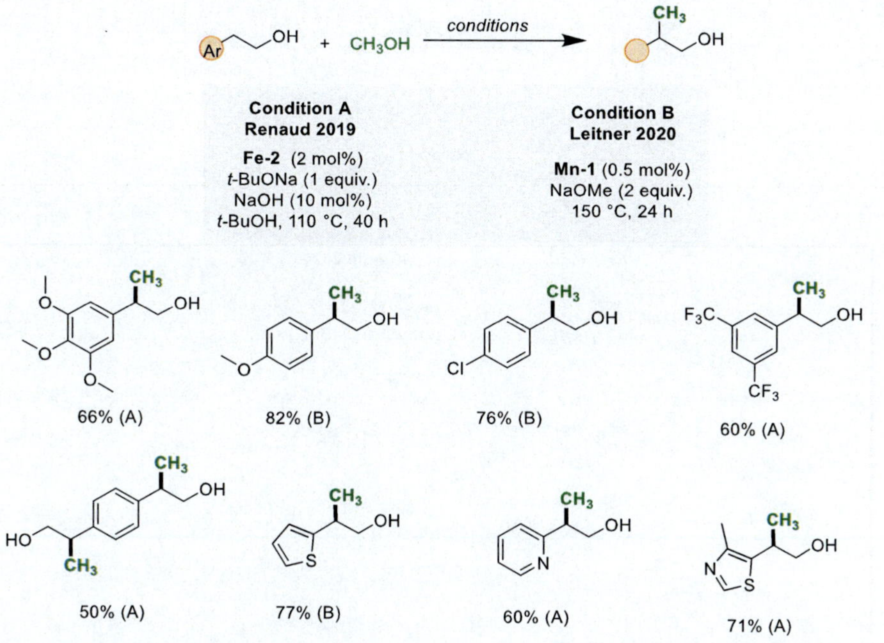

Scheme 10 β-Methylation of 2-arylethanol

3.7 α-Methylation of Nitriles

As presented in the previous section, most of the *C*-methylation protocols involve direct α-methylation of ketones or in situ formation of ketones from alcohols, allylic alcohols, and unsaturated ketones. In addition, most of these transformations require the use of equivalent amounts of base.

In 2019, Rueping/El-Sepelgy et al. demonstrated a manganese-catalyzed approach for α-methylation of nitriles under relatively mild conditions. This method utilized 5 mol% of the PNP manganese complex **(Mn-2)**, combined with 10 mol% of the mild base Cs_2CO_3. The reaction was conducted at 135°C in the presence of 1,4-dioxane as a cosolvent [42]. The key to success of this transformation even using only 10% of carbonate salt is the critical role of the bifunctional manganese catalyst for the activation of both methanol and the nitrile substrate. The proposed mechanism is shown in Scheme 12 below.

The cobalt version of this transformation was also reported by Liu et al. Employing conditions similar to those employed in the α-methylation of ketones, this method involved the use of a mixture of 1–2.5 mol% $Co(BF_4)_2{\cdot}6H_2O$ and tetradentate phosphine ligand (PP_3). The reaction was carried out in the presence of 1 equivalent of K_3PO_4, serving as the base, at 100°C for 24 to 48 h [36].

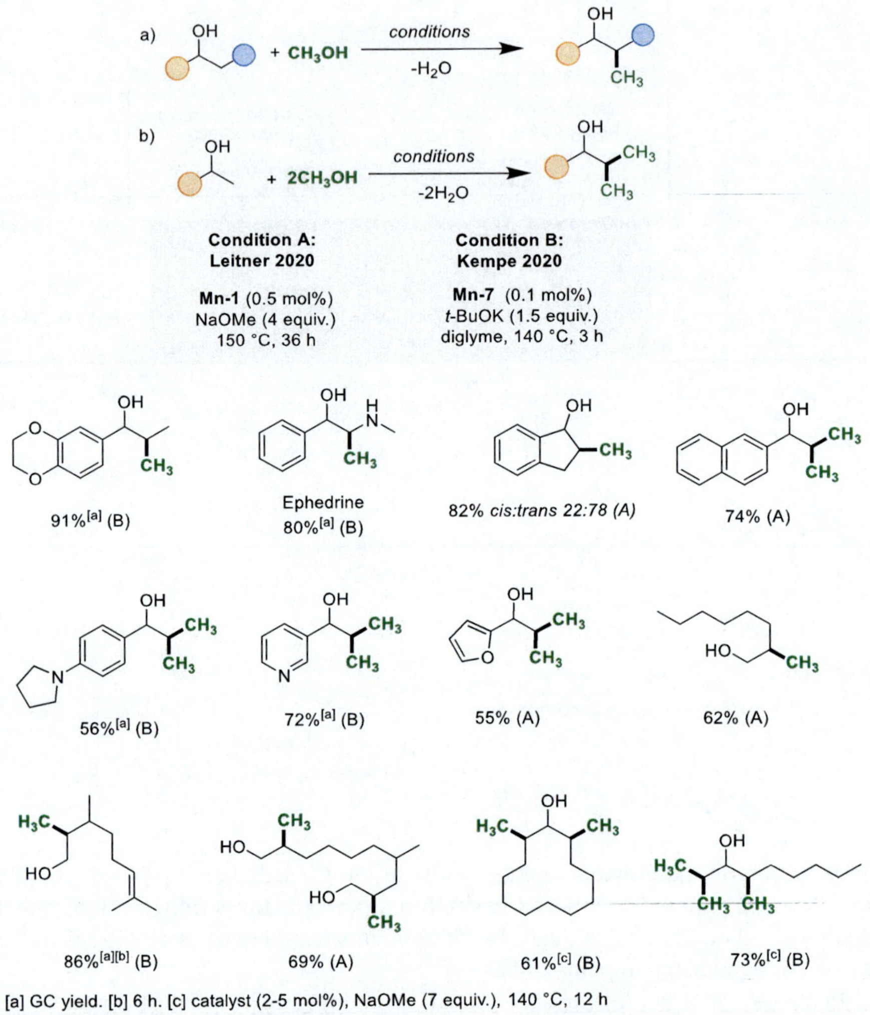

Scheme 11 β-Methylation of alcohols

3.8 *Methylation of N-Heterocycles*

Liu et al. also extended their cobalt-catalyzed methylation system ($Co(BF_4)_2{\cdot}6H_2O$/PP_3) to C(3)-methylation indoles. The reaction was carried out in the presence of 1 equivalent of K_3PO_4, serving as the base at 100°C (Scheme 13, Condition A) [36]. Also, Morill et al. have shown that iron could be used for the same transformation (Scheme 13, Condition B) [32].

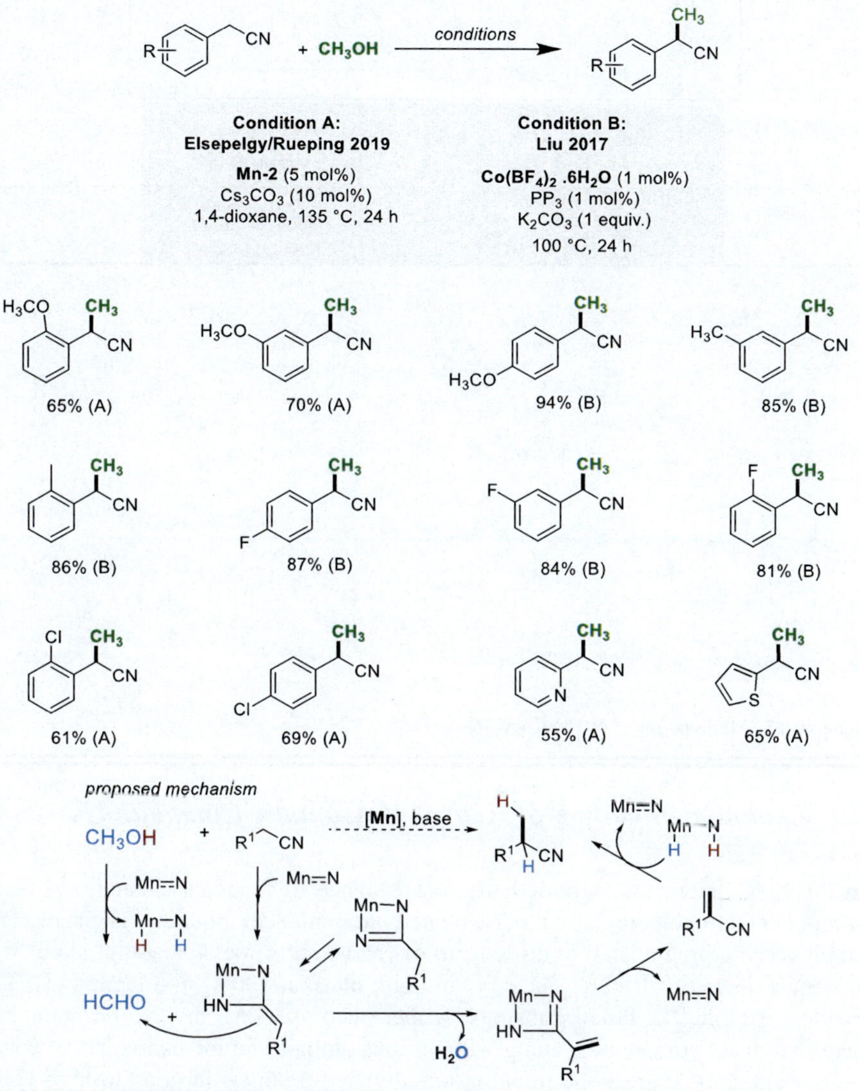

Scheme 12 α-Methylation of Nitriles

4 C- and Heteroatom Methylation

The chemistry of C- and N-methylation using hydrogen borrowing can be further applied to more complex transformations. In this regard, the hydrogen borrowing process is interrupted by the addition of nucleophile such as amines or alcohols.

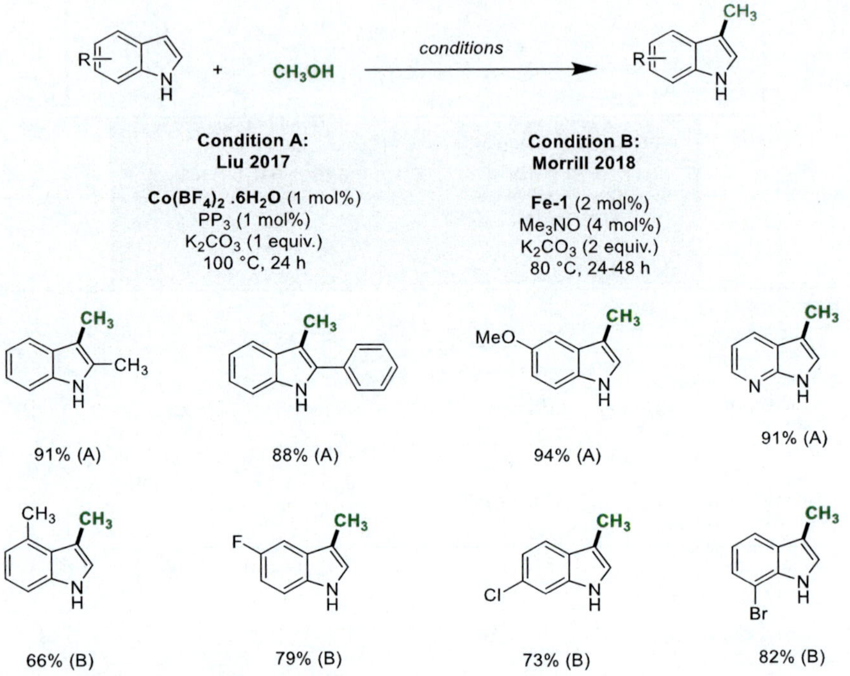

Scheme 13 Methylation of N-Heterocycles

4.1 *Aminomethylation of Activated Aromatic Compounds*

In 2017, Kichner et al. reported the first instance of manganese-catalyzed three-component aminomethylation of activated aromatic compounds. Their research employed the preprepared hydride form of (**Mn-5**) at 4 mol% loading, along with 1.3 equivalents of t-BuOK. The reaction took place at 130°C in toluene [43]. This method exhibited a broad substrate scope, encompassing diverse aromatic and heteroaromatic compounds, along with various amines and methanol. These conditions yielded the corresponding amino-methylated products in good-to-high yields (Scheme 14).

4.2 *α-Aminomethylation of Ketones*

In 2018, Xiao et al. introduced a successful cobalt-catalyzed approach for α-aminomethylation of ketones using methanol as the C1 source. This method involved a one-pot, two-step process for the three-component α-aminomethylation of ketones. The used catalyst was $CoCl_2{\cdot}6H_2O$ (3 mol%), along with 2 equivalents

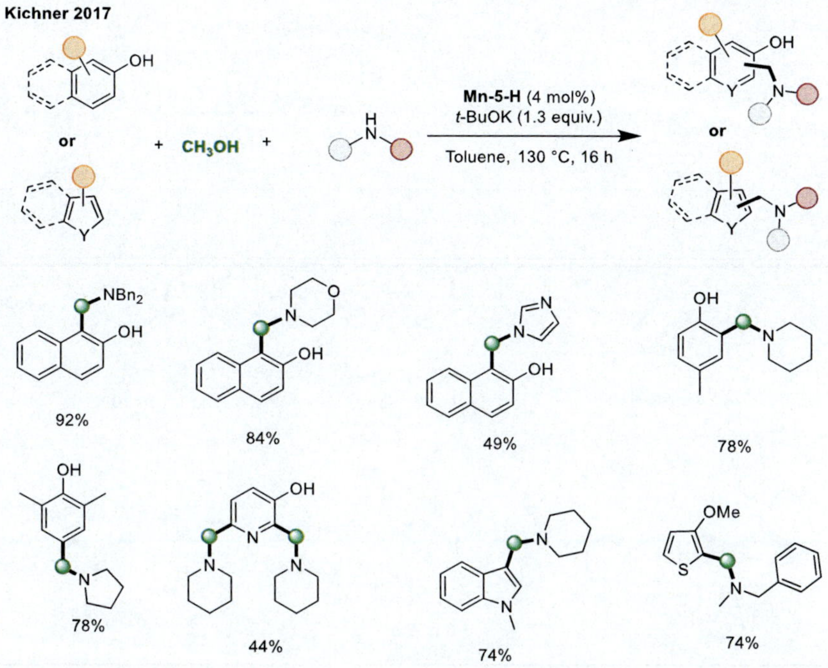

Scheme 14 Aminomethylation of activated aromatic compounds

of Cs_2CO_3 as the base and 1.5 equivalents of *t*-butyl hydroperoxide (TBHP) as the oxidizing agent. The reaction pathway, outlined in Scheme 15, emphasizes the amination of the in situ formed unsaturated ketone as the key step. TBHP's presence is proposed to prevent the completion of the hydrogen borrowing process. This established technique exhibited wide applicability that facilitate the synthesis of various α-aminomethylated ketones, including heterocyclic-based substrates [44].

4.3 α-Methoxymethylation of Ketones

The same group extended the established procedure for the aminomethylation of ketones to α-methoxymethylation of ketones. Interestingly, methanol plays a dual role as methylation and methoxylation source [44] (Scheme 16).

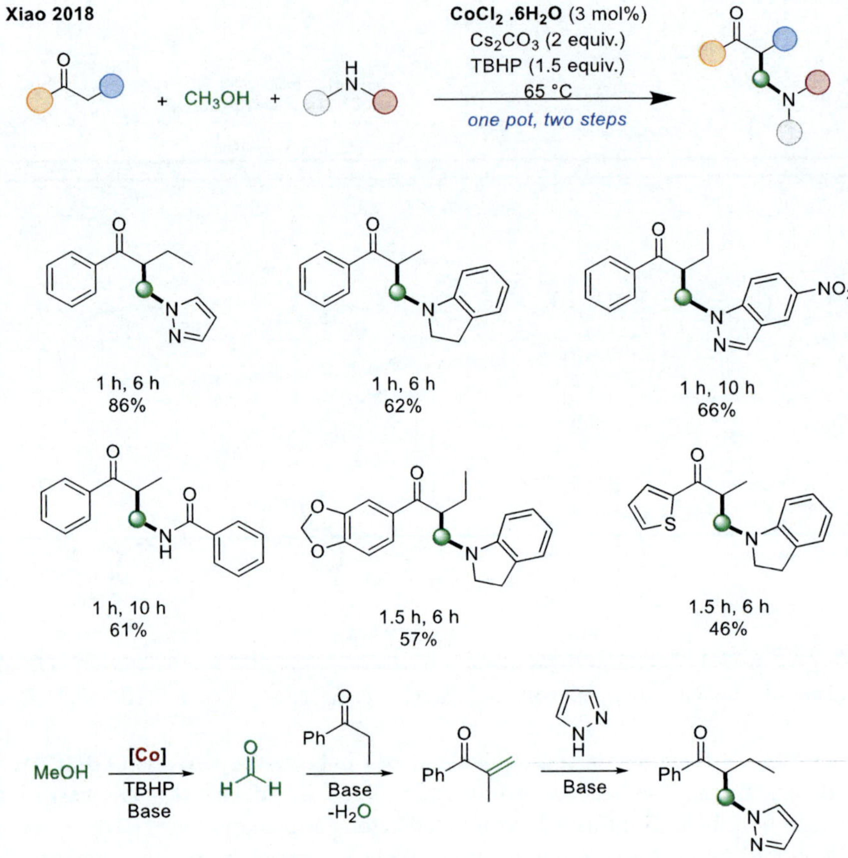

Scheme 15 α-aminomethylation of ketones

5 Two Carbon Methylation

In 2021, Chandrasekhar et al. presented a rare instance of utilizing methanol as a C1 source for synthesizing symmetrical 1,5-diketones. This innovative process employed a Co(II) porphyrin complex **(Co-1)** with *t*-butoxide salt at 135°C [45]. The sequence involves methanol dehydrogenation to formaldehyde, followed by base-catalyzed condensation, resulting in an α-methylenated intermediate. Within the described reaction conditions, this methylenated intermediate reacts with another ketone substrate molecule in the presence of the butoxide base, leading to Michael-type reactions. These steps culminate in the production of diverse 1,5-ketones with yields ranging from moderate to excellent (Scheme 17).

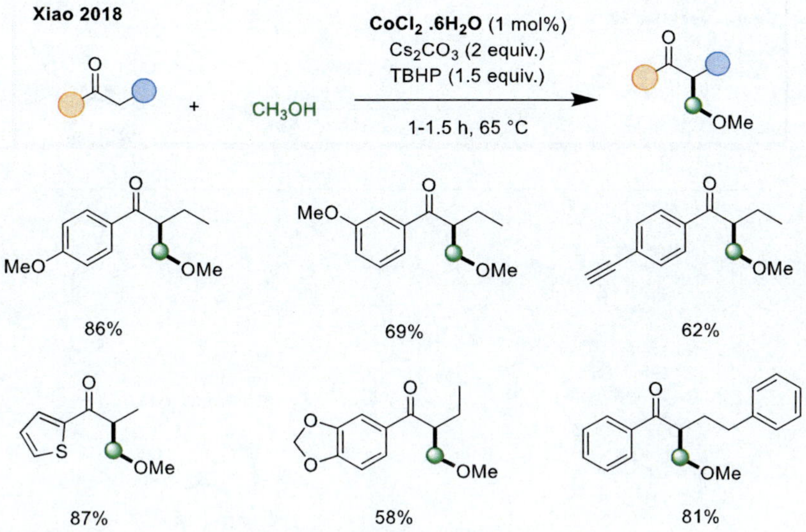

Scheme 16 α-methoxymethylation of ketones

6 Conclusion and Outlook

In summary, we have provided an overview of the chemistry involving methanol as a methylating reagent through the process of base-metal-catalyzed hydrogen borrowing. The inception of this chemistry dates back to 2016. It revolves around two primary processes: the hydrogen borrowing steps catalyzed by iron, cobalt, and manganese complexes, and the subsequent base-catalyzed condensation between the substrate and the in situ generated formaldehyde. Among these metals, manganese has exhibited the highest activity in most cases. Nonetheless, the cost-effective iron carbonyl complexes hold notable economic significance. Cobalt catalytic systems have demonstrated the advantage of not requiring pre-synthesized complexes.

Future endeavors might focus on the utilization of more affordable and stable homogeneous catalysis, specifically exploring combinations of readily available base metal salts such as Co(II), Fe(II), and Mn(II) with NHC or nitrogen-based ligands. Additionally, developing recyclable heterogeneous catalysis would mark a crucial advancement [46–49] for large-scale applications. Substituting high thermal energy with environmentally friendly visible light offers another avenue [50].

However, the central challenge lies in broadening the scope of this chemistry to encompass a wider range of substrates such as unactivated amides, esters, and other aromatic-aliphatic C-H containing compounds. This will entail devising alternatives to the base-catalyzed conditions that can be seamlessly integrated with hydrogen borrowing catalysis.

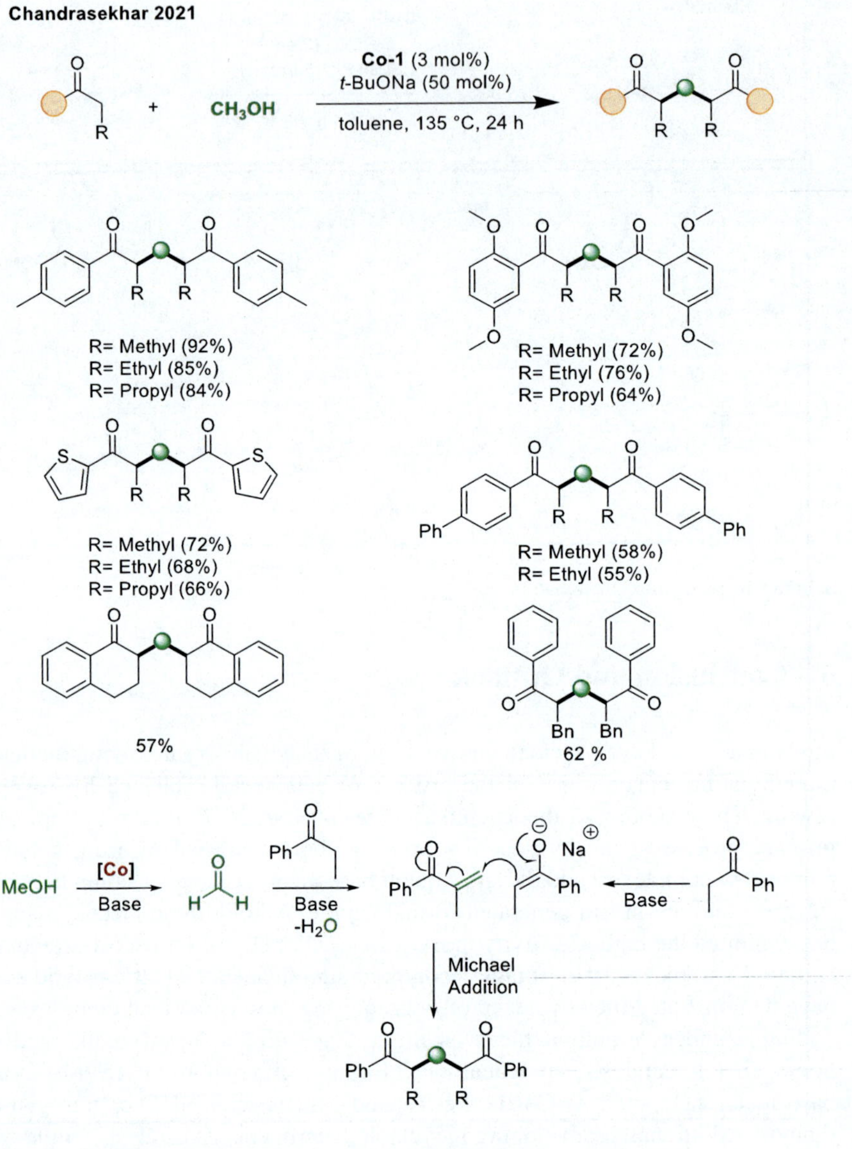

Scheme 17 Synthesis of symmetrical 1,5-diketons using methanol as C1 source

References

1. Aynetdinova D, Callens MC, Hicks HB, Poh CYX, Shennan BDA, Boyd AM, Lim ZH, Leitch JA, Dixon DJ (2021) Installing the "magic methyl" – C-H methylation in synthesis. Chem Soc Rev 50(9):5517-5563. https://doi.org/10.1039/D0CS00973C

2. Schönherr H, Cernak T (2013) Profound methyl effects in drug discovery and a call for new C–H methylation reactions. Angew Chem Int Ed 52(47):12256–12267. https://doi.org/10.1002/anie.201303207
3. Chen Y (2019) Recent advances in methylation: a guide for selecting methylation reagents. Chem A Eur J 25(14):3405–3439. https://doi.org/10.1002/chem.201803642
4. Pine SH, Sanchez BL (1971) Formic acid-formaldehyde methylation of amines. J Org Chem 36(6):829–832. https://doi.org/10.1021/jo00805a022
5. Selva M, Perosa A (2008) Green chemistry metrics: a comparative evaluation of dimethyl carbonate, methyl iodide, dimethyl sulfate and methanol as methylating agents. Green Chem 10(4):457–464. https://doi.org/10.1039/B713985C
6. Corma A, Navas J, Sabater MJ (2018) Advances in one-pot synthesis through borrowing hydrogen catalysis. Chem Rev 118(4):1410–1459. https://doi.org/10.1021/acs.chemrev.7b00340
7. Irrgang T, Kempe R (2019) 3d-metal catalyzed N- and C-alkylation reactions via borrowing hydrogen or hydrogen autotransfer. Chem Rev 119(4):2524–2549. https://doi.org/10.1021/acs.chemrev.8b00306
8. Casey CP, Guan H (2007) An efficient and chemoselective iron catalyst for the hydrogenation of ketones. J Am Chem Soc 129(18):5816–5817. https://doi.org/10.1021/ja071159f
9. Knölker H-J, Baum E, Goesmann H, Klauss R (1999) Demetalation of tricarbonyl (cyclopentadienone)iron complexes initiated by a ligand exchange reaction with NaOH – X-ray analysis of a complex with nearly square-planar coordinated sodium. Angew Chem Int Ed 38(13–14):2064–2066. https://doi.org/10.1002/(SICI)1521-3773(19990712)38:13/14<2064::AID-ANIE2064>3.0.CO;2-W
10. Pagnoux-Ozherelyeva A, Pannetier N, Mbaye MD, Gaillard S, Renaud J-L (2012) Knoelker′s iron complex: an efficient in situ generated catalyst for reductive amination of alkyl aldehydes an amines. Angew Chem Int Ed 51(20):4976-4980. https://doi.org/10.1002/anie.201201360
11. Fleischer S, Zhou S, Junge K, Beller M (2013) General and highly efficient iron-catalyzed hydrogenation of aldehydes, ketones, and α,β-unsaturated aldehydes. Angew Chem Int Ed 52(19):5120–5124. https://doi.org/10.1002/anie.201301239
12. Berkessel A, Reichau S, von der Höh A, Leconte N, Neudörfl J-M (2011) Light-induced enantioselective hydrogenation using chiral derivatives of Casey's iron–cyclopentadienone catalyst. Organometallics 30(14):3880–3887. https://doi.org/10.1021/om200459s
13. Quintard A, Rodriguez J (2014) Iron cyclopentadienone complexes: discovery, properties, and catalytic reactivity. Angew Chem Int Ed 53(16):4044–4055. https://doi.org/10.1002/anie.201310788
14. Korstanje TJ, Ivar van der Vlugt J, Elsevier CJ, de Bruin B (2015) Hydrogenation of carboxylic acids with a homogeneous cobalt catalyst. Science 350(6258):298–302. https://doi.org/10.1126/science.aaa8938
15. Liu W, Sahoo B, Junge K, Beller M (2018) Cobalt complexes as an emerging class of catalysts for homogeneous hydrogenations. Acc Chem Res 51(8):1858–1869. https://doi.org/10.1021/acs.accounts.8b00262
16. Elangovan S, Topf C, Fischer S, Jiao H, Spannenberg A, Baumann W, Ludwig R, Junge K, Beller M (2016) Selective catalytic hydrogenations of nitriles, ketones, and aldehydes by well-defined manganese pincer complexes. J Am Chem Soc 138(28):8809–8814. https://doi.org/10.1021/jacs.6b03709
17. Mukherjee A, Nerush A, Leitus G, Shimon LJW, Ben David Y, Espinosa Jalapa NA, Milstein D (2016) Manganese-catalyzed environmentally benign dehydrogenative coupling of alcohols and amines to form aldimines and H2: a catalytic and mechanistic study. J Am Chem Soc 138(13):4298–4301. https://doi.org/10.1021/jacs.5b13519
18. Alig L, Fritz M, Schneider S (2019) First-row transition metal (de)hydrogenation catalysis based on functional pincer ligands. Chem Rev 119(4):2681–2751. https://doi.org/10.1021/acs.chemrev.8b00555

19. Filonenko GA, van Putten R, Hensen EJM, Pidko EA (2018) Catalytic (de)hydrogenation promoted by non-precious metals – Co, Fe and Mn: recent advances in an emerging field. Chem Soc Rev 47(4):1459–1483. https://doi.org/10.1039/C7CS00334J
20. Kallmeier F, Kempe R (2018) Manganese complexes for (de)hydrogenation catalysis: a comparison to cobalt and iron catalysts. Angew Chem Int Ed 57(1):46–60. https://doi.org/10.1002/anie.201709010
21. Goyal V, Sarki N, Narani A, Naik G, Natte K, Jagadeesh RV (2023) Recent advances in the catalytic N-methylation and N-trideuteromethylation reactions using methanol and deuterated methanol. Coord Chem Rev 474:214827. https://doi.org/10.1016/j.ccr.2022.214827
22. Natte K, Neumann H, Beller M, Jagadeesh RV (2017) Transition-metal-catalyzed utilization of methanol as a C_1 source in organic synthesis. Angew Chem Int Ed 56(23):6384–6394. https://doi.org/10.1002/anie.201612520
23. Sarki N, Goyal V, Natte K, Jagadeesh RV (2021) Base metal-catalyzed C-methylation reactions using methanol. Adv Synth Catal 363(22):5028–5046. https://doi.org/10.1002/adsc.202100762
24. Sheetal, Mehara P, Das P (2023) Methanol as a greener C1 synthon under non-noble transition metal-catalyzed conditions. Coord Chem Rev 475:214851. https://doi.org/10.1016/j.ccr.2022.214851
25. Elangovan S, Neumann J, Sortais J-B, Junge K, Darcel C, Beller M (2016) Efficient and selective N-alkylation of amines with alcohols catalysed by manganese pincer complexes. Nat Commun 7(1):12641. https://doi.org/10.1038/ncomms12641
26. Bruneau-Voisine A, Wang D, Dorcet V, Roisnel T, Darcel C, Sortais J-B (2017) Mono-N-methylation of anilines with methanol catalyzed by a manganese pincer-complex. J Catal 347:57–62. https://doi.org/10.1016/j.jcat.2017.01.004
27. Neumann J, Elangovan S, Spannenberg A, Junge K, Beller M (2017) Improved and general manganese-catalyzed N-methylation of aromatic amines using methanol. Chem A Eur J 23(23):5410–5413. https://doi.org/10.1002/chem.201605218
28. Huang M, Li Y, Li Y, Liu J, Shu S, Liu Y, Ke Z (2019) Room temperature N-heterocyclic carbene manganese catalyzed selective N-alkylation of anilines with alcohols. Chem Commun 55(44):6213–6216. https://doi.org/10.1039/C9CC02989C
29. Liu Z, Yang Z, Yu X, Zhang H, Yu B, Zhao Y, Liu Z (2017) Efficient cobalt-catalyzed methylation of amines using methanol. Adv Synth Catal 359(24):4278–4283. https://doi.org/10.1002/adsc.201701044
30. Lator A, Gaillard S, Poater A, Renaud J-L (2018) Well-defined phosphine-free iron-catalyzed N-ethylation and N-methylation of amines with ethanol and methanol. Org Lett 20(19):5985–5990. https://doi.org/10.1021/acs.orglett.8b02080
31. Reed-Berendt BG, Mast N, Morrill LC (2020) Manganese-catalyzed one-pot conversion of nitroarenes into N-methylarylamines using methanol. Eur J Org Chem 2020(9):1136–1140. https://doi.org/10.1002/ejoc.201901854
32. Polidano K, Allen BDW, Williams JMJ, Morrill LC (2018) Iron-catalyzed methylation using the borrowing hydrogen approach. ACS Catal 8(7):6440–6445. https://doi.org/10.1021/acscatal.8b02158
33. Sklyaruk J, Borghs JC, El-Sepelgy O, Rueping M (2019) Catalytic C1 alkylation with methanol and isotope-labeled methanol. Angew Chem Int Ed 58(3):775–779. https://doi.org/10.1002/anie.201810885
34. Bruneau-Voisine A, Pallova L, Bastin S, César V, Sortais J-B (2019) Manganese catalyzed α-methylation of ketones with methanol as a C1 source. Chem Commun 55(3):314–317. https://doi.org/10.1039/C8CC08064J
35. Emayavaramban B, Chakraborty P, Dahiya P, Sundararaju B (2022) Iron-catalyzed α-methylation of ketones using methanol as the C1 source under photoirradiation. Org Lett 24(33):6219–6223. https://doi.org/10.1021/acs.orglett.2c02545
36. Liu Z, Yang Z, Yu X, Zhang H, Yu B, Zhao Y, Liu Z (2017) Methylation of C(sp3)–H/C(sp2)–H bonds with methanol catalyzed by cobalt system. Org Lett 19(19):5228–5231. https://doi.org/10.1021/acs.orglett.7b02462

37. Latham DE, Polidano K, Williams JMJ, Morrill LC (2019) One-pot conversion of allylic alcohols to α-methyl ketones via iron-catalyzed isomerization–methylation. Org Lett 21(19): 7914–7918. https://doi.org/10.1021/acs.orglett.9b02900
38. Ganguli K, Mandal A, Kundu S (2022) Well-defined bis(NHC)Mn(I) complex catalyzed tandem transformation of α,β-unsaturated ketones to α-methylated ketones using methanol. ACS Catal 12(19):12444–12457. https://doi.org/10.1021/acscatal.2c04131
39. Bettoni L, Gaillard S, Renaud J-L (2019) Iron-catalyzed β-alkylation of alcohols. Org Lett 21(20):8404–8408. https://doi.org/10.1021/acs.orglett.9b03171
40. Kaithal A, van Bonn P, Hölscher M, Leitner W (2020) Manganese(I)-catalyzed β-methylation of alcohols using methanol as C1 source. Angew Chem Int Ed 59(1):215–220. https://doi.org/10.1002/anie.201909035
41. Schlagbauer M, Kallmeier F, Irrgang T, Kempe R (2020) Manganese-catalyzed β-methylation of alcohols by methanol. Angew Chem Int Ed 59(4):1485–1490. https://doi.org/10.1002/anie.201912055
42. Borghs JC, Tran MA, Sklyaruk J, Rueping M, El-Sepelgy O (2019) Sustainable alkylation of nitriles with alcohols by manganese catalysis. J Org Chem 84(12):7927–7935. https://doi.org/10.1021/acs.joc.9b00792
43. Mastalir M, Pittenauer E, Allmaier G, Kirchner K (2017) Manganese-catalyzed aminomethylation of aromatic compounds with methanol as a sustainable C1 building block. J Am Chem Soc 139(26):8812–8815. https://doi.org/10.1021/jacs.7b05253
44. Yang J, Chen S, Zhou H, Wu C, Ma B, Xiao J (2018) Cobalt-catalyzed α-methoxymethylation and aminomethylation of ketones with methanol as a C1 source. Org Lett 20(21):6774–6779. https://doi.org/10.1021/acs.orglett.8b02892
45. Biswal P, Samser S, Nayak P, Chandrasekhar V, Venkatasubbaiah K (2021) Cobalt(II)-porphyrin-mediated selective synthesis of 1,5-diketones via an interrupted-borrowing hydrogen strategy using methanol as a C1 source. J Org Chem 86(9):6744–6754. https://doi.org/10.1021/acs.joc.1c00476
46. Das J, Singh K, Vellakkaran M, Banerjee D (2018) Nickel-catalyzed hydrogen-borrowing strategy for α-alkylation of ketones with alcohols: a new route to branched gem-Bis(alkyl) ketones. Org Lett 20(18):5587–5591. https://doi.org/10.1021/acs.orglett.8b02256
47. Charvieux A, Duguet N, Métay E (2019) α-Methylation of ketones with methanol catalyzed by Ni/SiO2-Al2O3. Eur J Org Chem 2019(22):3694–3698. https://doi.org/10.1002/ejoc.201900602
48. Zhang H, Wang J, Liu M, Liu D (2020) Cu/Al2O3 catalyst prepared by a double hydrolysis method for a green, continuous and controlled N-methylation reaction of nitroarenes with methanol. Appl Surf Sci 526:146708. https://doi.org/10.1016/j.apsusc.2020.146708
49. Ma Z, Zhou B, Li X, Kadam RG, Gawande MB, Petr M, Zbořil R, Beller M, Jagadeesh RV (2022) Reusable Co-nanoparticles for general and selective N-alkylation of amines and ammonia with alcohols. Chem Sci 13(1):111–117. https://doi.org/10.1039/D1SC05913K
50. Li H, Li C, Liu W, Yao Y, Li Y, Zhang B, Qiu C Photo-induced C1 substitution using methanol as a C1 source. ChemSusChem:e202300377. https://doi.org/10.1002/cssc.202300377

Top Organomet Chem (2024) 73: 199–226
https://doi.org/10.1007/3418_2023_95

Published online: 12 October 2023

Recent Advancement of 3d Metal-Catalyzed Ethanol Upgradation via the Guerbet Reaction

Tanmoy Mandal, Manisha Pal, and Joyanta Choudhury

Contents

Abstract The catalytic transformation of (bio)ethanol to butanol is a tempting method for improving the modest fuel qualities of this readily accessible bioderived feedstock into a molecule that possesses attributes substantially closer to standard gasoline. Through the use of borrowing hydrogen chemistry, longer-chain alcohols can be manufactured from their shorter-chain analogues via the Guerbet reaction, with the goal of using these long-chain alcohols as synthetic lubricants and sustainable fuels. Although this reaction for ethanol appears to be straightforward, it presents significant difficulties, particularly when trying to achieve high selectivity. Over the past century, there have been improvements in both heterogeneous and homogeneous catalysis by using noble metals, but there has been very little research conducted on 3d transition metal-based catalyst systems. This chapter emphasizes the findings of homogeneous and heterogeneous 3d metal-based catalyst systems that show promising results in terms of generating high activity and selectivity. While only Mn-based catalyst systems are known to catalyze this reaction under

T. Mandal, M. Pal, and J. Choudhury (✉)
Organometallics and Smart Materials Laboratory, Department of Chemistry, Indian Institute of Science Education and Research Bhopal, Bhopal, India
e-mail: joyanta@iiserb.ac.in

homogeneous conditions, metal oxide-supported systems dominate the field of heterogeneous catalysts.

Keywords 3d metal · Biofuel · Butanol · Catalysis · Ethanol upgrading · Guerbet reaction

Abbreviations

AC	Activated carbon
AES	Atomic emission spectroscopy
Ar	Aromatic
atm	Atmospheric pressure
BET	Brunauer, Emmett, and Teller
BH	Borrowing hydrogen
Bu	Butyl
CNT	Carbon nanotube
Cy	Cyclohexyl
Et	Ethyl
HAS	High surface area
ICP	Inductively coupled plasma
^{i}Pr	Isopropyl
LA	Lewis acid
LB	Lewis base
LDH	Layered double hydroxides
Me	Methyl
MLC	Metal-ligand cooperation
MMO	Mixed metal oxide
NMR	Nuclear magnetic resonance
OAc	Acetate
OES	Optical emission spectroscopy
Ph	Phenyl
PMO	Porous metal oxide
SEM	Scanning electron microscopy
^{t}Bu	Tert-butyl
TEM	Transmission electron microscopy
TM	Transition metal
TON	Turn over number
wt%	Weight percentage
XPS	X-ray photoelectron spectroscopy
XRD	X-ray diffraction

1 Introduction

1.1 Biofuels and Ethanol

Due to the fact that no renewable energy source currently exists that can match fossil fuels with regard to accessibility or generation of energy, there is still a lot of study being done on how to make them more effective and useful. In modern times, due to concerns related to environmental safety, energy preservation, and sustainability, people are looking for alternative energy sources [1]. The primary goals of each renewable energy technology are to identify the source of clean energy that can be continuously regenerated while also reducing the quantity of CO_2 that is discharged into the environment. After electricity and heating, the transport sector produces the second highest amount of CO_2 globally by burning fuels like petrol and diesel. The transportation sector is starting to take into account the wide range of potential fuel alternatives to gasoline given the importance of automobiles in modern lifestyle. Crude oil, a petroleum-based product that is used to make gasoline, is anticipated to be exhausted very soon, which emphasizes the urgency of searching for alternatives toward sustainable fuel. Out of the various popular petrol substitutes, currently, biofuels are considered one of the most popular ones and are being investigated as potential replacements for dwindling fossil fuels [2].

As biofuels are low in cost and environment-friendly also, there is a need to create a systematic process and technology for their production. Research in academia and industry is focusing more on biofuels from sustainable biomass sources [3]. At present, bioethanol is being investigated as a very good substituent (or fuel additive) for conventional gasoline. Bioethanol, the first-generation biofuel is primarily made by fermenting crops rich in sugar or starch using microbial or yeast catalysts. Despite its benefits, bioethanol has certain disadvantages as well. For example, multiple processes must be carried out before a usable fuel can be made, including pre-treatment, hydrolysis, fermentation, and distillation. In addition to having a low energy density (70%) and being extremely soluble in water, this substance makes separation and shipping challenging. Additionally, pure ethanol has a low vapor pressure that makes it difficult to start an engine when it is cold. Also, it is corrosive to fuel tanks, making prolonged storage troublesome. Due to these problems, it is debatable if ethanol is capable of being considered as an advanced biofuel, since it can only be used in small amounts of petrol mixtures.

1.2 Butanol as the Advanced Biofuel

Advanced biofuels should be drop-in substitutes for petrol that require little to no adjustment. Butanol (or related longer-chain alcohols) can be used to reduce the issues caused by ethanol. Butanol has an energy density by volume that is 91% over that of gasoline, compared to ethanol's value of 67%. As a result, a vehicle could go further on just 1 liter of butanol than it could on an equivalent amount of ethanol, and it could travel as far as it could on petrol. Additionally, butanol is substantially less prone than ethanol to soak up water, probably as a result of the enhanced effect of the hydrophobic carbon network. Butanol only has a 7.7 wt% water solubility, but ethanol and water are totally miscible, allowing for the safe blending of butanol with petrol prior to distribution. Additionally, butanol is less corrosive than ethanol, making it safe to store and utilize within the existing facilities. A rise in the content of moisture along with other harmful contaminants in engines can result from the unintentional evaporation of ethanol from gasoline-ethanol mixtures, lowering engine performance and lifespan [4]. This problem is solved as butanol-gasoline combinations have a lower vapor pressure than pure petrol. Transport is made safer by lower vapor pressures, which also lower the possibility of accidental explosion or fire. Finally, butanol is capable of igniting engines in cold conditions as it has a lesser autoignition temperature than ethanol while having a lower vapor pressure. All of these attributes make butanol as a promising advanced biofuel candidate in the automobile sector.

1.3 Butanol Formation via the Guerbet Coupling

The known process of butanol synthesis is ABE (acetone, butanol, ethanol) fermentation but it has concerns with selectivity, separation, and poor yields, making it difficult to produce large quantities of pure butanol [5]. Butanol can also be made from propylene through hydroformylation or hydrogenation reaction but that depends on non-renewable feedstock [6]. So, the synthesis of butanol from ethanol via the Guerbet reaction has been found to overcome all these problems [7]. This reaction goes through the hydrogen borrowing pathway with three steps (Fig. 1). The first step is the dehydrogenation of alcohol to aldehyde, then aldol condensation where water is the only by-product and the final step is the hydrogenation of the unsaturated aldehyde to alcohol.

But the production of butanol with the help of this process is challenging too because acetaldehyde is highly reactive; so selectivity becomes very less. In addition, the first step of dehydrogenation is thermodynamically not feasible. The

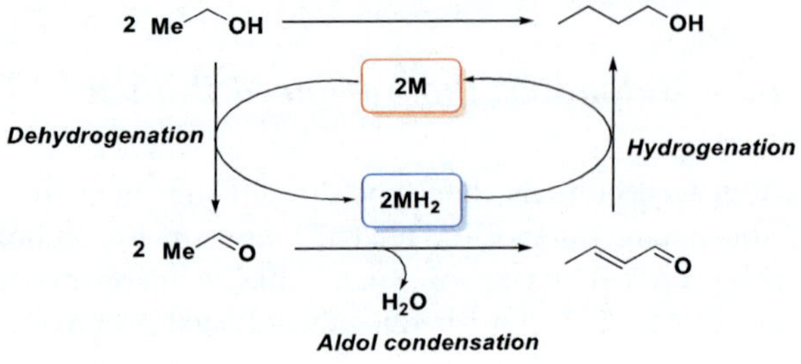

Fig. 1 General mechanism of ethanol upgradation to butanol via the Guerbet reaction [4]

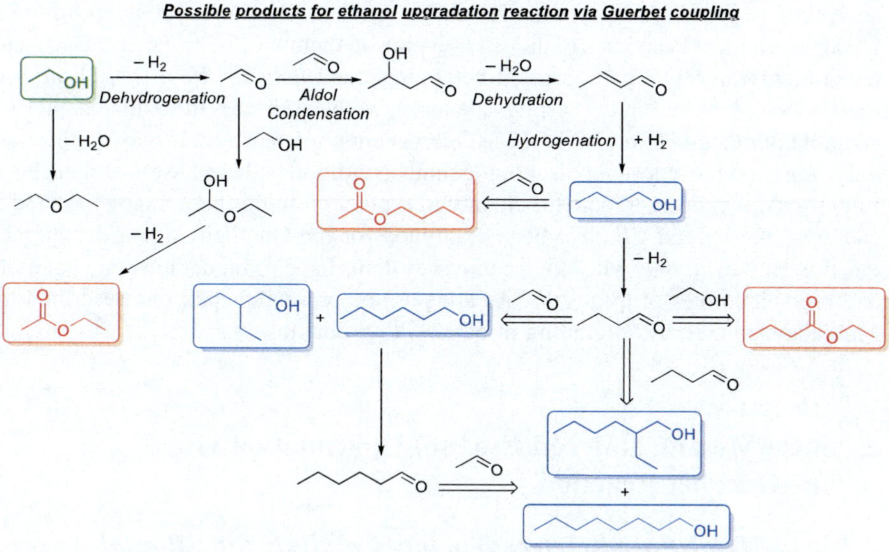

Fig. 2 Possible products for ethanol upgradation reaction via the Guerbet coupling [4]

following figure summarizes the possible products that can be formed during the upgradation of ethanol to 1-butanol via the Guerbet reaction (Fig. 2).

Homogeneous Ir and Ru-based catalysts have already been reported for performing the Guerbet reactions with high selectivity [4]. But yield using these catalysts is very low. One of the main objectives of modern chemistry has been to substitute precious metal catalysts for selective transformations with non-precious and less harmful metals. To enable practical applications, scientists and chemical firms are compelled to seek out fresh sustainable setups due to the scarce supply and high cost of these precious metal salts. The usage of 3d metals is environmentally less hazardous and adheres to various green chemistry standards because they typically have lesser toxicity when compared to 4d and 5d metal complexes [8]. As noble metal precursors are becoming more expensive and in demand, it ultimately has increased interest in finding more affordable homogeneous catalysts to replace them. In recognition of this, first-row transition metal catalysts are widely used in a diversity of manufacturing processes, which include catalytic hydrogenation, dehydrogenation, and related reactions [9]. So, in the Guerbet reaction also if the noble metal catalyst is replaced by 3d metals, it will be more beneficial. But there has also been a significant advancement in the Guerbet chemistry of ethanol using heterogeneous catalytic systems. For that pathway, basic metal oxides and transition-metal catalysts have been used substantially. Although several innovative heterogeneous catalysts and methods have been developed for the ethanol Guerbet reaction, the challenging reaction conditions, low conversions, and a poor selectivity profile are the major drawbacks of this field of study [10]. There are several first-row transition-metal catalysts that effectively perform dehydrogenation reactions in acceptorless pathways to produce aldehyde from alcohol under basic conditions. Under equivalent reaction conditions, several of them can hydrogenate C=C and C=O bonds. In this regard, several heterogeneous catalyst systems based on base metals were explored in the last decade and have been discussed in this chapter for ethanol upgradation to butanol via the Guerbet coupling reaction. However, there are only finger count reports under homogenous conditions where base metal successfully uses this process to convert ethanol to *n*-butanol. In this case, many base metal catalysts like Fe and Co have been examined for the Guerbet reaction to upgrade ethanol. However, only Mn-based catalyst systems have so far demonstrated activity and produced noteworthy results. As a result, the present chapter has described the Mn-catalyzed Guerbet upgrading of ethanol to *n*-butanol.

2 Base-Metal Catalyzed Ethanol Upgradation via the Guerbet Reaction

2.1 3d Metal-Based Heterogeneous Catalysts for Ethanol Upgradation

In 2015, the Poliakoff group showed *n*-BuOH production from EtOH in a flow reactor by using Cu-based catalysts supported on diverse solid platforms like Al_2O_3, TiO_2, Si/Al, ZSM-5 zeolite, CeO_2 with high surface area ($HSACeO_2$), and one regular surface area CeO_2 [11]. Out of all these catalysts, the best catalytic result was obtained with **Cu/HSACeO₂** catalyst which provided 67% ethanol conversion with 30% butanol yield followed by the **Cu/CeO₂** catalyst (39% conversion and 13% yield) at 260 °C (Fig. 3). The **Cu/TiO₂** catalyst showed similar butanol yields where for the other catalyst systems the yield falls down to the range of 0–6%. Further increasing the temperature to 330 °C, the ethanol conversion was increased but the butanol yield remained constant or decreased. On decreasing the temperature to 190 °C, they observed a reduction in ethanol conversion as well as butanol yield. At this temperature also the **Cu/HSACeO₂** catalyst provided the maximum butanol yield (10%) with a maximum ethanol conversion of 16%. They proposed that the $HSACeO_2$ having the higher surface area showed better activity for the aldol step and thus the **Cu/ HSACeO₂** catalyst showed superior activity.

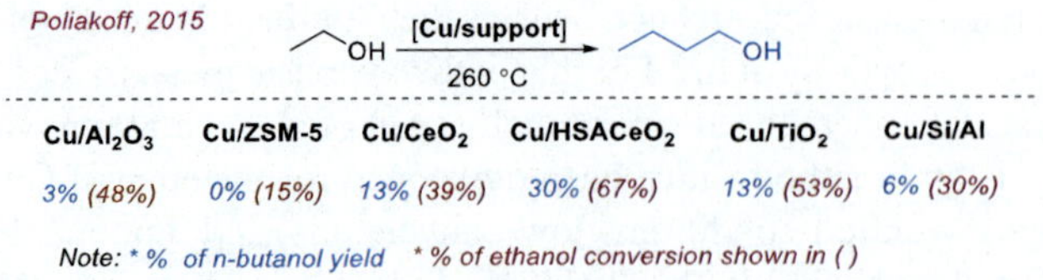

Fig. 3 Catalytic conversion of ethanol to *n*-butanol as reported by the Poliakoff group [11]

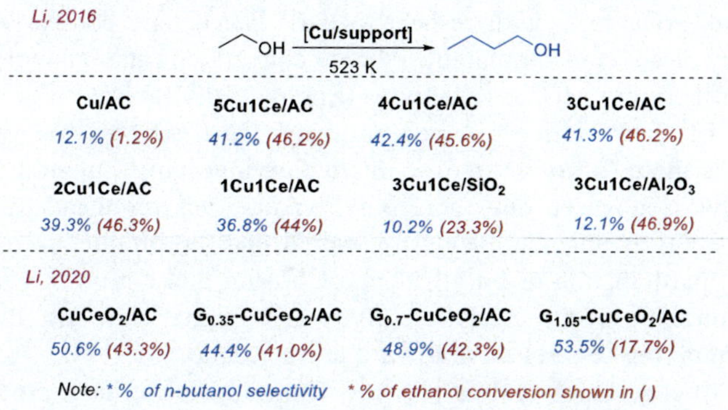

Fig. 4 Supported Cu catalysts for the conversion of ethanol to *n*-butanol reported by the Li group [12, 13]

In 2016, the Li group reported **Cu–CeO$_2$/AC** catalysts for *n*-butanol formation from ethanol via the Guerbet reaction in a flow reactor [12]. They prepared **Cu–CeO$_2$/AC** catalysts with diverse Cu/Ce molar ratios and compared the catalytic performance with other supported catalysts like **Cu/AC**, **3Cu1Ce/SiO$_2$**, and **3Cu1Ce/Al$_2$O$_3$**. They observed a dramatic improvement in the catalytic activity by the incorporation of Ce, as all the **Cu–CeO$_2$/AC** catalysts, irrespective of various molar ratios (Cu/Ce) showed ethanol conversion in the range of 44.0–46.2% with *n*-butanol selectivity of 36.8–42.4% (Fig. 4). In comparison, the **Cu/AC** catalyst showed only trace amount of ethanol conversion (1.2%). Instead of AC, when SiO$_2$ was used as the Cu–CeO$_2$ support, the ethanol conversion decreased to 23.3% with 10.2% butanol selectivity whereas the Al$_2$O$_3$ supported Cu–CeO$_2$ provided the highest ethanol conversion of 46.9% but the selectivity of butanol decreased to only 12.1%. They also tested the catalysis in a batch reactor by taking **3Cu1CeO$_2$/AC** as a catalyst for 48 h at 523 K with 250 mL ethanol and found 39.1% ethanol consumption with 55.2% selectivity toward n-butanol. To gain some mechanistic insights, they calculated the average diameter of Cu nanoparticles. They found that for **Cu/AC** and **Cu–CeO$_2$/AC** catalysts the average diameter was 4–6 nm which increased to 22 nm after the reaction for **Cu/AC** catalyst; however for Cu–**CeO$_2$/AC** catalysts, these values were less than 10 nm. They proposed that the presence of Ce restricted the growth of Cu nanoparticles which preserved the high dispersion of Cu over the **Cu–CeO$_2$/AC** catalysts. The high specific surface area of AC and the effective interaction that exist between Cu and CeO$_2$ ought to operate together to provide a significant dispersion of Cu and CeO$_2$ over **Cu-CeO$_2$/AC** catalysts. They further concluded that high ethanol conversion was facilitated by the significant dispersion of Cu metals while the aldol step benefited from the presence of extremely dispersive Ce species as basic sites, which shifted the reaction equilibrium toward the formation of *n*-butanol.

Next to adjust the interaction, they found that glucose appeared to be capable of physically stabilizing metal particles [13]. So, to regulate the Cu-CeO$_2$ interaction via glucose, they constructed a variety of **Gx-CuCeO$_2$/AC** catalysts (G = glucose and x is the molar ratio with respect to Cu). When they compared the catalytic activity, they discovered comparable ethanol conversions and butanol selectivity but variation in the ethyl acetate selectivity (Fig. 4). Ethyl acetate was the substantial by-product of their earlier research utilizing **Cu-CeO_2/AC** catalysts, which is thought to be the result of an intense interaction involving Cu and Ce species. The interaction was enhanced and more ethyl acetate was produced by elevating the loading of CeO_2 against Cu. They discovered that the production of ethyl acetate requires the participation of both Cu^0 and Cu^+ and that a ratio of Cu^0 to Cu^+ may serve as a measure of the catalytic activity. The higher the ratio, the more active toward butanol, but less selective toward ethyl acetate.

In order to create high-carbon alcohols from ethanol, in 2016, the Zheng group employed Ni-based catalysts supported on MgAlO in a fixed-bed reactor [14]. They initially optimized the Mg/Al ratio in the **Ni/MgAlO** catalysts and discovered that at an Mg/Al proportion of 4:1, the selectivity of *n*-butanol improved to 50.2% which suggests that the Mg/Al quantity in oxide catalysts strongly impacts the *n*-butanol selectivity and ethanol conversion. Both the conversion of ethanol and the selectivity for *n*-butanol dropped with a reduction in this proportion, and without the use of any Mg, a simple **Ni/Al_2O_3** catalyst resulted in 25% conversion of ethanol and a 35% selectivity for *n*-butanol. Additionally, they looked into how the mixed-oxide catalyst's Ni content affects ethanol conversion and butanol selectivity. Only 0.5% of the ethanol was converted when Ni was not present. The increase in Ni content from 0% to 19.5 wt% (**Ni4MgAlO**) resulted in an upsurge in ethanol conversion to 18.8% with 58% selectivity. However, the alcohol selectivity was dramatically reduced when the Ni concentration was raised further to 45.2 wt%. Using the **Ni4MgAlO** catalyst, they eventually investigated how temperature affects the conversion and selectivity of alcohols. They discovered that when the reaction temperature was raised from 473 to 548 K, the ethanol conversion raised gradually from 7.2% to 22.1% while the selectivity for C_4-C_8 alcohols stayed beyond 80%.

In 2017, the Barta group demonstrated the transformation of ethanol to 1-butanol using a variety of innovative porous metal oxide (PMO) compositions that are devoid of precious metals [15]. A number of hydrotalcite precursors were synthesized employing modular processes that were subjected to calcination to produce these distinct PMO catalysts. The Mg^{2+} in the catalyst structure of hydrotalcite, which is a mixed metal oxide of Mg/Al, was replaced with a minor amount of Cu and Ni to produce the final catalyst components. With the objective of figuring out the structure-activity patterns, they carried out a comprehensive characterization using a variety of methods including powder XRD, SEM, TEM, BET, etc. Since the product 1-butanol formation continues with several side reactions, the mixing of acidic and basic sites on the catalysts can affect both catalytic activity and product selectivity and thus the ratio of Mg to Al in hydrotalcite also can influence the product selectivity. Along with that, by adding other metal additives, the composition of Mg-Al mixed-oxides can be easily altered which further tunes their catalytic performance. The effectiveness of **MgAl-PMO** in encouraging the self-coupling of ethanol to 1-butanol has previously been investigated. The effects of Cu- and Ni-doping on the **MgAl-PMO** for the Guerbet reaction of ethanol were investigated (Fig. 5). At

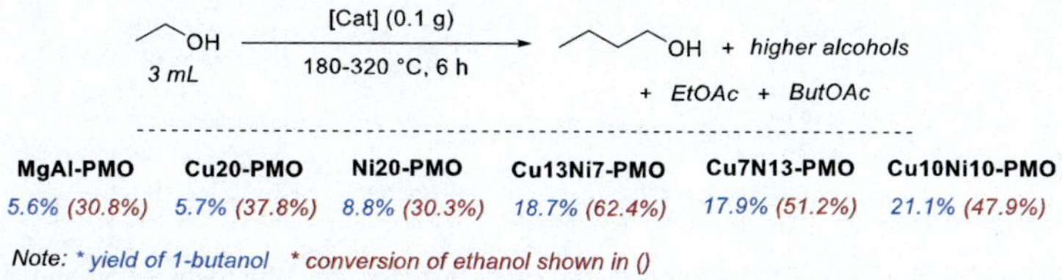

Fig. 5 PMO-based heterogeneous catalysts for ethanol upgradation to 1-butanol shown by the Barta group [15]

first, hydrotalcites (HTCs) were designed each of which had a little quantity of Cu and/or Ni dopant in place of a portion of Mg^{2+} while maintaining a constant 3:1 M^{2+}/M^{3+} ratio, and based on it, multiple materials were manufactured and tested for Guerbet reaction of ethanol. The catalytic results showed that a 5.6% yield of 1-butanol along with higher alcohols was obtained in 6 h with **MgAl-PMO**, which was free of any extra dopants. With **Cu20-PMO** ($Cu_{0.6}Mg_{2.4}Al_{1.0}$) as a catalyst, the ethanol conversion enhanced but the yield of 1-butanol (5.7%) was nearly comparable to that obtained with **MgAl-PMO**, giving ethyl acetate as its major product (18.1% yield). The yield of 1-butanol increased to 8.8% using **Ni20-PMO** ($Ni_{0.6}Mg_{2.4}Al_{1.0}$) as a catalyst where nickel was used simply as a dopant, but other long-chain alcohols continued to be the predominant output. Copper and nickel-containing catalysts produced the most impressive catalytic performance, with 1-butanol yields of 18.7% using **Cu13Ni7-PMO** catalyst, 17.9% with **Cu7Ni13-PMO** catalyst, and 21.1% with **Cu10Ni10-PMO** catalyst. The relationship between conversion and yield values with reaction temperature (between 180 and 320 °C) was also studied where the conversion steadily increased from 2.3% at 180 °C to 56.5% at 320 °C, with a maximum yield of 22.2% for 1-butanol. Then they tested the reusability of the **Cu10Ni10-PMO** catalyst where they discovered good resilience even after seven cycles. For the first two cycles, the yield of 1-butanol varied (21% vs. 14%), but it stayed steady for the next cycles and persisted within the range of 10–14%. After the catalysis reaction, they observed BET surface area drop from 255.7 m^2/g to 120.5 m^2/g along with a decrease in pore volume from 1.06 to 0.67 cm^3/g. They explained that there was a significant aggregation of smaller nanoparticles to create many larger nanoparticles with a diameter up to 100 nm which was the main difference between the fresh catalyst and recovered catalyst which was also the reason for its decreased catalytic activity. Furthermore, ICP analysis revealed nearly no leach of the Cu and Ni integrated into the catalyst structure, although a minor quantity of Mg and Al loss was noticed after the first cycle but almost no quantities were lost in the following cycles. They further proved that the coexistence of Cu and Ni caused the creation of a Cu-Ni alloy phase during the reaction which favored the dehydrogenation of ethanol and improved the stability of the catalyst.

In 2019, the Siqueira group also tested the ethanol conversion reaction to make 1-butanol by using a catalyst containing Cu which was supported on mixed metal oxide (MMO) [16]. To make the hydrotalcite, the layered double hydroxides (LDH),

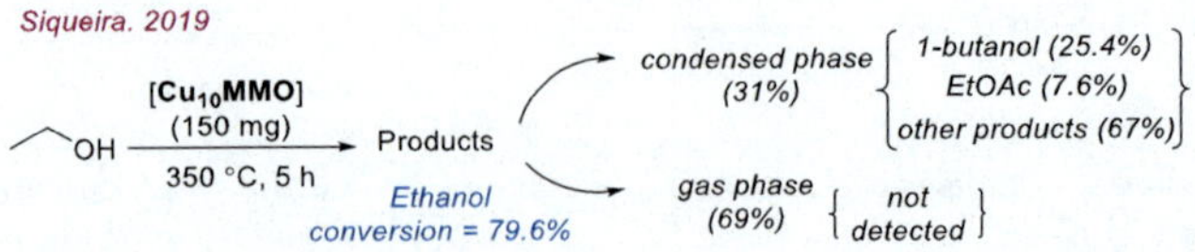

Fig. 6 Siqueira's report on MMO-supported Cu catalysts for the conversion of ethanol to *n*-butanol [16]

they used Mg^{2+} and Al^{3+} which served as the support, and by replacing 10 mol% of Mg^{2+} with Cu^{2+} they made the final version of the catalyst **Cu_{10}MMO**. They found that the surface area did not change significantly after substitution by Cu^{2+} as it decreased slightly from 178 to 171 m^2g^{-1}. In this study, their main focus was to optimize the reaction parameters like time, temperature, and catalyst loading and their impact on catalysis. They started the catalysis by taking 30 mg of **Cu_{10}MMO** catalyst at 280 °C where they found 20% ethanol conversion in 2 h with only 2% of 1-butanol. Then after performing several reactions at different conditions, they observed the highest conversion of 79.6%, by using 150 mg catalyst and 5 h of reaction at 350 °C. The composition of the products was divided into the condensed phase (31%) and the gas phase (69%) (Fig. 6). They did not identify the products in the gas phase whereas the condensed phase contained the products like 1-butanol (25.4%), ethyl acetate (7.6%), and other Guerbet condensation products (67%). They also proposed the importance of Cu^{+2} in the present catalysis as it favored the hydride formation and its transfer where the overall hydrogenation activity improved resulting in an increase of ethanol transformation. They showed three times reusability of the catalysts where ethanol conversion was in the range of 79% to 73% but butanol formation reduced to 14% only which suggested some structural changes on the catalyst surface.

In 2019 the Martins group showed the Guerbet condensation of ethanol in a flow reactor by using AlMgO/AlCaO-supported Cu catalysts [17]. In an H_2 environment, the impact of changing system pressure on ethanol coupling over **Cu/AlMgO** catalyst was investigated. At 1 atm pressure, the catalyst showed 17.9% ethanol conversion with acetaldehyde (56% selectivity) as the major product along with the other coupling products (C_{4+} alcohols and aldehydes) with 22.1% selectivity (Fig. 7). They noticed that the product selectivity changed as the H_2 pressure increased. For instance, at 31 bars of H_2 pressure, acetaldehyde selectivity dropped to 12.8% while C_{4+} alcohol selectivity rose to 49.2% with a 19.7% ethanol conversion. They observed a reduced activity in the absence of Cu as with **AlMgO** and **AlMgO-P** converting at just 7.1 and 11.1% ethanol whereas **AlCaO** and **AlCaO-P** showed nearly zero activity. However, with the addition of Cu, activity increased for all the catalysts. They observed similar activities by the **Cu/AlMgO** and **Cu/AlMgO-P** catalysts, with conversions of 19.7 and 23.3% with alcohol selectivity of 21.8 and 23.5%, respectively. With the **Cu/AlCaO** and **Cu/AlCaO-P** catalysts, though more selective desired products (54.8 and 48.8%, respectively) were observed but ethanol conversion was limited to only 8.7 and 8.4%.

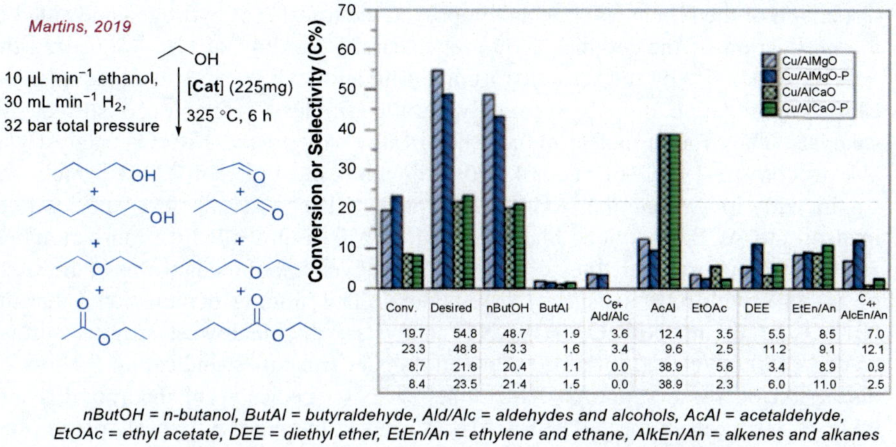

Fig. 7 Catalytic conversion of ethanol to *n*-butanol as reported by the Martins group [17] (The plot corresponding to the ethanol conversion with different product selectivity was reused with permission from [17])

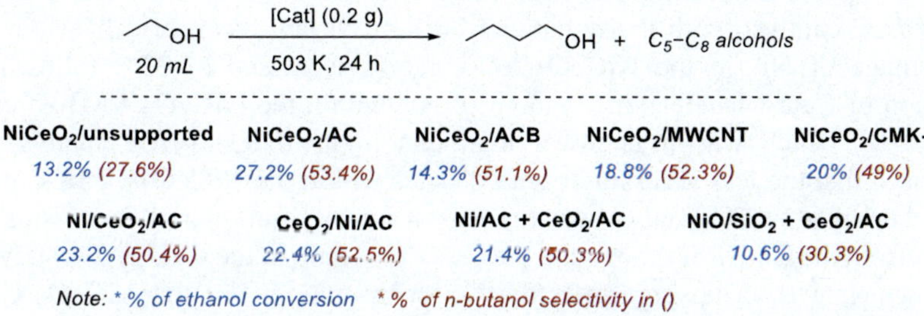

Fig. 8 A series of $NiCeO_2$ catalysts utilized by the Pang group for the catalytic conversion of ethanol to *n*-butanol [18]

To demonstrate the impact of metal dispersion and interactions on the transformation of ethanol to *n*-butanol, the Pang group synthesized a variety of $\mathbf{NiCeO_2}$ catalysts in 2020, illustrating the distribution of Ni and CeO_2 on various carbon substrates [18]. First, they used several $\mathbf{NiCeO_2/AC}$ catalysts at 503 K for 24 h to test the impact of Ni and CeO_2 loading on activated carbon (AC) for the selective conversion of ethanol to *n*-butanol (Fig. 8). Both Ni and CeO_2 loading varied in the range of 5–20 wt%. First maintaining the CeO_2 loading fixed at 15 wt%, they discovered the highest activity with 10 wt% Ni loading, which indicated ethanol conversion of 27.2% with 54% *n*-butanol selectivity. Then, they varied the concentration of CeO_2 while fixing 10% Ni on AC and discovered that in the absence of CeO_2 with 10% **Ni/AC** catalyst, only 8.4% of ethanol could be converted with 31.7% *n*-butanol selectivity. They explained that the CeO_2 present in the material enhanced the aldol condensation process, favoring the equilibrium toward *n*-butanol and thus having an effect on the product selectivity. In addition, CeO_2 enhanced the dispersion of the Ni particles, which improved the transfer of hydrogen and aided in the production of the product. However excessive loading of CeO_2 squeezed the product selectivity by further transforming *n*-butanol, while excessive loading of Ni caused aggregation, which eventually resulted in less available Ni surface for catalysis. They found that in the absence of any support the **$NiCeO_2$** catalyst was able to convert 13.2% of ethanol with *n*-butanol selectivity of 27.6% which was significantly lower than the **$NiCeO_2$/AC** catalyst. Then, Ni and CeO_2 species were loaded onto ACB (acetylene black) and MWCNT (multiwalled carbon nanotubes, 30–50 nm) support where they discovered the conversion of ethanol was in the range of 14–19% with 52% selectivity to *n*-butanol. This finding demonstrated that the **$NiCeO_2$/ACB** and **$NiCeO_2$/MWCNT** catalysts with the lowest specific surface areas (≈150 m^2/g) had inferior hydrogen transfer capacities, indicating the low Ni dispersion on these supports. After that, $NiCeO_2$ species were incorporated into CMK-3, resulting in the **$NiCeO_2$/CMK-3** catalyst having the highest specific surface area of 786 m^2/g. However, this catalyst also showed lower activity than the **$NiCeO_2$/AC** catalyst with ethanol conversion of 20.0% and 49.0% *n*-butanol selectivity. Then they performed the XPS investigation to analyze the surface states of Ni and Ce species. From the Ni-XPS analysis, they detected the presence of both Ni^0 and Ni^{2+} species inside the materials. In the absence of CeO_2, Ni^0 made up 30% of the **Ni/AC** catalyst, with the remaining 70% of Ni being conserved for Ni^{2+}. The percentage of Ni^0 in the **$NiCeO_2$/AC** catalyst decreased to 22% following the addition of CeO_2, whereas the amount of Ni^0 within the **$NiCeO_2$/CMK-3** catalyst was 17%, which was little lower than that in the **$NiCeO_2$/AC** catalyst. They proposed that the less accessibility of Ni^0 sites within the **$NiCeO_2$/CMK-3** catalyst than the **$NiCeO_2$/AC** catalyst was the reason for inferior ethanol conversion. Similarly, from the Ce-XPS, they found the Ce^{3+}/Ce^{4+} ratio in the **CeO_2/AC** catalyst was 0.49 which was an upsurge after the Ni incorporation. For the **$NiCeO_2$/CMK-3** catalyst, this value was 0.55 whereas, for **$NiCeO_2$/AC** catalyst, it increased to 0.63 which suggested that more Ce presented as Ce^{3+} inside the **$NiCeO_2$/AC** catalyst. The larger ionic size of Ce^{3+} in comparison with Ce^{4+} created more defects inside the **$NiCeO_2$/AC** catalyst which ultimately accounted for higher catalytic efficiency toward the formation of *n*-butanol. They were also able to demonstrate four times reusability of the **$NiCeO_2$/AC** catalyst where they found ethanol conversion in the range of 20–24% with *n*-butanol selectivity of 58% which indicated the robustness of the material.

The Zheng group, in 2020, demonstrated the conversion of ethanol to butanol using a binary catalytic system comprising **$FeNiO_x$** and LiOH [19]. Based on the XRD and XRF data, they speculated that the **$FeNiO_x$** catalyst is made up of FeNi oxides with a Fe/Ni ratio of 4:1. With the goal of ethanol conversion, initially, they tested a range of metal catalysts in the presence of $NaHCO_3$.They discovered that over the **NiO_x** and **CuO_x** catalysts, only 4.4%–4.6% ethanol conversions were achieved with butanol selectivity of 53.0%, but with the **$FeNiO_x$** catalyst, the butanol selectivity increased to 68.3% with the same ethanol conversion rate. A maximum of 3% ethanol conversions were observed with other supported metal catalysts including **Ru/C**, **Pd/C**, **Rh/TiO_2**, **NiIr/MC**, and **Ni/CeO_2** (Fig. 9). In the presence of

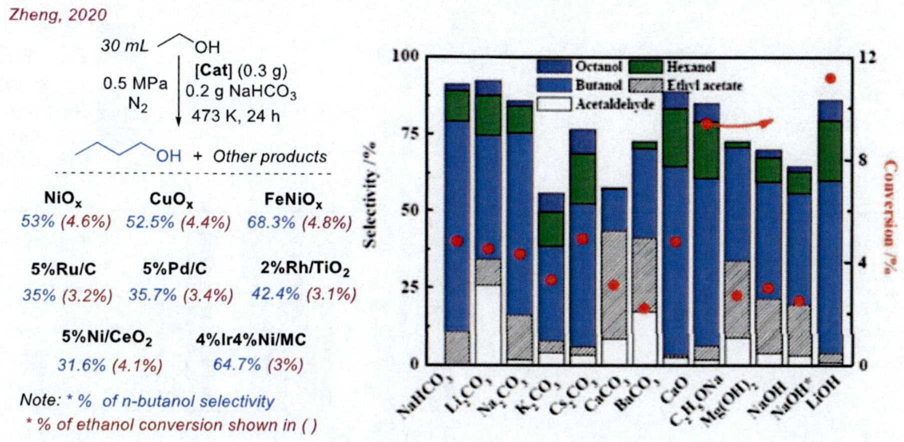

Fig. 9 Zheng's report on catalytic conversion of ethanol to *n*-butanol using a binary catalytic system [19] (The plot corresponding to the ethanol conversion with different product selectivity was reused with permission from [19])

$FeNiO_x$ catalysts, various salts or bases with the same weight ratio were evaluated for ethanol conversion in substitution of $NaHCO_3$ (Fig. 9).

With the rest of the alkali metal carbonates, the butanol selectivity was lower than 60%. On the contrary, when alkaline earth metal carbonates like $CaCO_3$ and $BaCO_3$ were used in the process, they observed the major product changed to ethyl acetate with a reduced ethanol conversion. When NaOEt was evaluated for ethanol conversion, it demonstrated 54.9% butanol selectivity at 9.4% ethanol conversion. Hydroxide bases were next put to the test in the conversion of ethanol. Low ethanol conversion (3%) with 30% butanol selectivity was achieved over $Mg(OH)_2$ and NaOH; however, LiOH considerably increased the ethanol conversion to 11.2% with a butanol selectivity of 56.1%. Next, they adjusted the base concentration and discovered that, as LiOH content rose from 0.05 g to 0.8 g, the ethanol conversion first climbed, then remained constant, before it declined, with the maximum conversion of 28% and the highest butanol selectivity of 71% observed at 0.2 g. Here, they proposed that LiOH catalyzed the aldol condensation reaction and also improved ethanol conversion as a result of the synergistic interaction with **$FeNiO_x$**, which played a significant role in the hydrogen transfer reaction. At last, they showed seven times recyclability of the **$FeNiO_x$ catalyst** which further indicated its stability.

Similarly, in 2021, the Metzker group also utilized Mg^{2+} and Al^{3+} containing mixed metal oxide-based catalysts where 20 mol% Mg^{2+} was replaced by various metals like Fe^{2+}, Co^{2+}, Ni^{2+}, Cu^{2+}, and Zn^{2+} to make a series of heterogeneous catalysts for the Guerbet reaction to transform ethanol (Fig. 10) [20]. After synthesis, they checked the catalytic activity of these materials in a fixed-bed flow reactor at 450 °C where they found a maximum ethanol consumption was achieved after 1 h of reaction for all the catalysts. Out of all these catalysts, highest activity was noticed with **Zn_{20}MMO** which showed 85% ethanol conversion whereas **Fe_{20}MMO** provided the lowest ethanol conversion of 47%. They detected the presence of alcohols

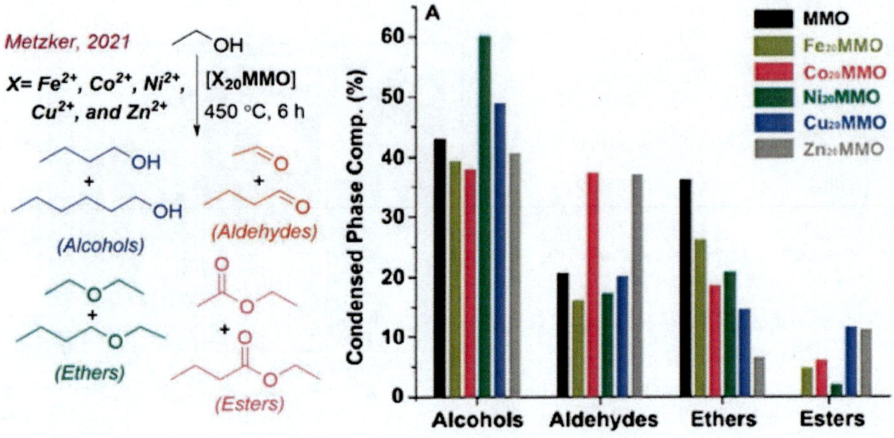

Fig. 10 Metzker's report on various 3d metal-supported catalytic systems for the conversion of ethanol to *n*-butanol by [20] (The plot corresponding to the ethanol conversion with different product selectivity was reused with permission from [20])

(95% *n*-butanol +5% *n*-hexanol), aldehydes, ethers, and esters as the products of ethanol conversion. In terms of alcohol selectivity, the **Ni_{20}MMO** catalyst provided the best result with 60% selectivity.

In 2021, the Ni group used **Cu@UiO-66** catalysts to effectively upgrade ethanol to *n*-butanol [21]. They synthesized a variety of UiO-66 encapsulated Cu (**Cu@UiO-66**) catalysts by adjusting the loading of Cu and were able to regulate the interaction between Cu metal and Zr6 nodes. They discovered the close connection between internal Cu metal sites and Lewis acid-oxygen vacancy pairs in Zr nodes, which are active sites, dominating the critical step reactions in the upgrading of ethanol to *n*-butanol. Then, they tested the catalytic activity of these materials at 250 °C by taking 0.5 g catalyst in a flow reactor. An ethanol conversion of 2.6% was obtained by using **UiO-66** as the catalyst, indicating that the **UiO-66** support is virtually inert for ethanol dehydrogenation (Fig. 11). The addition of Cu metal greatly boosted the efficiency of ethanol conversion as with the **1% Cu@UiO-66** catalyst, 23.8% ethanol conversion was observed with an *n*-butanol selectivity of 62.1%. Increasing the Cu loading, by using **2% Cu@UiO-66** as the catalyst, provided the highest selectivity of *n*-butanol with a yield of 16.2% with 60.3% selectivity. With additional Cu loading, even though ethanol conversion increased slightly, selectivity also decreased a bit, resulting in a yield of butanol that is within the same range. The **Cu@UiO-66** catalysts were then tested at higher temperatures (280 °C) and for longer periods, yielding up to 22.2% of *n*-butanol. They speculated that the active sites for upgrading ethanol to *n*-butanol are Cu metal sites and Lewis acid-oxygen vacancy pairs of Zr nodes within the pores of **UiO-66**, as opposed to Cu metal sites and Zr nodes on the outside of **UiO-66**. From the catalytic results, they suggested that the catalyst's structure only allowed for a maximum of 2% Cu to be

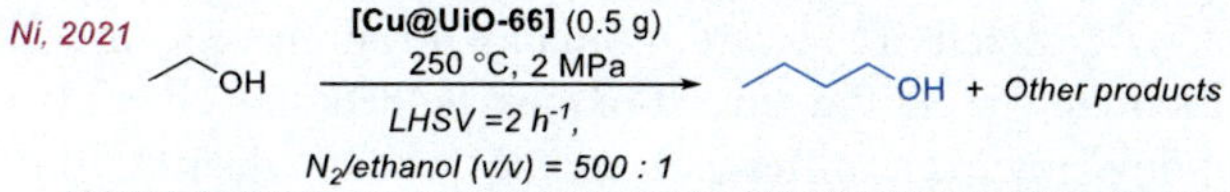

Catalyst	Ethanol conversion	n-butanol selectivity	n-butanol Yield
UiO-66	2.6%	0.3%	-
1% Cu@UiO-66	23.8%	62.1%	14.8%
2% Cu@UiO-66	26.8%	60.3%	16.2%
3% Cu@UiO-66	28.6%	56.7%	16.2%
5% Cu@UiO-66	28.3%	58.2%	16.5%

Fig. 11 MOF-based Cu catalyst for the catalytic conversion of ethanol to *n*-butanol showed by the Ni group [21]

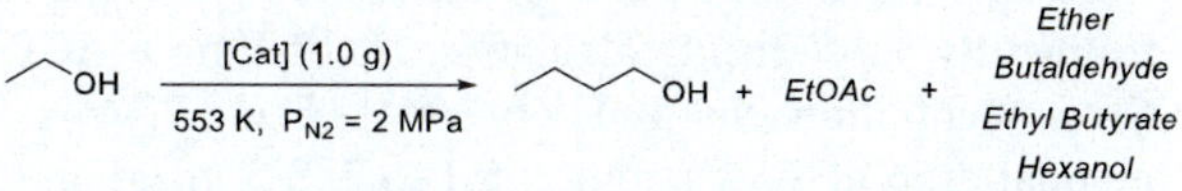

Catalyst	ethanol conversion	n-butanol selectivity	EtOAc selectivity
$MgAlO_x$	4.4%	0%	0.7%
$ZnAlO_x$	8.5%	0.2%	38.6%
$Cu/MgAlO_x$	43.1%	33.2%	11.8%
$Cu/ZnAlO_x$	33.9%	7.7%	42.2%

Fig. 12 Ding's report on catalytic conversion of ethanol to *n*-butanol by mixed-oxides supported Cu catalysts [22]

present on the internal surface of **UiO-66** and that a larger Cu loading only caused the development of larger Cu metal particles on the catalyst's external surface which were less involved in catalysis.

In 2022, the Ding group reported the conversion of bioethanol through the Guerbet coupling process for $\mathbf{MgAlO_x}$ and $\mathbf{ZnAlO_x}$ mixed-oxides supported Cu catalysts [22]. From the ICP-AES, they found similar Cu content inside the materials with values of 1.11 wt.% and 1.35 wt.% respectively. They also calculated the surface area of the $\mathbf{Cu/MgAlO_x}$ and $\mathbf{Cu/ZnAlO_x}$ catalysts which were 101.5 and 115.7 m^2/g, slightly less than its precursors (119.7 and 155.9 m^2/g, respectively). From the XPS investigations, they found the presence of Cu^+ and Cu^0 species inside both materials; however, for the $\mathbf{Cu/ZnAlO_x}$ catalyst they detected the Cu^{2+} species also which was not present in the $\mathbf{Cu/MgAlO_x}$ catalyst. Then using these catalysts, the reaction for ethanol upgradation was performed in a fixed-bed reactor at 553 K and 2 MPa of N_2. In the absence of Cu support, only using $\mathbf{MgAlO_x}$ and $\mathbf{ZnAlO_x}$ as catalysts showed very less ethanol conversion (4.4% and 8.5%) with little selectivity toward butanol (Fig. 12). The addition of Cu in these materials led to enhance the ethanol conversion as for $\mathbf{Cu/MgAlO_x}$ catalyst ethanol conversion was increased to 43.1% with *n*-butanol selectivity of 33.2% along with several other by-products. Similarly, $\mathbf{Cu/ZnAlO_x}$ catalyst showed 33.9% ethanol conversion but, in this case, ethyl acetate was observed as the major product with a selectivity of 42.2%, providing just 7.7% selectivity toward *n*-butanol. From these results, they concluded that the addition of Cu in the mixed-oxides helped the ethanol conversion via Guerbet reaction. However, difference in the product distribution suggested the role of the solid support, implying a synergistic interaction among Cu and mixed-oxides. After catalysis, they detected the larger Cu particles for both materials which formed via aggregation and resulted in a decrease in the catalytic efficiency. They proposed that the presence of Cu^0/Cu^+ species inside the materials promoted the hydride transfer and improved the ethanol conversion.

Later in a work in 2022, the Ding group prepared a number of **Cu/NiAlO$_x$** catalysts that were formed from NiAl-hydrotalcite and demonstrated improved activity in the continuous catalytic conversion of ethanol to produce butanol [23]. The hydrothermal deposition precipitation technique was used to create the set of **Cu/NiAlO$_x$** catalysts with various Cu loadings. Well-dispersed Cu species were contained in the pores of the HT-derived material, designated as **Cu/NiAlO$_x$**, formed by subsequent calcination. According to ICP-AES results, all samples had Cu concentrations that fall between 0.15 and 7.36%. As the Cu loading increased, a gradual drop in S_{BET} (from 255.1 to 229.5 m^2/g) and Vp (from 0.33 to 0.28 cm^3/g) was shown in **Cu/NiAlO$_x$** catalysts. This was explained by the accumulation of the Cu species within the small holes of **NiAlO$_x$**. Next, the catalytic transformation of ethanol to form butanol was evaluated with various **Cu/NiAlO$_x$** catalysts in a continuous fixed-bed reactor operating at a reaction temperature between 200 and 300 °C. Both butanol selectivity (about 12%) and ethanol conversion (around 15%) for **NiAlO$_x$** were comparatively poor. The ethanol conversion along with butanol selectivity appeared to have improved with the addition of Cu, as the **0.15%Cu/NiAlO$_x$** catalyst showed 22% conversion with 33% butanol selectivity. With a butanol selectivity of 45% and an ethanol conversion of about 25%, the **0.75% Cu/NiAlO$_x$** offered the best reactivity at 250 °C. Further raising the temperature to 300 °C, though ethanol conversion was increased to 55%, a concurrent decline in butanol selectivity to 18% was observed. The catalytic performance was marginally improved by raising the Cu loadings further to 1.5% and 7.5%, with an ideal balance between butanol selectivity (45%) and ethanol conversion (30%) occurring at 250 °C. According to additional characterization tests, they proved that the addition of Cu to the catalysts might not only offer redox sites to speed up the dehydrogenation and hydrogenation processes but could also control the base/acid characteristic to reduce the likelihood of side reactions. Further from the EDX-mapping and H_2-TPR characteristic, they proved the improved interaction between Cu and Ni species which was the proposed reason for the good stability.

Recently, in 2023, the Zheng group demonstrated the use of **NiCeO$_2$@CNTs** catalysts to convert ethanol into *n*-butanol [24]. For that, carbon nanotubes (CNTs) with various outer diameters were employed in this study. Ethanol conversions were considerably improved by anchoring a comparable quantity of Ni (11.3%) along with CeO_2 (16.8%) within CNTs. Using the **NiCeO$_2$@CNTs 30–50 nm** catalysts, they achieved an ethanol conversion of 12.6% with 54.1% *n*-butanol selectivity whereas the ethanol conversion improved progressively as CNT pore size decreased, reaching a maximum of 23.2% with 58.7% *n*-butanol selectivity over the

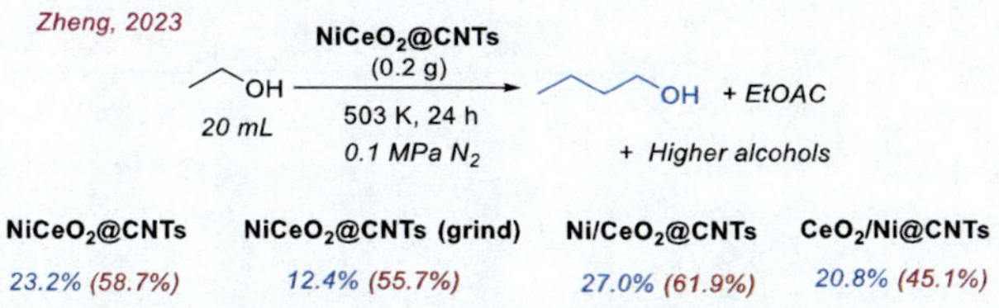

Fig. 13 Zheng's report for the ethanol conversion into *n*-butanol over $NiCeO_2$@CNTs catalyst [24]

$NiCeO_2$@CNTs 4–6 nm catalysts (Fig. 13). From these results, they speculated that the pore size of CNTs played a crucial role in regulating ethanol conversion and *n*-butanol selectivity. To confirm that the CNTs' pore structure was destroyed by ball milling the CNTs for 12 h, which resulted in a 12.4% reduction in ethanol conversion with 55.7% *n*-butanol selectivity. The disappearing promoting effect after destroying the nanotube structure showed that the 4–6 nm nanotube structure was essential for high ethanol conversions. For ethanol transformations, the location of the metal centers within or on the CNTs was also crucial. Using Ni particles on CNTs to bind CeO_2 species, **Ni/CeO_2@CNTs** catalyst was prepared for which the ethanol conversion was found to be 27.0% with *n*-butanol selectivity of 61.9%. However, after altering the loading method, they made **CeO_2/Ni@CNTs** catalyst for which the ethanol conversion and *n*-butanol selectivity decreased to 20.8 and 45.1%, respectively. The catalytic results for the **Ni/CeO_2@CNTs** and **CeO_2/Ni@CNTs** catalysts were different, suggesting that CeO_2 species were preferentially placed within the channels of CNTs for condensations and that this location subsequently increased both the ethanol conversion and *n*-butanol selectivity. The fact that the ethanol conversion and *n*-butanol selectivity remained at 22.7 and 56.2% after three cycles showed the great stability of this catalyst in the ethanol to *n*-butanol conversion process.

In 2023, the Lu group demonstrated the catalytic use of the **$Co_nMg_{3-n}AlO_x$** catalysts for the Guerbet reaction for the conversion of biomass-derived ethanol to C_{4+} molecules in a fixed-bed reactor. They created the catalysts by dispersing Co^{2+} on **$MgAlO_x$**, where the Co molar ratio had a crucial impact on the product selectivity as well as ethanol conversion (Fig. 14). The ethanol conversion improved from 11.5% to 32.9% with an increase in the cobalt molar ratio, which ranged from 0.015 to 0.15; but after that, the ethanol conversion appeared to decline with further increase in the cobalt molar ratio to 0.6. At 523 K and 0.1 MPa, the ethanol conversion was achieved over the most optimal catalyst, **$Co_{0.15}Mg_{2.85}AlO_x$**, with a 95.4% selectivity for C_{4-10} alcohols. In contrast, the **Mg_3AlO_x** catalyst without any Co, demonstrated 14.0% ethanol conversion and an 83.3% selectivity for C_{4-10} alcohols. Further, they monitored the influence of reaction temperature on catalysis and found that when the reaction temperature was lowered to 473 K, both the generation of alcohol-selective products and the conversion of ethanol declined. In addition, when the reaction temperature was raised to 623 K, the ethanol conversion climbed to 41.4%, but the entire selectivity for the C_{4-10} alcohols dropped to 49.2%.

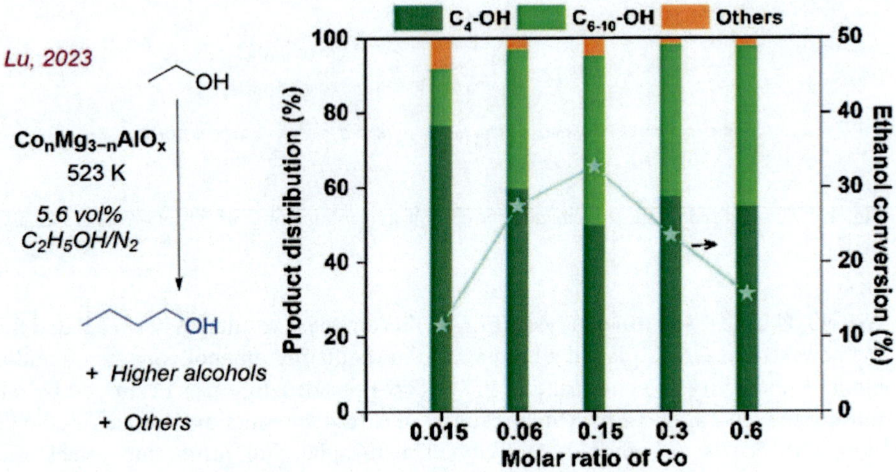

Fig. 14 Lu's report on the ethanol conversion to higher alcohols using $Co_nMg_{3-n}AlO_x$ catalysts [25] The plot corresponding to the ethanol conversion with different product selectivity was reused with permission from [25])

For comparing the role of the Co oxidation states, the $\mathbf{Co_{0.15}Mg_{2.85}AlO_x}$ catalyst was reduced in order to acquire more metallic Co for this catalysis, where they found that the conversion of ethanol dropped by about 7% while the selectivity remained the same. From this result, they concluded that metallic Co can also promote this reaction, though with somewhat less activity.

In 2023, the Wang group reported an amphiphilic **NiSn@C-MgO** catalyst to use the Guerbet pathway to produce higher alcohols (C_{4+} alcohols) from ethanol [26]. When the **NiSn@C-MgO** catalyst, consisting of NiSn nanoparticles (NPs) encased with carbon shell and MgO nanoflakes support, was used to upgrade aqueous ethanol, it was discovered that the carbonization temperature (400 to 700 °C) had an impact on the catalytic performance. The ethanol conversion increased from 62.5% to 73.3% and subsequently decreased to 60.6% when the carbonization temperature was increased from 400 to 700 °C. With an ethanol conversion of 73.3%, the optimum catalyst **NiSn@C-MgO-550** exhibited excellent activity with selectivity for C_{6+} alcohols reached up to 60.9% along with a 36.0% yield. Then, they investigated on how reaction temperature affected the upgrading of ethanol over the **NiSn@C-MgO-550** catalyst and found that as the reaction temperature increased from 230 to 260 °C, the ethanol conversion steadily climbed from 56.0% to 76.1%, showing that the elevated reaction temperature was advantageous for the conversion of ethanol.

2.2 3d Metal-Based Homogeneous Catalysts for Ethanol Upgradation

In 2017, the Liu group demonstrated the Guerbet-type condensation reaction to show the formation of 1-butanol from ethanol by using Mn-based molecular homogeneous catalyst systems [27]. This was the first report where instead of using a precious noble metal-based catalyst system, simply using base metal-based catalysts, the upgradation of ethanol into higher alcohols under homogeneous conditions was successfully demonstrated. For that, they used PNP and NNP-type pincer ligands and synthesized a series of well-defined pincer complexes of Mn (Fig. 15). When they checked the catalytic activity of these Mn-catalysts, they found similar activity for bis- and tris-carbonyl PNP-based Mn(I) complexes (**Mn-1a**, **Mn-1b**, **Mn-2a**, **Mn-2b**). It was ascribed to the formation of Mn(I) bis-carbonyl complexes from the corresponding tricarbonyl Mn(I) complexes under the catalysis condition of high temperature. However, when the bulkier ligand ($^{t\mathrm{Bu}}$PNP)-containing complex $^{t\mathrm{Bu}}$PNP-Mn(I) was used as a catalyst for the same reaction, they did not observe any catalytic activity. The same applied to ^{t}Bu group-containing NNP-Mn (I) catalysts also whereas the ^{i}Pr group-containing NNP-Mn(I) catalysts showed low activity for this catalysis. Additionally, the PNP-based Mn(II) catalyst and the precursor complex $Mn(CO)_5Br$ led to no product formation signifying the important role of auxiliary carbonyl and pincer ligands together.

After complete optimization of the reaction condition, they were able to achieve a TON of 114,120 with 92% selectivity for the upgradation of ethanol (1,275 mmol) into 1-butanol by using **Mn-1** (0.00125 mmol) as the catalyst and EtONa (150 mmol) as a base. Though high TON was achieved, yield was found in the

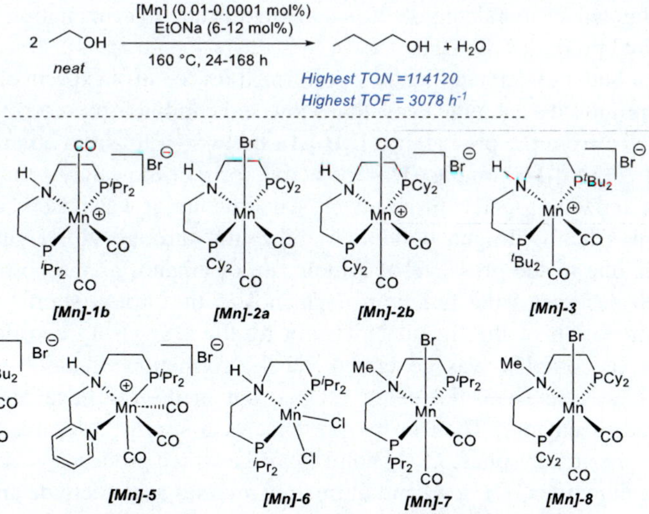

Fig. 15 Pincer Mn-complexes utilized by the Liu group for ethanol upgradation to 1-butanol [27]

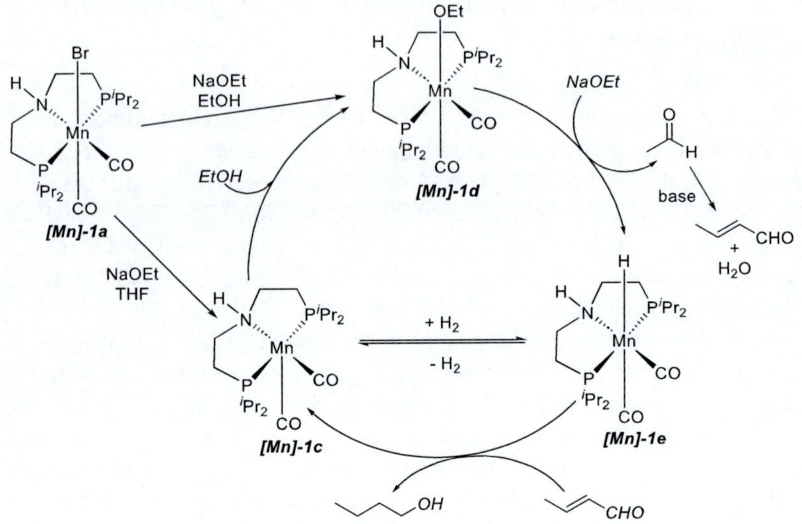

Fig. 16 Proposed catalytic cycle for Guerbet reaction of ethanol as shown by the Liu group [27]

range of 10% only. The addition of Hg or PMe_3 ligand did not hamper the catalytic efficiency of the catalyst suggesting the homogeneous nature of the catalysis process. To show the bifunctional nature of the catalysts, they prepared the corresponding N–Me analogue of the PN^HP-Mn catalysts (**Mn-7** and **Mn-8**) and found much lesser reactivity for this transformation with these catalysts. So, mechanistically, under the catalytic condition, at first, ethanol dehydrogenation took place to form acetaldehyde which underwent aldol condensation reaction to form the unsaturated crotonaldehyde from where further hydrogenation occurred to form the final product 1-BuOH.

For better understanding they performed few control experiments based on which they proposed a catalytic cycle that showed the initial formation of ethoxide complex **[Mn]-1d** from the pre-catalyst **[Mn]-1a** by using EtONa in ethanol (Fig. 16).

This **[Mn]-1d** complex was the active form of the catalyst that they isolated and characterized properly. From the crystal structure, it was observed that the **[Mn]-1d** complex has hydrogen bonding stabilization through N···H and O···H hydrogen bonds due to the presence of a molecule of ethanol as the co-crystallized solvent which designates the function of ethanol as the proton shuttle and facilitates the deprotonation of the ligand N–H unit by the OEt group coordinated to Mn. The **[Mn]-1d** complex was generated via the formation of amido complex **[Mn]-1c** which was also isolated using EtONa, but in the presence of THF as a solvent instead of ethanol. Then in the presence of a strong base, this **[Mn]-1d** complex undergoes inner-sphere C–H bond cleavage which proceeds via coordination of Mn to the ethoxide C–H bond and ultimately releases acetaldehyde and itself transforms into Mn–H species **[Mn]-1e**. Under the basic condition, acetaldehyde undergoes aldol condensation, which is an uncatalyzed process, and is transformed into

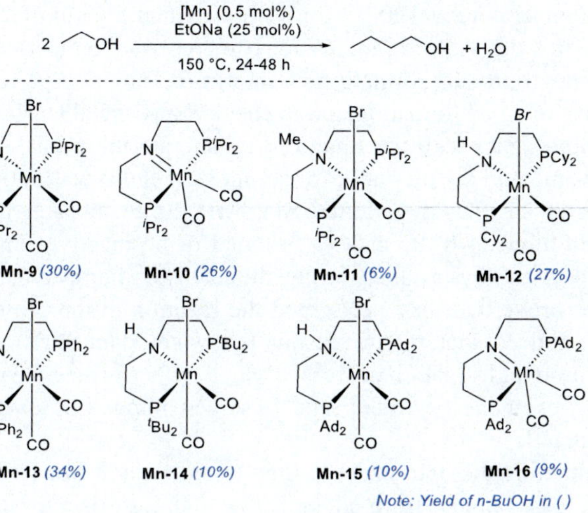

Fig. 17 Pincer Mn-complexes utilized by the Jones group for ethanol upgradation to 1-butanol [28]

crotonaldehyde. Then, this unsaturated aldehyde is hydrogenated by the Mn–H species **[Mn]-1e** to form 1-BuOH as the end product along with the formation of amido complex **[Mn]-1c** which undergoes alcoholysis to regenerate the active catalyst **[Mn]-1d** and restarts the catalytic cycle.

Nearly at the same time the Jones group also reported the utility of PNP-based Mn pincer catalysts toward the ethanol upgradation to *n*-butanol (Fig. 17) [28]. Here, they performed a detailed investigation based on reaction solvent, base, and temperature as well as the role of ligand substituents on the catalyst backbone and demonstrated its impact on the catalytic efficiency. They also began their investigation by taking $[HN\text{-}(CH_2CH_2P^iPr_2)_2]Mn(CO)_2Br$ (**Mn-9**) complex as the catalyst and started with base optimization for the Guerbet reaction of neat ethanol. They tested a series of bases like KOH, NaOEt, KO^{t}Bu, NaOMe, PhONa, and NaOAc, out of which the best activity was observed with NaOEt. They found that under 0.5 mol % loading of catalyst **Mn-9**, with 25 mol% NaOEt as the base, ethanol (6.85 mmol) can be upgraded to *n*-butanol with 30% yield along with the formation of 10% higher alcohols when the reaction was carried out at 150 °C for 24 h. The presence of NaOAc did not provide any product formation while a weak base like PhONa furnished only 5% of the desired product. In the absence of a base, no catalysis was observed to proceed, whereas the rest of the bases provided moderate reactivity with yields in the range of 10–20%. They also tried to investigate the impact of catalyst/base ratio on the product activity and selectivity, but the best result was observed with only 0.5 mol% catalyst loading in the presence of 25 mol% NaOEt. Then keeping the catalyst and base loading fixed, they checked the impact of reaction time and temperature for this reaction where they found that increasing the reaction time to 48 h did not improve the yield significantly. By decreasing the reaction temperature to 120 °C, they observed a decrease in product yield; however, keeping the temperature at 120 °C for 48 h provided a yield of 28% of the desired product. On the other hand, increasing the reaction temperature to 180 °C, a significant drop of product formation was noticed. They also performed the reaction in the presence of an additional ligand to check if any metal leached during catalysis that would bind again with the ligand to regenerate the catalyst; however, only a slight improvement in the product formation (31% yield) was noticed. After observing the reaction profile, they found that with time, the activity of the catalyst decreased and after 24 h the catalyst seemed to be inactive. They suspected that water formed as the by-product during the catalysis hampered the catalyst's efficiency and to prove that they performed the reaction in the presence of additional water. They noticed that by increasing the water concentration in the reaction mixture, the product formation decreased. So, they performed the reaction by adding molecular sieves at the beginning and observed minor improvement with a 33% yield of *n*-butanol.

After completing the optimization, they tested several Mn-PNP catalysts toward the formation of *n*-butanol from ethanol (Fig. 17). Surprisingly the amido complex **Mn-10**, which was supposed to be the active form of the catalyst **Mn-9**, provided less reactivity which they ascribed to its poor solubility. Complex **Mn-11**, which was the N–methylated analogue of complex **Mn-9** offered only 6% product formation and supported the importance of bifunctional cooperativity among the metal and ligand backbone; however, the product formation, even in low amount, also suggested an alternate route of the reaction. Then to monitor the steric and electronic impact of the ligand substituents, they compared the catalytic activity with various [(^{R}PNP)-MnBr(CO)$_2$] catalysts where R = ^{i}Pr, Cy, ^{t}Bu, Ph, or Ad. Complex **Mn-12** with cyclohexyl substituents which offered more bulkiness afforded slightly less activity than the ^{i}Pr substituted catalyst **Mn-9**. The complexes with the larger substituents ^{t}Bu or Ad group provided further less activity and both offered only 10% yield of *n*-butanol. Out of all these complexes, the best result was shown by catalyst **Mn-13** comprising the Ph substituent on the ligand backbone which showed a 34% yield of desired product along with 8% of higher alcohol.

In 2020, the Wass group utilized a series of chelating diphosphine and mixed-donor phosphinoamine-based Mn-catalysts for catalytic conversion of ethanol and methanol to higher alcohol *i*-BuOH [29] (Fig. 18). In this case, they wanted to explore whether the simple bidentate chelating ligands containing P – P and P – N donor sites can also lead to catalytic activity for this Guerbet-type condensation reaction or if it is essential to have a pincer-based catalyst backbone, as demonstrated earlier. For that, they synthesized a variety of mono- and bis-chelate complexes of Mn and utilized them successfully to demonstrate the formation of *i*-BuOH by using methanol and ethanol as the starting materials. Here first ethanol was coupled with methanol to form n-propanol which reacted with another molecule of methanol to make *i*-BuOH as the end product.

After synthesizing the Mn-complexes when they utilized them as catalysts toward *i*-PrOH formation, the highest activity was achieved with catalyst **Mn-18** with around 11% yield associated with a TON of 113 (Fig. 18). The rest of the catalysts provided poor to moderate reactivity with TON values in the range of 12–87. Then

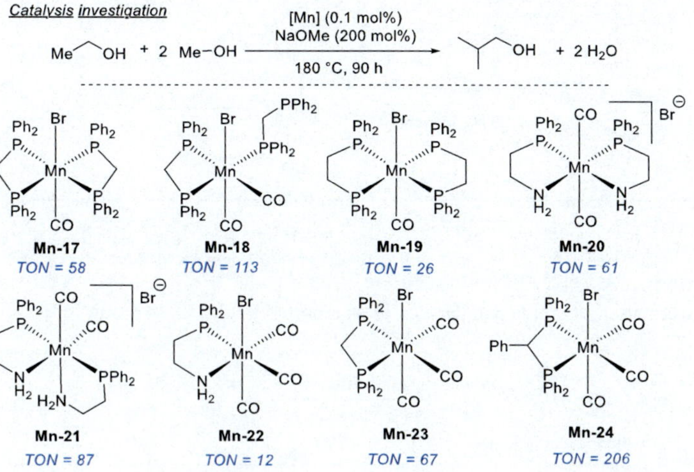

Fig. 18 Chelated Mn-complexes utilized by the Wass group for Guerbet reaction [29]

taking the most active catalysts **Mn-18** and **Mn-21**, they also checked the homocoupling reaction of ethanol to make *n*-butanol as the end product. But surprisingly, the catalyst **Mn-18** which showed the highest activity toward *i*-PrOH formation showed very less activity in this case whereas catalyst **Mn-21** showed good activity and a TON of 100 was achieved. In this case, after the catalysis, they observed a dark brown-colored solution for catalyst **Mn-18**, whereas catalyst **Mn-21** remained bright yellow which indicated that the catalyst **Mn-18** was decomposed under the reaction condition but not the catalyst **Mn-21** and that suggested their catalytic efficiencies.

To check the mechanistic aspect of the *i*-BuOH formation, the most active catalyst **Mn-18** was treated with NaOMe where they found the phosphine ligand dissociation and formation of the mono-chelate Mn-complex **Mn-23**. However, when this complex was utilized as a catalyst, it showed lesser reactivity. Based on the literature survey they found that treatment of complex **Mn-23** with KHMDS could make complex **Mn-23b** which could be converted to the hydride complex **Mn-23c** under hydrogen pressure. They also found that this hydride complex formation was more facile by substituting the $–CH_2–$ backbone with the –CH(Ph)– backbone (Fig. 19). Based on that they prepared complex **Mn-24**, an analogue of complex **Mn-23** with just phenyl substituted backbone and when this complex was tested for forming *i*-BuOH, they observed the highest yield of 21% with a TON value of 206. Thus, they explained a ligand-assisted mechanism where the Ph substitution on the ligand backbone accelerated the catalytic activity.

In 2023, the Shao group demonstrated the cross-coupling of ethanol with primary alcohols, for its upgradation to linear alcohols via the Guerbet reaction [30]. At first, they compared the catalytic activity between the various Mn-catalysts containing PNP-based $[HN(CH_2CH_2PR_2)_2]$ ligands and observed that the ^{cy}PNP-Mn (I) complex showed the best activity at 140 °C under the solvent-free condition

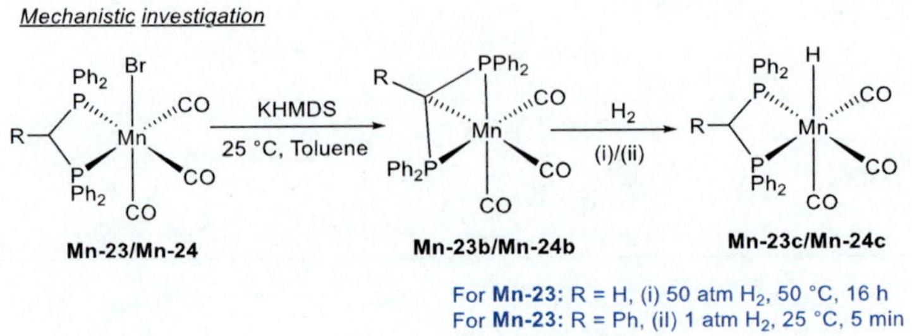

Fig. 19 Formation of active Mn–H species as reported by the Wass group [29]

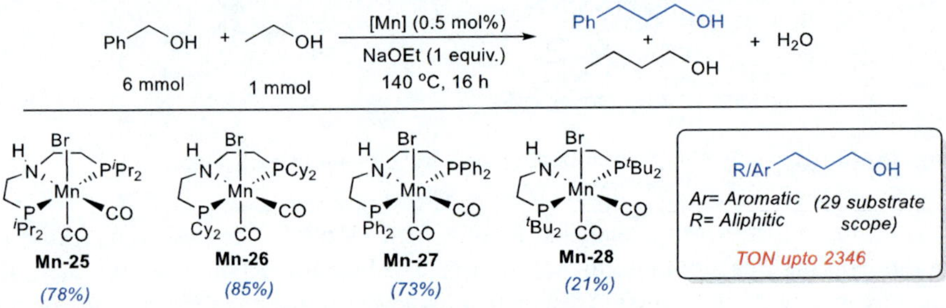

Fig. 20 Shao's report on the cross-coupling of ethanol with primary alcohols [30]

with a TON of 2,336 (Fig. 20). The reaction was optimized with 0.5 mol% Mn-catalyst with excess amounts of benzylic alcohol (6 mmol) and ethanol (1 mmol) as the model substrates in the presence of 1 equiv. of NaOEt as the base, and they discovered high-yield production of phenyl propanol (82%) with 4% by-product produced by ethanol homocoupling. Further, increasing the reaction temperature did not lead to any improvement whereas decreasing the temperature resulted in decreased activity. There were no appreciable alterations in the catalytic outcome when the benzylic alcohol/ethanol ratio was raised, while a lower yield was noted when the benzylic alcohol amount was lowered. For this reaction, the ^{cy}PNP-Mn(I) complex outperformed the ^{iPr}PNP-Mn(I) and ^{Ph}PNP-Mn(I) catalysts with only a minor decrease in efficiency. By contrast, the activity of the ^{tBu}PNP-Mn(I) catalyst was significantly lower (with 21% yield of phenyl propanol). When $MnCl_2$ or Mn $(CO)_5Br$ was employed as the catalyst, they did not observe any activity implying the important role played by the pincer ligands. At last, a large variety of substrates were demonstrated by cross-coupling of ethanol with aromatic primary alcohols. 1-hexanol, the aliphatic primary alcohol generated trace amounts of the desired cross-coupling product along with ethanol in the presence of the ^{cy}PNP-Mn (I) catalyst. But using the ^{Ph}PNP-Mn catalyst at an elevated temperature of 165 °C, some good activity (40% yield) was observed for aliphatic primary alcohols.

In 2019, the Pathak group conducted theoretical analyses based on an Mn-PNP complex to comprehend the molecular mechanisms for ethanol conversion to

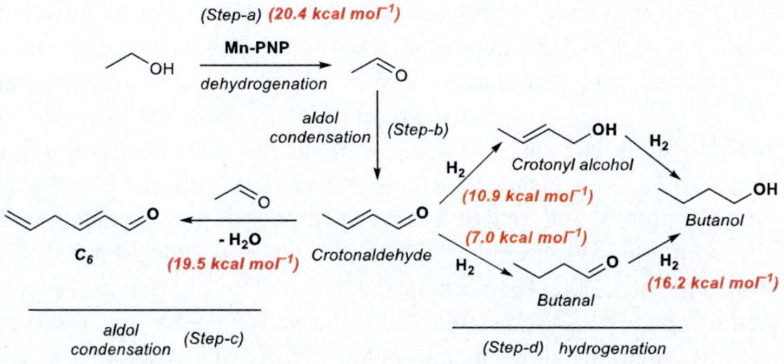

Fig. 21 Pathak's report on theoretical investigation for upgrading ethanol to *n*-butanol [31]

n-butanol [31]. They discovered that the conversion of ethanol to acetaldehyde regulates the substrate conversion, whereas the aldol condensation and hydrogenation of crotonaldehyde directly impact the product selectivity. Their research demonstrates the significance of N–H functionality for illuminating the MLC mechanism within the Mn-PNP complex. Overall, the catalytic process is completed in three steps: (i) the dehydrogenation process, which converts ethanol to acetaldehyde (Step-a), (ii) the C–C coupling process that converts acetaldehyde to crotonaldehyde (C_4 product) (Step-b), and (iii) the subsequent hydrogenation of crotonaldehyde to the desired product, *n*-butanol (Step-d). They found that, in comparison with the other steps, ethanol dehydrogenation (Step-a) step is energetically the most expensive (20.4 kcal mol^{-1}) (Fig. 21). Next, the rate associated with the aldol condensation process (Step-c) and hydrogenation of the C_4 product (Step-d) determines how selectively butanol is produced in comparison with C_{4+} products. In comparison with the aldol condensation of the C_4 product (19.5 kcal mol^{-1}), the determined energy range for the hydrogenation of C_4 is 7.0 kcal mol^{-1} for C=C and 10.9 kcal mol^{-1} for C=O. Consequently, it was consistent with the earlier investigations that revealed butanal and crotyl alcohol are produced as a result of hydrogenation of C_4 products. Next, butanal to butanol hydrogenation is 16.2 kcal mol^{-1} easier than crotyl alcohol hydrogenation. This indicates that two distinct products (butanol and crotyl alcohol) were produced throughout the hydrogenation of crotonaldehyde; however, only butanol was produced from butanal, as the conversion of crotyl alcohol to butanol was unfavorable. Because of the efficient hydrogenation performed by the Mn-PNP catalysts due to the N–H functionality's dual activity, butanol is formed more selectively than higher C-based products.

3 Conclusions

The topic of this chapter covered the Guerbet coupling reaction for the formation of butanol and higher alcohols from ethanol utilizing both homogeneous and heterogeneous catalysts based on 3d metals. The reaction proceeded with the dehydrogenation of ethanol followed by aldol condensation to form unsaturated aldehyde and the last hydrogenation of unsaturated aldehyde to make the saturated alcohol. With that, some side reactions were also observed to proceed which led to the formation of a diverse range of products along with the desired one. In this regard, both the homogeneous and heterogeneous reactions had their unique advantages. For heterogeneous catalyst systems, it was observed that based on the variation of catalyst support, both reactivity and selectivity of product changed. Similarly, keeping the supported system fixed, with varying the catalytic 3d metal also affected the reaction. The interaction between the catalyst and the support played a significant role in governing the catalyst's stability. Additionally, researchers observed how the oxidation state of the catalytic metal center affected the reaction's efficiency and selectivity. Generally, for using the heterogeneous catalysts, high-temperature conditions were required than the homogeneous catalyst systems. Researchers highlighted the significant function of the ligand backbone in homogeneous catalyst systems for this reaction. In this regard, metal-ligand cooperation within the catalyst was a crucial factor. Interestingly, though the Guerbet reaction for benzyl alcohol was known to be catalyzed by several base metals like Ni, Co, Mn, Fe, etc., however, for upgrading ethanol to butanol, only Mn-based catalyst systems were known to provide any reactivity. Though the Guerbet reaction can be a useful technique for ethanol upgradation, still it has several drawbacks and restrictions. As a number of products are formed in this reaction, the separation of the desired product can be difficult, and specific methods may be necessary to separate the intended product from the by-products. Managing selectivity remains difficult so far. Ethanol backbone can dimerize or trimerize as a result of the Guerbet reaction, which can decrease the selectivity and yield of the desired product. Condensation and isomerization are two related side reactions that can happen and can result in the production of undesirable by-products. These unwanted side effects might be challenging to manage and can reduce the reaction's efficacy. Additionally, the Guerbet reaction is carried out on a laboratory scale; however, scaling it up to an industrial level can present additional challenges. The economic viability of the Guerbet reaction is influenced by several variables, including the cost of the chemicals, catalysts, and apparatus used, as well as the total yield and selectivity attained. It can be difficult to balance these parameters in a way that makes the reaction economically viable. Combining experimental and theoretical methods with a thorough knowledge of organic chemistry and reaction mechanisms will be necessary to address these practical problems. Researchers are still working on creating methods to increase the Guerbet reaction's effectiveness, selectivity, and usefulness for upgrading ethanol. Improvements in process intensification methods like continuous flow chemistry may improve the effectiveness and control of the Guerbet reaction of ethanol. Continuous flow systems make it possible to precisely adjust reaction parameters, thus improving yields and selectivity in the process. Joint efforts between professionals in engineering, organic chemistry, catalysis, and other fields could develop creative ways of the ethanol Guerbet reaction.

Acknowledgments The authors thank IISER Bhopal and SERB for generous funding. T.M. and M.P. acknowledge SERB and UGC for fellowship.

References

1. Höök M, Tang X (2013) Energy Policy 52:797–809
2. Kiehbadroudinezhad M, Merabet A, Ghenai C, Abo-Khalil AG, Heliyon ST (2023) Heliyon 9: e13407
3. Hui Liew W, Hassim MH, Ng DKS (2014) J Clean Prod 71:11–29
4. Aitchison H, Wingad RL, Wass DF (2016) ACS Catal 6:7125–7132
5. Anbarasan P, Baer ZC, Sreekumar S, Gross E, Binder JB, Blanch HW, Clark DS, Toste FD (2012) Nature 491:235
6. Pruett RL (1979) Adv Organomet Chem 17:1–60
7. Guerbet MCR (1899) Acad Sci Paris 128:1002–1004
8. Filonenko GA, Putten RV, Hensen EJM, Pidko EA (2018) Chem Soc Rev 47:1459–1483
9. Alig L, Fritz M, Schneider S (2019) Chem Rev 119:2681–2751
10. Kozlowski JT, Davis RJ (2013) ACS Catal 3:1588–1600
11. Earley JH, Bourne RA, Watsonb MJ, Poliakoff M (2015) Green Chem 17:3018–3025
12. Jiang D, Wu X, Mao J, Ni J, Li X (2016) Chem Commun 52:13749–13752
13. Tong Y, Zhou J, He Y, Tu P, Xue B, Cheng Y, Cen J, Zheng Y, Ni J, Li X (2020) ChemistrySelect 5:7714–7719
14. Pang J, Zheng M, He L, Li L, Pan X, Wang A, Wang X, Zhang T (2016) J Catal 344:184–193
15. Sun Z, Vasconcelos AC, Bottari G, Stuart MCA, Bonura G, Cannilla C, Frusteri F, Barta K (2017) ACS Sustain Chem Eng 5:1738–1746
16. Siqueira MR, Perrone OM, Metzker G, Lisboa DCO, Thoméo JC, Boscolo M (2019) Mol Catal 476:110516
17. Petrolini DD, Eagan N, Ball MR, Burt SP, Hermans I, Huber GW, Dumesic JA, Martins L (2019) Cat Sci Technol 9:2032–2042
18. Wang Z, Pang J, Song L, Li X, Yuan Q, Li X, Liu S, Zheng M (2020) Ind Eng Chem Res 59: 22057–22067
19. Pang J, Zheng M, Wang Z, Li S, Li X, Li X, Wang J, Zhang T (2020) Chinese J Catal 41:672–678
20. Metzker G, Vargas JAM, de Lima PL, Perrone OM, Siqueira MR, Varanda LC, Boscolo M (2021) Appl Catal Gen 623:118272
21. Zhou J, He Y, Xue B, Cheng Y, Zhou D, Wang D, He Y, Guan W, Fang K, Zhang L, Ni J, Li X (2021) Sustain Energy Fuels 5:4628–4636
22. Liu Z, Li J, Tan Y, Guo L, Ding Y (2022) Catalysts 12:1170
23. Li J, Lin L, Tan Y, Wang S, Yang W, Chen X, Luo W, Ding Y (2022) ChemCatChem 14: e202200539
24. Wang Z, Yin M, Pang J, Wu P, Song L, Li X, Zheng M (2023) Ind Eng Chem Res 62:2594–2604
25. Lv W, He L, Li W, Zhou B, Lv S, Lu A (2023) Green Chem 25:2653–2662
26. Cai X, Li X, Dang M, Huang H, Liu A, Zhang Q, Pi Y, Meng Q, Wu X, Wang T (2023) Fuel 335:126971
27. Fu S, Shao Z, Wang Y, Liu Q (2017) J Am Chem Soc 139:11941–11948
28. Kulkarni NV, Brennessel WW, Jones WD (2018) ACS Catal 8:997–1002
29. King AM, Sparkes HA, Wingad RL, Wass DF (2020) Organometallics 39:3873–3878
30. Shao Z, Li X, Zhang X, Zhao M (2023) Environ Chem Lett 21:1271–1279
31. Rawat KS, Mandal SC, Bhauriyal P, Garg P, Pathak B (2019) Cat Sci Technol 9:2794–2805

Top Organomet Chem (2024) 73: 227–256
https://doi.org/10.1007/3418_2023_96

Published online: 28 September 2023

Reformation of Alcohols to Esters, Acids, Amides, Ureas, Polyureas and Polyethyleneimine by 3d-Metal Catalysts

Claire Brodie and Amit Kumar

Contents

Abstract Dehydrogenative processes for the catalytic up-conversion of alcohols into higher value products have continued to receive sustained attention over the past few years with a drive towards the application of more sustainable transition metals (i.e. earth-abundant metals) within these processes. This chapter discusses the recent developments in the field of metal-ligand cooperative (MLC) and 3d transition metal catalysts for alcohol reformation to various products such as esters, amines, acids, ureas, polyureas and polyethyleneimines. These MLC facilitated dehydrogenative processes are cost-effective and atom-economic routes to alcohol reformation products, often only producing hydrogen as the only by-product.

Keywords Polyureas · Acids · Amides · Dehydrogenation · Esters · Iron · Manganese · Pincer · Polyethyleneimine · Ureas

---

C. Brodie and A. Kumar (✉)
Department of Chemistry, University of St Andrews, St Andrews, UK
e-mail: ak336@st-andrews.ac.uk

1 Introduction

Alcohols as feedstocks have several advantages; for example, they are cost-effective, commercially available and have low toxicity. Additionally, many alcohols can be made from renewable feedstock, a fact that makes them attractive choices for further valorisation to valuable organic compounds. For instance, the large-scale production of methanol from the hydrogenation of CO_2 has been demonstrated at an industrial level by a number of companies such as Lurgi AG (Germany), NIRE and RITE (Japan), Mitsui Chemicals (Japan), Carbon Recycling International (Japan), and Air Company (USA) under 40–87 bar pressure and 225–260 °C temperature [1]. The production of ethanol from microbial fermentation of sugars [2] has also been demonstrated at the industrial level. Enzymatic routes to produce *n*-propanol, *n*-butanol, isopropanol and *iso*butanol by the fermentation of sugars from lignocellulosic feedstock have been reported and recently reviewed by Schubert [3]. Methods for the upgradation of ethanol to higher alcohols such as butanol, *iso*butanol and C_4–C_{16} alcohols have also been studied [4, 5]. Additionally, several diols and polyols can also be sourced from biomass, e.g. ethylene glycol [6], propylene glycol [7], 1,4-butanediol [8], 1,5-pentanediol [9], 1,6-hexanediol [10], 2,5-*bis*(-hydroxymethyl)furan [11] and isosorbide [12]. As a by-product of biodiesel industry, glycerol is currently produced in surplus quantity, making it one of the most widely produced polyols [13]. The production of polyethercarbonate polyol from CO_2 has also been commercialised [14, 15]. Considering these examples, it becomes clear that alcohols are excellent choices of renewable feedstock and their transformation to valuable organic compounds will support a circular economy.

Processes based on catalytic dehydrogenation reactions involving alcohols are atom-economic routes for the synthesis of various organic compounds such as esters, acids, amides and urea derivatives. These reactions were first discovered using precious metal-based homogeneous catalysts such as ruthenium-pincer catalysts [16, 17]. However, in the past 8 years, organometallic catalysts involving earth-abundant metals have emerged as attractive options to achieve such transformations in a cost-effective manner. This chapter will discuss the reformation of alcohols to esters, acids, amides, ureas, polyureas and polyethyleneimine using well-defined molecular complexes of 3d-metals.

2 Ester Synthesis from the Transformation of Alcohols

Esterification is one of the most important synthetic transformations in modern industry, with the annual market estimated at $89 billion USD in 2022 [18]. This huge valuation results from the diverse range of applications esters are found in, including, but not limited to: biodiesel [19], pharmaceuticals [20], plastics and coatings [21, 22], solvents [23], flavourings [24], cosmetics [25], preservatives [26] and perfumes [27]. Traditional Fischer esterification involves the condensation

Scheme 1 Traditional (**a**) and dehydrogenative coupling (**b**) routes for ester synthesis

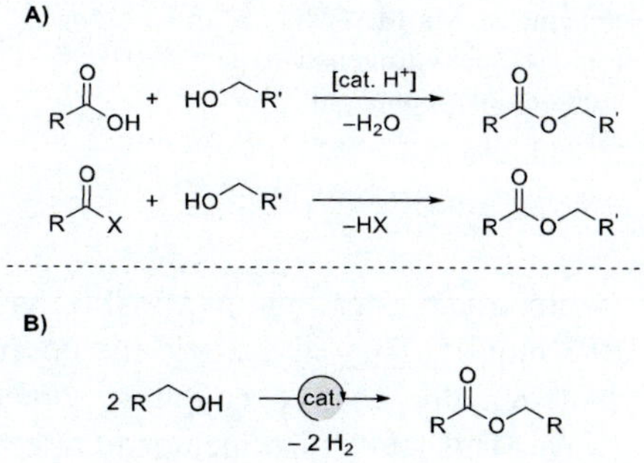

Scheme 2 Ester synthesis from Fe-MACHO catalysed acceptorless dehydrogenative coupling [34]

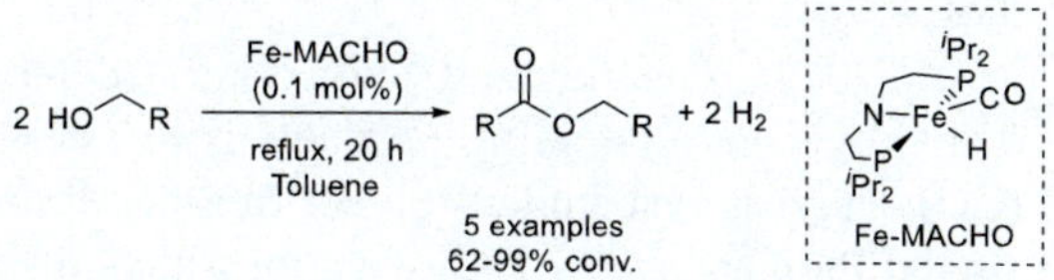

of an alcohol with a carboxylic acid, liberating water. However, this reaction is typically slow and requires heating the mixture at reflux for hours in the presence of an acid catalyst, typically conc. H_2SO_4, to achieve satisfactory yields [28]. Additionally, to produce the required carboxylic acid substrates for this typically requires use of a stoichiometric strong oxidising agent, such as $KMnO_4$ [29]. A popular alternative method for alcohol esterification involves the application of acyl-halide species, such as acyl chlorides, liberating, e.g., HCl (Scheme 1a). However, acyl chlorides require stoichiometric reaction of the corresponding acid with thionyl chloride or phosphorus trichloride [30].

An attractive alternative method for esterification is the catalytic acceptorless dehydrogenative coupling (ADC) of alcohols (Scheme 1b) [31]. This is a simple and elegant method for the synthesis of esters, without the need for external oxidant and where the only by-product is hydrogen gas [31–33], rather than stoichiometric metallic or corrosive waste as produced in traditional esterification routes. The dehydrogenative coupling of alcohols is accepted to proceed through a three-step process and is thermodynamically driven by the release of H_2 gas: (1) alcohol dehydrogenation to aldehyde; (2) reaction of aldehyde with a second equivalent of alcohol to produce a hemiacetal intermediate; and (3) hemiacetal dehydrogenation to yield ester.

In 2014, Schneider reported the first example of an earth-abundant 3d metal-catalysed acceptorless dehydrogenative coupling of alcohols to esters, using an Fe-MACHO catalyst, Fe(PNP-^{i}Pr)(CO)(H), (PNP-^{i}Pr = η^3-P,N,P-[*N*,*N′*-*bis*{diisopropylphoshinoethyl}amido]) [34]. While this study was largely focused on acceptorless alcohol dehydrogenation (AAD) of secondary alcohols to ketones, the authors also included some primary aliphatic alcohols within their study (Scheme 2). They showed the successful dehydrogenative coupling of a few aliphatic alcohols such as *n*-butan-1-ol, *n*-heptan-1-ol and 2-cyclohexylethanol to corresponding

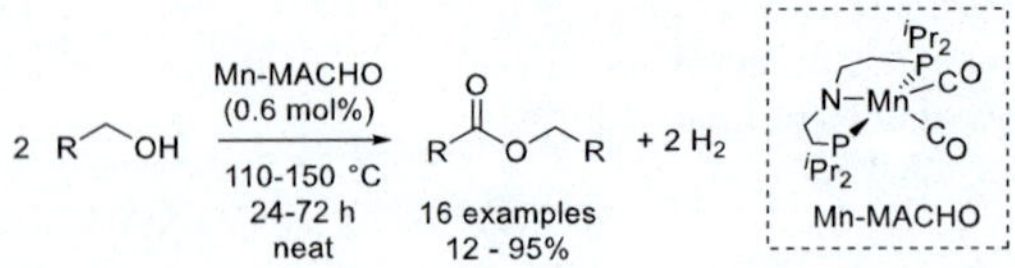

Scheme 3 Mn-MACHO catalysed dehydrogenative coupling of alcohols to esters [35]

symmetrical esters in up to 90% yield with remarkably low catalyst loadings (0.1 mol%). They also noted the observation of heptan-1-al in the case of heptan-1-ol dehydrogenative coupling, which supports the three-step process outlined above. This study also included the generation of γ-butyrolactone from dehydrogenation and subsequent intramolecular cyclisation and dehydrogenation of butan-1,4-diol.

In 2017, Gauvin and co-workers demonstrated the manganese-catalysed dehydrogenative coupling of alcohols to esters using Mn-MACHO [Mn(PNP-^{i}Pr)(CO)$_2$] precatalyst under solvent-free conditions (Scheme 3) [35]. They demonstrated the base-free ADC of various alcohols to corresponding symmetric esters, with scope including various linear and 2° branched aliphatic substrates alongside variously substituted benzylic alcohols using low catalyst loading (0.6 mol%) with reasonable reaction conditions for this class of catalyst (150°C for 24 h). Gauvin noted that for low boiling alcohols such as *n*-propanol and *n*-butanol, longer reaction times (*c.f.* 72 and 24 h) were required to achieve acceptable yields. This is possibly due to the lower temperature available for these reactions owing to the boiling point of the substrates as the reactions were conducted under open flow of inert gas. Indeed, the attempted dehydrogenative coupling of methanol returned only 3% conversion under reflux conditions for 72 h. Since the reactions were carried out under a stream of inert gas to allow for continuous hydrogen removal, it is possible that higher conversions in shorter reaction times may be achieved through use of a sealed system. This would allow higher temperatures to be used during reaction for these low boiling alcohols; a strategy that has been employed in, for instance, the dehydrogenative coupling of methanol [36]. In this study by Gauvin, the authors also noted that increasing steric bulk corresponds to a reduction in the observed alcohol to ester conversion. Very bulky substrates, such as neopentyl alcohol, remained unchanged under Gauvin's reaction conditions.

Importantly, the authors also examined the tolerance of this Mn-MACHO alcohol esterification system towards impurities. For instance, bio-sourced alcohols may contain water, methanol and acid impurities [37, 38], and common routes to Mn-MACHO may result in salt impurities, i.e. KBr [39]. The authors found that the presence of 5 mol% water or methanol (>8 equivalents vs [Mn]) has little effect on the catalytic activity; however, the presence of KBr (2 mol%) did drop the activity significantly (*c.f.* TON 83 and 153 with and without KBr, respectively) [35]. They also found that the presence of butyric acid significantly reduced activity (TON 10), which the authors attribute to the formation of a catalytically inactive carboxylate manganese complex, formed through activation of the acid hydroxyl group [40].

The derivative complex, Fe-MACHO-BH (Fe(PN^{H}P-iPr)(CO)(H)(η^1-H-BH$_3$) where PN^{H}P-iPr = η^3-P,N,P-[*N,N′*-*bis*{di*iso*propylphosphinoethyl}amine]) was

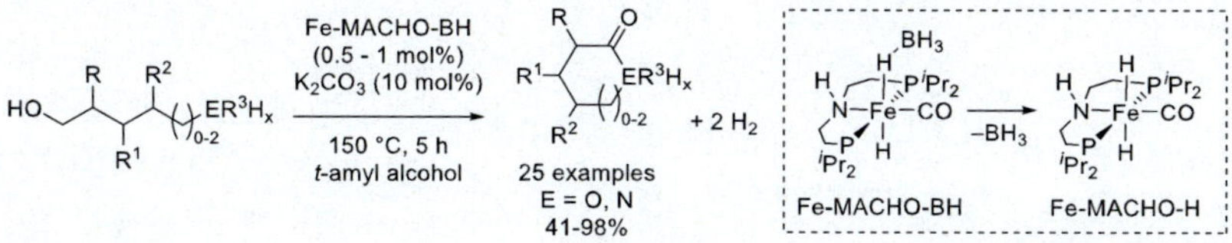

Scheme 4 Lactone and lactam syntheses from Fe-MACHO-BH catalysed diol/amino alcohol dehydrogenation [41]

demonstrated by the Beller group (Scheme 4) to promote acceptorless alcohol dehydrogenation to produce substituted γ-butyrolactones and varying ring-size/ substituted lactones from symmetric and asymmetric diols as well as lactams (cyclic amides) from the acceptorless dehydrogenation of amino alcohols [41]. This study significantly increased the substrate scope for these reactions, including variously substituted alkyl and aryl γ-butyrolactone derivatives; aliphatic and aromatic lactones with various substitution patterns, spirocyclic lactones and multiring lactones; diversely substituted lactams of different ring size as well as the production of some cyclic imines through an additional dehydration process. While the optimised conditions reported by Beller for these transformations (Scheme 4) do include the addition of catalytic quantities of base (10 mol%), they do report that this Fe--MACHO-BH precatalyst is still active without base (loss of BH_3 from Fe-MACHO-BH generates a catalytically active iron-hydride species, Fe-MACHO-H), but with slightly reduced activity: *c.f.* 99% and 82% yield obtained for phthalide synthesis from 1,2-benzenedimethanol with and without K_2CO_3 after 18 h at 150 °C, respectively [41].

Of note, the authors reported successful lactone synthesis from an unsaturated diol substrate (cyclohex-4-ene-1,2-diyldimethanol) where the alkene moiety retains unsaturation under their conditions, despite molecular hydrogen build up within the sealed system used within these reactions. This chemoselectivity is in contrast to what has been reported for ruthenium catalysed alcohol dehydrogenation reactions, where alkenes are readily hydrogenated [42, 43].

In 2020 Funk et al. reported lactone synthesis from diol dehydrogenation catalysed by a (cyclopentadienone)iron precatalyst in a transfer dehydrogenation process using acetone in a dual role as both solvent and hydrogen acceptor [44]. This class of catalyst requires trimethylamine *N*-oxide for pre-activation through oxidative removal of a carbonyl ligand (Scheme 5). While these reactions do require a hydrogen acceptor in place, the previous examples release molecular hydrogen as the only by-product, lactone formation here is accessible at lower temperature (90 °C). Funk and co-workers show five-, six- and seven-membered lactone formation from a variety of symmetric and asymmetric diol substrates in high yield. Furthemore, they expanded the scope of this transformation to include a variety of mixed primary and secondary diols bearing aromatic and aliphatic substituents in good yield (69–91%). Prior to this report, there was only a single previous example for lactone synthesis using base-metal catalyst: Beller's Fe-MACHO-BH catalysed

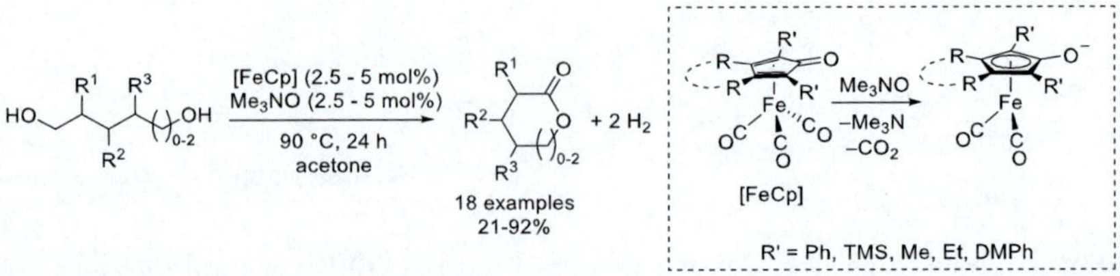

Scheme 5 Lactone synthesis from [FeCp] catalysed diol dehydrogenation [44]. DMPh = 3,5-dimethylphenyl

lactone formation via acceptorless alcohol dehydrogenation of 1-phenylbutane-1,4-diol [41].

To date, there remain no reports of the successful synthesis of four-membered β-lactones or β-lactams from the dehydrogenative cyclisation of diols or amino alcohols, respectively [41]. This is likely due to the high ring strain in these compounds and their susceptibility towards hydrolysis [45, 46]. β-Lactones are important synthetic intermediates with a wide application of uses [47].

3 Amide Synthesis from the Dehydrogenative Coupling of Alcohols and Amines

The formation of amide/formamides is one of the most important reactions in organic chemistry, not least due to their widespread occurrence in pharmaceuticals and biologically active compounds [48]. Indeed, amide linkages are found in 25% of all available drugs [49]. Amide preparation is such an important transformation that the ACS Green Chemistry Institute placed "*General methods for catalytic/sustainable (direct) amide or peptide formation*" in the second place of their list of key green chemistry research areas from an industrial perspective [50]. Vital to the progress of sustainable amide bond formation is an atom-economic transformation, as traditional amide syntheses employ coupling agents, generating stoichiometric waste (Scheme 6a). A 2016 article from the Stahl group does demonstrate the generation of amides from alcohols and amines applying a copper catalyst; however,

Scheme 6 Traditional (**a**) and dehydrogenative coupling (**b**) approaches to amide bond formation

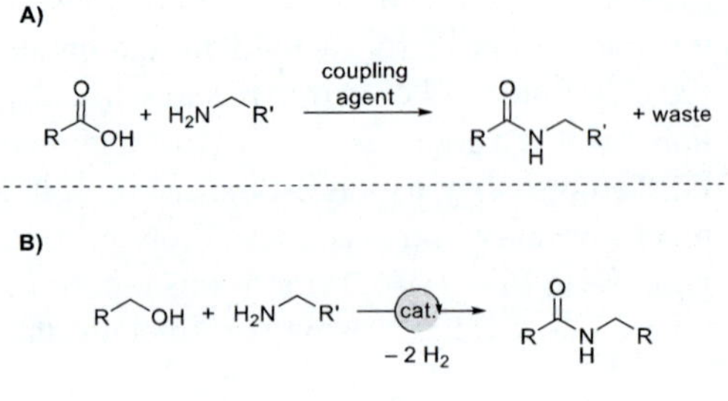

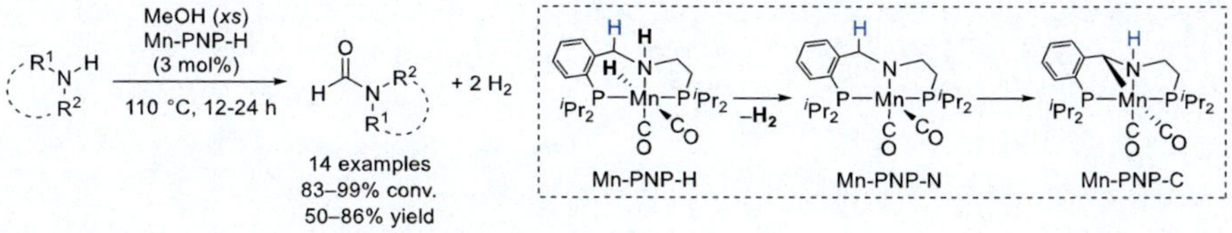

Scheme 7 Mn-catalysed dehydrogenative coupling of methanol and amines to formamides [52]

this oxidative amination process requires an atmosphere of oxygen and a radical co-catalyst (e.g. ABNO) for amide selectivity [51].

A more elegant approach for the preparation of amides or formamides is the dehydrogenative C-N coupling of amines and alcohols. This process follows a similar stepwise mechanism as seen in ester bond formation: (1) dehydrogenation of an alcohol to aldehyde; (2) reaction of aldehyde with one equivalent of amine to generate a hemiaminal intermediate; (3) dehydrogenation of hemiaminal to produce amide [52]. There is abundant literature on the homogeneous dehydrogenative coupling of amines and alcohols to amides using precious metal catalysts, in particular, ruthenium [53–62], along with examples using rhodium [63] and rhenium [64]. However, reports on this transformation using 3d metals are more limited as described below.

Using a manganese hydride pincer precatalyst, Mn(PNP-H)$(CO)_2$(H) (PNP-H = 2-(di*iso*propylphosphino)-*N*-(2-(di*iso*propylphosphino)benzyl)ethanamine, Milstein and co-workers demonstrated the first example of a 3d metal-catalysed formamide synthesis from methanol and amine (Scheme 7) [52]. This Mn-PNP-H precatalyst promotes the coupling of methanol and amine through successive dehydrogenation steps, in line with those outlined above. The authors tested a variety of amine substrates, primary and secondary amines, including cyclic secondary amines (piperidine, pyrrolidine, morpholine), and achieved excellent conversion (83–99%) in all cases under their optimised catalytic conditions. These reactions were carried out in sealed pressure tubes to allow for higher temperatures (relative to the boiling point of methanol) to be used; however, no additional hydrogen acceptor was required despite build-up of molecular hydrogen gas within the reaction vessel. Indeed, molecular hydrogen within the reaction headspace was detected by GC gas analysis, supporting the proposed dehydrogenative pathway.

Here, the authors propose dihydrogen loss from Mn-PNP-H (Scheme 7; insert) to generate an intermediate Mn-amido species, Mn-PNP-N, that then undergoes intramolecular C-H activation to produce the thermodynamically stable metalated amino complex, Mn-PNP-C. Both dihydrogen loss organometallic products from Mn-PNP-H were demonstrated to activate methanol to a Mn-alkoxy-amino complex through MLC amino-amido functionality [65], which can then undergo dehydrogenation to formaldehyde [52].

At a similar time, Bernskoetter and co-workers published a report detailing Fe-MACHO catalysed dehydrogenative coupling of secondary amines with alcohols to form amides using low catalyst loadings (0.1 mol%) and relatively mild

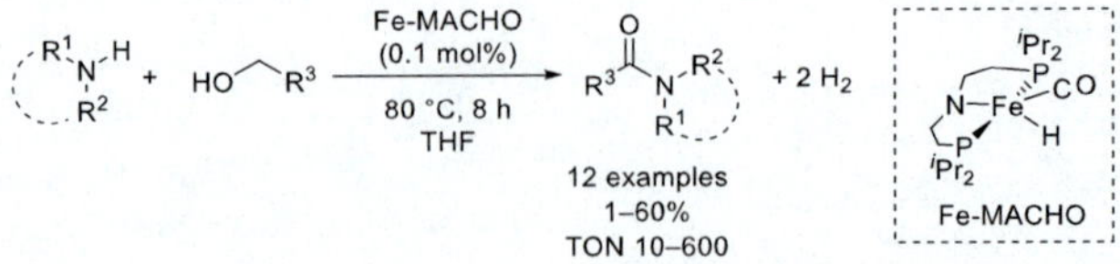

Scheme 8 Fe-catalysed dehydrogenative coupling of secondary alcohols to amides [66]

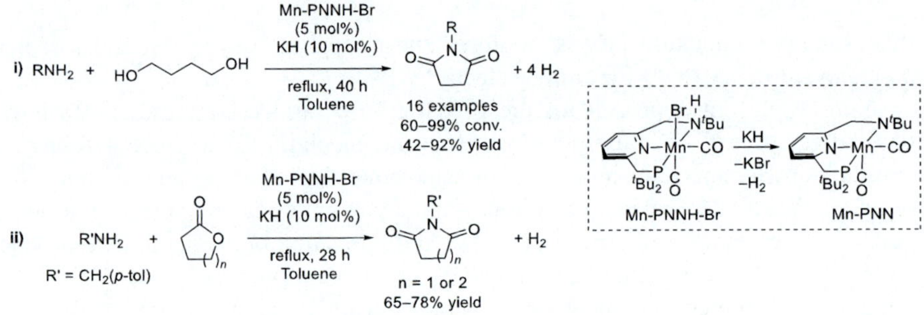

Scheme 9 Mn-catalysed dehydrogenative coupling of diols and amines to cyclic imides [68]

conditions (Scheme 8) [66]. The scope of both secondary amine and alcohol substrates was explored. Of note, a relationship between the steric bulk of the amine substrate with the relative TON calculated was determined, where increasingly sterically congested amines strongly correlated to decreasing TON. For instance, a TON of 600 was calculated for 1,2,3,4-tetrahydro*iso*quinoline, where the TON calculated for *N,N′-bis*(phenylmethyl)amine is 126 under the same reaction conditions. Attempts to generate diamides from cyclic amide substrates were unsuccessful under these conditions. Furthermore, the authors also note that attempts to generate amides from the dehydrogenative coupling of primary amines with methanol initially formed formamides and then ureas, a reaction they exploit (vide infra) [67].

Variations to the alcohol substrate within these reactions revealed a similar trend as Bernskoetter observed amine variation. Increasing steric profiles of the alcohol also results in a dramatic decrease to the observed TON while using this Fe-MACHO catalyst [66]. These observations support their claims that the activity of this Fe-MACHO catalyst is strongly constrained by steric effects at both sides of the product carbonyl moiety, a fact they attribute to steric congestion around the Fe-centre imposed by MACHO phosphine isopropyl moieties.

In the direction of the dehydrogenative amidation reactions, Milstein in 2017 reported the first example of the synthesis of cyclic amides using a catalyst of earth-abundant metals. A manganese precatalyst in the presence of catalytic quantities of base was used for the preparation of various succinimide derivatives from reaction of primary amine with butan-1,4-diol (Scheme 9i) [68]. This reaction requires long reaction times (40 h), but can be carried out under toluene at reflux conditions, i.e. no sealed vessel required and dihydrogen is continuously removed. The catalyst applied

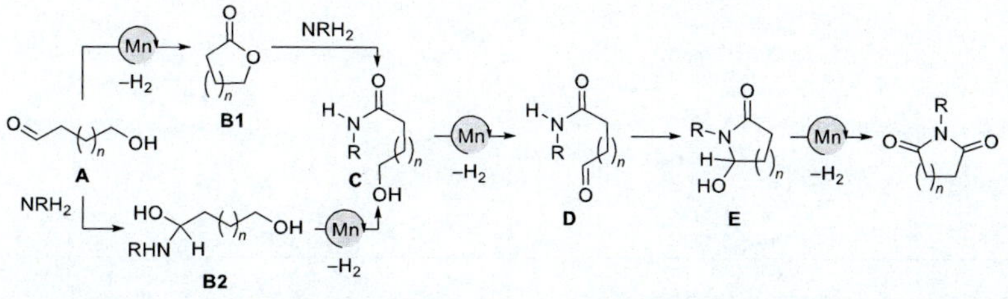

Scheme 10 Proposed pathway for the formation of cyclic imides from diols and amines [68]

here, $(Mn(PNNH)(Br)(CO)_2$ [(Mn-PNNH-Br), where PNNH = *N*-(6-{(di*tert*butylphosphiomethyl)pyridin-2-yl}-methyl(*tert*butyl)amine)], requires activation with base to generate the catalytically active dehydrohalogenated product, Mn-PNN (Scheme 9; insert). Worth mentioning, mechanistic studies by the authors determined this catalyst requires amine/amido-MLC to successfully mediate alcohol dehydrogenation and subsequent cyclic imide formation. This is in contrast to related Ru-PNN systems, which have been demonstrated to facilitate amine and alcohol coupling to amides proceeding via aromatisation/dearomatisation of the methylpyridinyl moiety [53].

The precatalyst, Mn-PNNH-Br, mediates the dehydrogenative synthesis of succinimide derivatives bearing a myriad of substituted methylphenyl primary amines in high yield when used at 5 mol% loading in the presence of 2 equivalents (to [Mn]) of potassium hydride [68]. However, it was noted that no reaction was observed when either aniline or *tert*butylaniline was tested in this reaction. Interestingly, γ-butyrolactone and δ-lactones were also converted to cyclic imides under the same reaction conditions (Scheme 9 ii). In fact, lactones were observed during in situ reaction monitoring, suggesting that these could be an intermediate on the reaction pathway to cyclic imides. However, as hydroxyamide species were present from the early stages of these reactions, the authors propose there are competing mechanisms for initial hydroxyamide formation after initial diol dehydrogenation to a hydroxyaldehyde (Scheme 10).

The fundamental mechanistic steps proposed for cyclic imide formation are thus as follows: 1) alcohol dehydrogenation to hydroxyaldehyde **A**; then either of: 2a) dehydrogenation and intramolecular cyclisation of **A** to lactone **B1** and subsequent reaction with amine to generate hydroxyamide **C**, or 2b) amine reaction with **A** to form a hydroxyhemiaminal intermediate **B2** and subsequent dehydrogenation to **C**; then 3) dehydrogenation of **C** to aldehyde **D**; intermolecular cyclisation to hemiaminal **E**; 4) dehydrogenation of **E** to generate cyclic imide product [68]. Overall, this reaction releases 4 mol of hydrogen per mole cyclic imide produced.

Amide bond formation and the reverse reaction, amide hydrogenation to regenerate amine and alcohol, is a particularly attractive method for hydrogen storage materials; liquid organic hydrogen carriers (LOHCs) [69]. Several groups have realised the potential of these (de)hydrogenative process for prospective LOHCs;

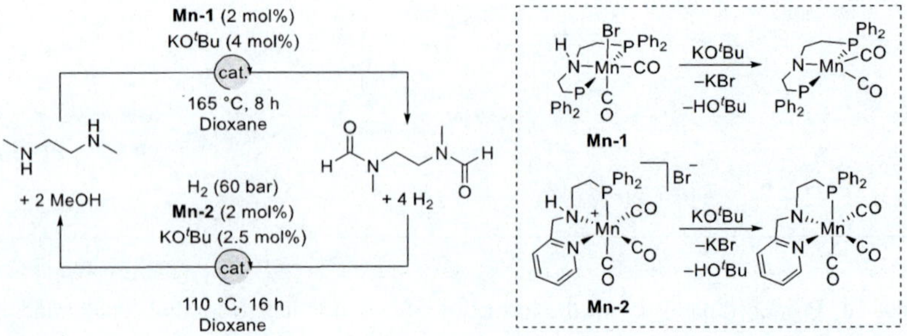

Scheme 11 Optimised conditions for formation and hydrogenation of amides for hydrogen storage [69]

however, these typically apply precious metal catalysts based on Ru-pincer systems [70–73].

In 2020 Liu reported the first use of 3d metal (manganese) precatalysts for the reversible (de)hydrogenative coupling of *N,N′*-dimethylethylenediamine (DMEDA) with methanol to form *N,N′*-(ethane-1,2-diyl)*bis*(*N*-methylformamide) [69]. This remarkable work achieves 97% maximum selectivity for DMEDA/MeOH (de)-hydrogenation demonstrating a system with 5.3 wt% H_2 storage capacity (Scheme 11). Reaction of DMEDA with methanol (6 equivalents to DMEDA; 3 × excess) in the presence of catalytic Mn(PN^{H}P-Ph)$(CO)_2$ (**Mn-1**; PN^{H}P-Ph = η^3-P,N,P-[*N,N′*-*bis*(diphenylphosphinoethyl)amine]) and base at elevated temperature (165 °C) generates the desired diformamide in 98% yield via acceptorless dehydrogenative coupling through amino/amido-MLC at manganese. This dehydrogenation is performed in a sealed vessel, allowing dihydrogen to build up during the reaction. The authors note that the build-up of dihydrogen within the vessel decreases the selectivity of the reaction, resulting in the observation of decarbonylation products, *N*-methyl-*N*-[2-(methylamino)ethyl]formamide and 3-methyl-1-methyleneimidazolidin-1-ium. However, this issue is overcome through stopping the reaction after 2 h and removing the dihydrogen atmosphere, before continuing the reaction for 6 further hours. This hydrogen release during the reaction course also had the benefit of increasing the yield obtained from 92% (after 16 h reaction) to 98% (after 8 h) as dihydrogen inhibition of the dehydrogenation reaction is decreased [69].

The reverse reaction, hydrogenative cleavage of diformamide, *N,N′*-(ethane-1,2-diyl)*bis*(*N*-methylformamide), was most successful when a different precatalyst, [Mn(PNN-Ph)$(CO)_3$]Br (**Mn-2**, PNN-Ph = η^3-P,N,N-[*N*-{diphenylphosphinoethyl}-*N′*-(2-{methylpyridyl})amine]) was used, releasing DMEDA and MeOH products in a 99% yield [69]. This hydrogenation reaction is performed with 60 bar H_2 overpressure along with temperature of 110 °C. Importantly, the activity of this hydrogenation reaction is unaffected by the presence of **Mn-1**, as determined by the performance of sequential dehydrogenation then hydrogenation reactions under optimised conditions. Finally, this reaction can also be carried out using **Mn-1** precatalyst, but this requires harsher conditions (180 °C, 80 bar H_2) to attain 94% yield. This work on Mn-based LOHC systems is an important step with regard to realising a hydrogen economy, however, as it stands, these 3d metal systems do require harsher reaction conditions and higher catalytic loading when compared to analogous Ru-systems [73]. As such, there remains work to be done to improve the efficiency of these systems.

4 Acid Synthesis from the Transformation of Alcohols

Carboxylic acid derivatives such as butyric acid (BA) and lactic acid (LA) are useful feedstock for chemical industry. Butyric acid is an industrially important chemical, with ~80,000 metric tons produced each year [74]. It has uses in the chemical [75], food [76], pharmaceutical [77], perfume [78] and animal feed [79] industries as well as increasing interest in the possibility of its use as a feedstock for biobutanol fuel [77]. Lactic acid has a considerably larger demand, with annual production volume ~1.4 million metric tons and has various applications including within food [80], pharmaceutical [81], cosmetic [82], textile [83] and fine chemical [84] industries.

Butyric acid is produced chemically from hydroformylation of propylene (originating from fossil fuels) to butyraldehyde and subsequent oxidation to BA (Scheme 12a) [85]. While the hydroformylation step is 100% atom economic, it typically requires a rhodium catalyst (although cobalt can also be used) along with harsh reaction conditions (high temperatures and pressures) [86]. From a chemical process perspective, the most economically feasible route to LA production begins with reaction of acetal aldehyde (a by-product of the acrylonitrile industry) with hydrogen cyanide to generate a lactonitrile intermediate, subsequent acid hydrolysis produces lactic acid alongside ammonium salt waste (Scheme 12b) [87]. However, over 90% of lactic acid production comes from the microbial fermentation of sugars [88]. The

Scheme 12 Industrial (**a**, **b**) and AAD (**c**) routes to butyric acid (BA) and lactic acid (LA)

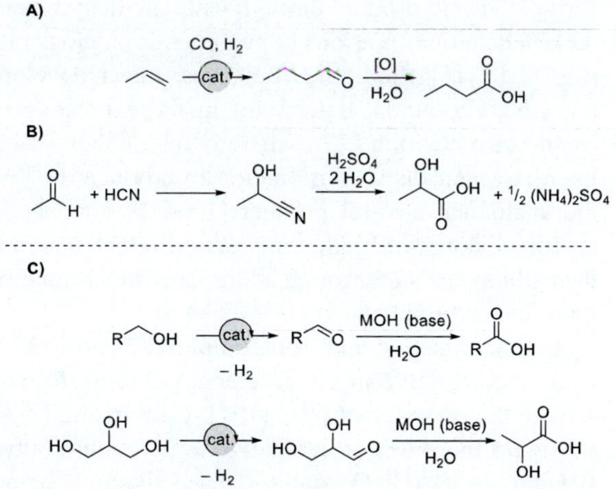

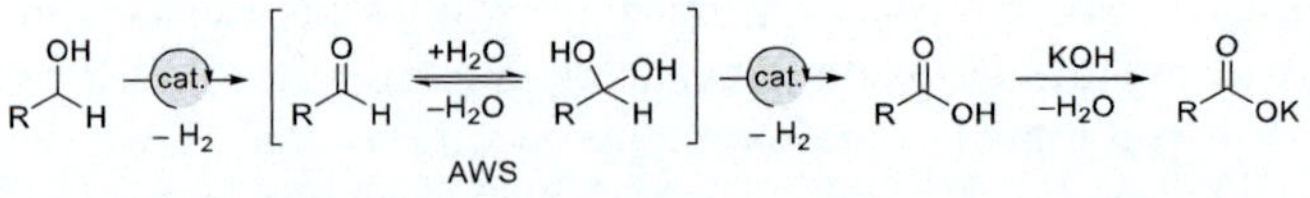

Scheme 13 Proposed mechanism of AAD preparative route to carboxylic acids. [93] AWS = aldehyde-water shift

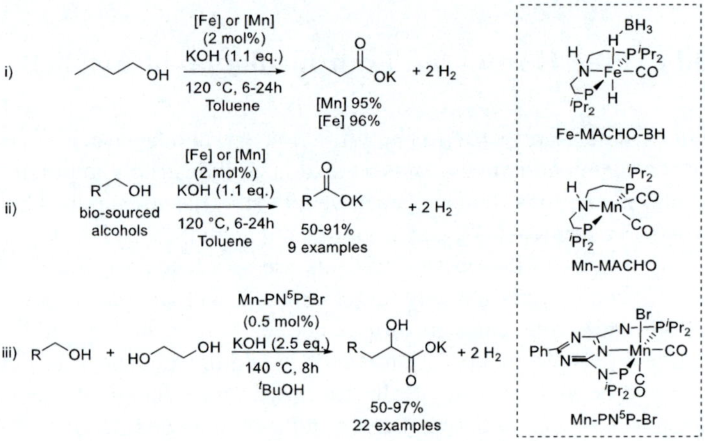

Scheme 14 AAD for Mn- or Fe-catalysed transformation of butan-1-ol and analogues to derivative acids [101, 103]

high cost of the refined sugar feedstocks required for fermentation routes to LA has resulted in a renewed drive to advance LA production technologies [88].

A general approach for the preparation of carboxylic acids from sustainable alcohol sources is found in metal-catalysed acceptorless alcohol dehydrogenation (AAD) and base-initiated aldehyde-water shift (AWS) [89] to generate the desired carboxylic acid product through catalytic dehydrogenation of the intermediate *gem*-diol (Scheme 13). Various homogeneous precious metal catalyst systems (ruthenium [63, 90–95] and rhodium [96–98]) have been developed for the oxidation of alcohols to carboxylic acids. Selectivity in these processes can suffer from a competing Cannizzaro reaction [99], whereby an aldehyde with no α-hydrogens can undergo intermolecular disproportionation to produce the desired carboxylic acid alongside the analogous alcohol product [100]. A good approach to avoid unwanted side reactions along with allowing easy product separation is to isolate the product as the carboxylate salt through addition of stoichiometric base [101, 102] before work-up to the carboxylic acid.

Both manganese and iron complexes bearing MACHO ligand motifs have been shown to be active in the efficient AAD transformation of alcohols to carboxylate salts in the presence of base [101]. Gauvin and co-workers obtained 95% and 96% yields for the AAD of butan-1-ol to potassium butyrate mediated by Fe-MACHO-BH and Mn-MACHO catalysts, respectively (Scheme 14i). Interestingly, the authors

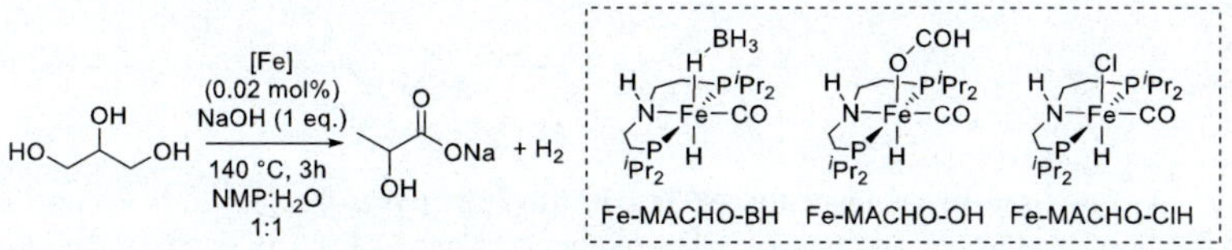

Scheme 15 Fe-catalysed glycerol to lactic acid AAD [102]

note that carrying out this reaction in the presence of water (2 equivalents to alcohol) resulted in a decrease in observed yield (*c.f.* 12% with water and 95% in the absence of water).

The substrate scope of this transformation was also extended to include several bio-sourced alcohols, including fatty acid terpenoic derivative alcohols, of which the authors report NMR yields of 50–91% with Mn- and Fe-precatalysts [101]. Typically, Mn-MACHO catalysed reactions gave better yields which is attributed to the higher robustness of Mn complexes versus analogous Fe complexes under the conditions used. Also of note, these AAD transformations are tolerant of unsaturation within the alcoholic substrate. This concept has been recently expanded by Maji and co-workers (Scheme 14 iii) for the synthesis of α-hydroxycarboxylic acids from the dehydrogenative coupling of ethylene glycol and primary alcohols using Mn-PN^{5}P-Br (Mn(PN^{5}P)$(CO)_2$Br, where PN^{5}P = κ^3-P,N,P-(2,4-[*bis*(-di*iso*propylphosphino)-λ^2-azenyl]-6-phenyl-1,3,5-triazine) precatalyst [103]. Various α-hydroxycarboxylic acids including lactic acid, α-amino acid, α-thiocarboxylic acid, and several pharmaceutically relevant molecules were synthesised with a TON in the range of 200–10,000.

Hazari and co-workers have further extended this chemistry to the impressive AAD transformation of glycerol to sodium lactate for lactic acid preparation using a series of Fe-MACHO analogues [102] in a rare example of glycerol upgrading using a base-metal catalyst. Optimising the reaction using Fe-MACHO-BH, the efficiency of this transformation was determined to be conditional upon temperature, base loading, and solvent, with the optimal conditions shown in Scheme 15, achieving a TON of 770 for modest yield, ~17% using 1:1 NMP:water after 6 h. Interestingly, the authors also report a base initiated side reaction where NMP (solvent) undergoes ring-opening to give sodium-4-*N*-methylaminobutanoate via hydroxide nucleophilic attack. The reported TON does decrease as the concentration of NMP is increased: TON 360 with 2:1 NMP:water. However, no reactions with >1 equivalent of base are reported, therefore the effect this side reaction has on the obtainable yield cannot be commented upon. Under these conditions, catalyst deactivation is significant after 0.5 h (as indicated by hydrogen evolution monitoring) and is complete after 3 h. This is in line with previous reports of Fe-MACHO complex instability at elevated temperatures [101].

The proposed mechanism for lactic acid formation from glycerol dehydrogenation is shown in Scheme 16. There are 4 major steps to generate lactic acid from glycerol: 1) AAD of glycerol to glyceraldehyde; 2) base catalysed dehydration; 3)

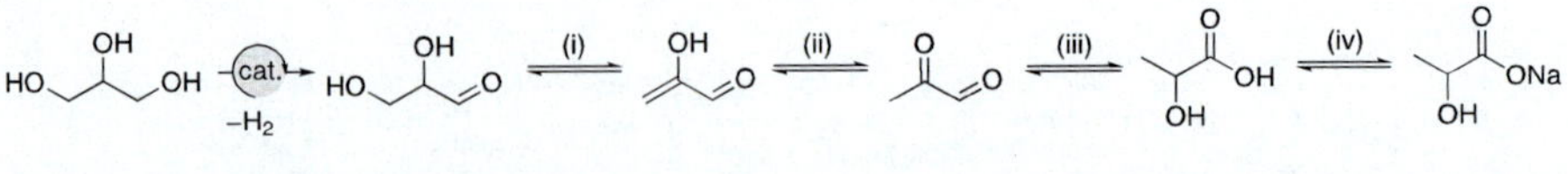

Scheme 16 Proposed mechanism for AAD pathway for glycerol to lactic acid transformation. (i) dehydration; (ii) tautomerisation; (iii) intramolecular Cannizzaro; (iv) deprotonation/metalation [102]

tautomerisation to 2-oxopropanal; 4) an intramolecular Cannizzaro reaction of 2-oxopropanal to lactic acid. Under these basic conditions, lactic acid is then rapidly deprotonated to produce sodium lactate.

5 Urea Derivatives from Dehydrogenative Coupling of Methanol and Amines

Urea derivatives are ubiquitous organic compounds widely applied for use in pharmaceutical drugs [104–106], pesticides [107, 108], fertilisers [109–111], dyes [112] and resin precursors [113–115]. Currently, urea derivatives are produced industrially from the reaction of amines with phosgene gas [116] or isocyanates (Scheme 17a) [117]. However, these starting materials are both toxic, with phosgene gas falling into the extremely toxic category [118]. As such, there is a need to replace these substrates with greener, less toxic and more sustainable alternatives.

Several examples of the dehydrogenative coupling of amines directly with formamides have been reported; however, these typically employ precious metal catalysts, in particular, ruthenium-based systems [119–122]. While dehydrogenative coupling of amines with formamide is preferable to the use of isocyanates or phosgene, the preparation of formamide for use as a substrate in these reactions typically requires harsh formylation agents [123] or precious metal catalysts [124].

Better still is the metal-catalysed dehydrogenative coupling of amines with methanol to generate urea derivatives (Scheme 17b). Methanol can be produced through completely renewable processes, making it a particularly attractive feedstock [125]. This transformation was initially reported using ruthenium-based systems [42, 126], but has recently been expanded to include both iron [67] and

Scheme 17 Traditional (**a**) and dehydrogenative coupling (**b**) routes to urea derivatives

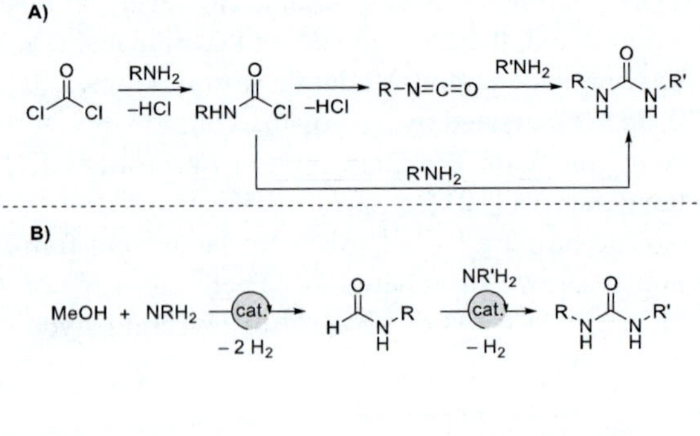

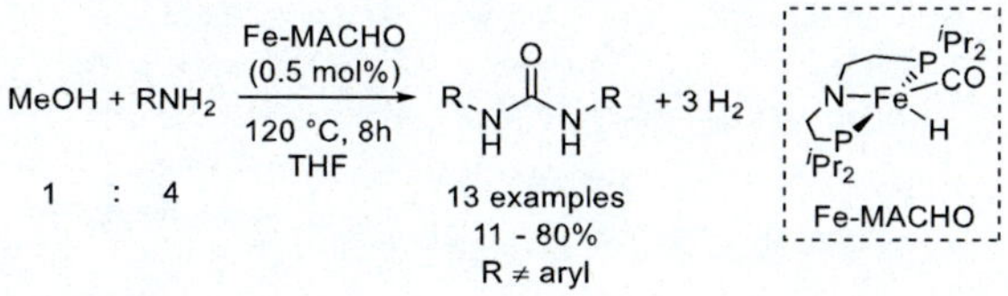

Scheme 18 Preparation of symmetrical urea derivatives mediated by Fe-MACHO [67]

manganese [36, 127]. The preparation of urea derivatives from methanol and amines is suggested to proceed via three steps, releasing three equivalents of hydrogen gas (relative to urea) as a by-product: (1) methanol dehydrogenation to formaldehyde; (2) dehydrogenative coupling of formaldehyde and amine to produce formamide in situ; and (3) dehydrogenative coupling of formamide with a second equivalent of amine to generate the urea product. The final step of the process has been postulated to proceed via two possible mechanisms, either proceeding via (a) reaction of formamide with amine to produce an aminal intermediate and subsequent dehydrogenation to urea [128], or (b) dehydrogenation of formamide to an isocyanate intermediate which produces urea upon reaction with amine [36, 67, 129].

The first report of base-metal catalysed dehydrogenative coupling between methanol and primary amines came from the group of Bernskoetter in 2018 using a base-free Fe-MACHO (Fe(PN^{H}P-iPr)(CO)H, PN^{H}P-iPr = η^3-P,N,P-[*N,N′-bis*{di*iso*propylphosphinoethyl}amine]) catalyst (Scheme 18) [67]. This Fe-MACHO catalyst was employed in the preparation of various symmetrical ureas, with the authors noting that the conditions used did not tolerate primary amines bearing an aryl substituent. The lack of reactivity observed towards aryl-amine dehydrogenative coupling was attributed to poor nucleophilicity, an observation that has been noted previously [59].

Bernskoetter and co-workers report isolated urea yields between 11 and 80% with TON calculated between 12 and 160. Low TON was typically observed for sterically demanding substrates, such as cyclohexylamine or 2-aminoheptane (TON 66 and 12, respectively), with the highest TON corresponding to terminal, sterically undemanding aliphatic amines. The electronic effect of the amine upon reactivity was also probed through a short series of benzyl-amine derivatives and found that, in line with previous reports [59], increasing the nucleophilicity of the amine corresponds to an increase in TON. The authors also attempted to prepare a cyclic urea from 1,2-diaminocyclohexane, and while successful, yielded little product (11% isolated yield), which they speculate may be due to additional amine coordination effects decreasing catalytic turnover.

In 2022, Kumar and co-workers expanded the catalytic space for the dehydrogenative coupling of methanol and amines to include various manganese pincer complexes (Scheme 19) [36]. They found that under the reaction conditions used, Mn-MACHO-Br, Mn(PN^{H}P-iPr)(CO)$_2$Br, was the most effective precatalyst trialled for the generation of urea derivatives (where R = *n*-octyl in Scheme 19). The authors also found it beneficial to use a higher reaction temperature (*c.f.* 120°C for Fe-MACHO and 150°C for Mn-MACHO-Br) to obtain quantitative conversion, resulting in the requirement for use of a sealed reaction vessel. Additionally, they

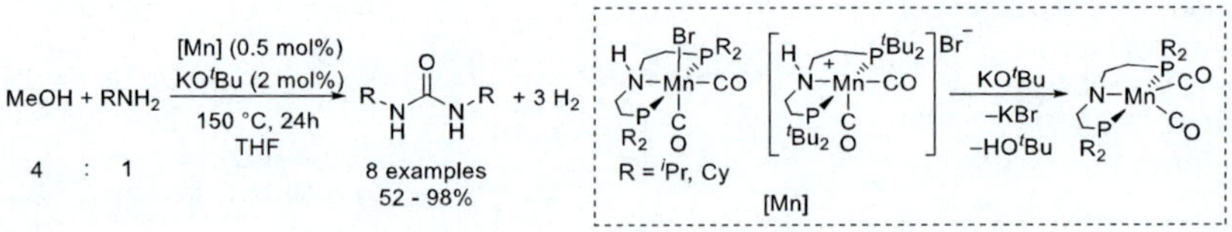

Scheme 19 Preparation of symmetrical urea derivatives mediated by Mn-pincer catalysts [36]

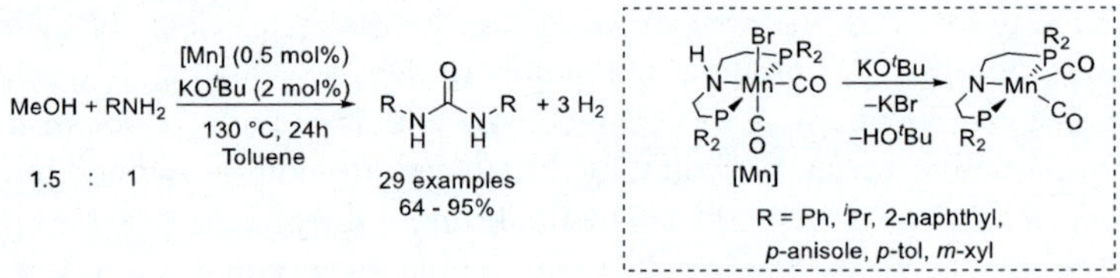

Scheme 20 Preparation of symmetrical urea derivatives mediated by Mn-pincer catalysts [127]

also reported a requirement to use methanol in excess to amine; when 0.5 equivalents of methanol to amine was used, they found that the yield dropped from 98% to 45%, a fact they attribute to the low boiling point of methanol (MeOH b.p. 64.7 °C at 1 atm).

In quick succession, a significant expansion to the substrate scope came from the group of Liu in 2022, demonstrating successful urea formation from 29 amine sources, including: aliphatic amines, substituted benzyl-amines and, notably, heteroaryl substituted methylamines such as pyridyl, furanyl and thienyl [127]. They applied various manganese catalysts including, and analogous to, Mn-MACHO-Br (see Scheme 20) and found the most productive precatalyst trialled under their conditions to be Mn(PN^{H}P-*m*-xyl)(CO)$_2$Br (where PN^{H}P-*m*-xyl = η^3-P, N,P-[*N*,*N*′-*bis*{di-*m*-xylylphosphinoethyl}amine]), obtaining an 84% conversion of benzyl-amine and methanol to *N*,*N*′-dibenzylurea, with analogous Mn-complexes generating the same product in 39–76% yields.

The optimised conditions determined by Liu and co-workers are similar to those used by Kumar et al.; however, they found they could achieve high yields at lower temperature (here, 130 °C). As in the report by Kumar and co., Liu also used methanol in excess (1.5–2 equivalents versus amine) and performed the reactions in sealed systems. The authors also reported that attempts to generate urea from secondary amines (e.g. *N*-methylbenzylamine) were unsuccessful, instead producing formamide. This observation supports the proposed mechanism of urea formation proceeding via formamide dehydrogenation to an isocyanate intermediate, in line with previous reports [36, 67, 129]. Also noted, substrates bearing a strong electron withdrawing group (e.g. CF_3, NO_2) failed to produce urea product.

The expansion of the chemical space to allow synthesis of urea derivatives from the dehydrogenative coupling of methanol and amines using earth-abundant catalysts is an important step towards a sustainable synthetic future. However, at the present time, the methodologies discussed do not currently allow for the preparation of asymmetric urea derivatives using a catalyst of earth-abundant metal, instead requiring the use of formamide and amine substrates for dehydrogenative coupling [36, 67]. As such, further development of this transformation to include asymmetric urea preparation remains sought after.

In addition, the current reports also typically require these reactions to be performed in sealed systems. This is to allow the reactions to reach the desired temperatures that are typically greater than the boiling point of the solvent used. While these batch reactions are suitable for small scale (reported scales are 2–10 mmol), it is likely that there will become a requirement for careful engineering upon scale up. Finally, there is also a need to expand this work to tolerate aromatic amine substituents to expand the substrate scope considerably, although this may prove difficult due to the poor nucleophilicity of these amines, for instance Ru-MACHO catalysed MeOH and aniline coupling produces 1,3-diphenylurea in 2% yield [122].

6 Polyurea Synthesis from Dehydrogenative Coupling of Methanol and Diamines

An important class of plastic, polyureas find applications in coatings [130], adhesives [130], construction materials [131] and drug delivery systems [132]. The annual global market for polyurea coatings in 2022 was \$1.2 billion USD [133]. The synthesis of polyureas is analogous to that of urea derivatives (see Sect. 5). As such, there exists the same requirement for the development of a greener, less toxic synthetic approach to prepare polyurea as industrially they are made from the reaction of diamines with diisocyanates that are made from phosgene gas (Scheme 21) [134].

Applying the same strategy as in urea preparation, Kumar [36] and Liu [127] have both reported the preparation of polyurea by the manganese-catalysed dehydrogenative coupling of methanol with diamines. These reactions proceed in an analogous manner to the formation of urea in a three-step process, each of which releases an equivalent of hydrogen gas as the by-product, making this approach

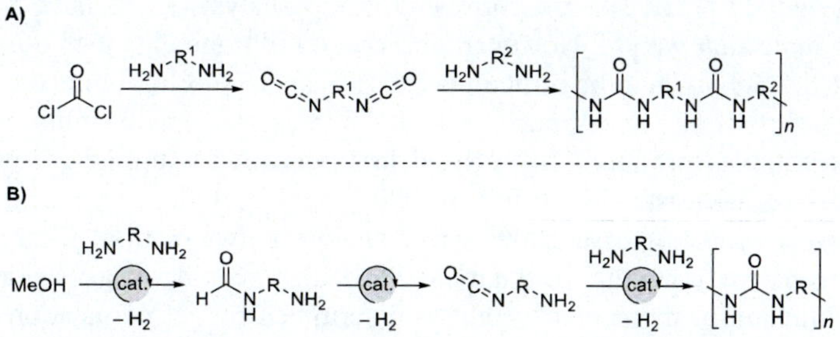

Scheme 21 Traditional (a) and dehydrogenative coupling (B) routes to polyurea

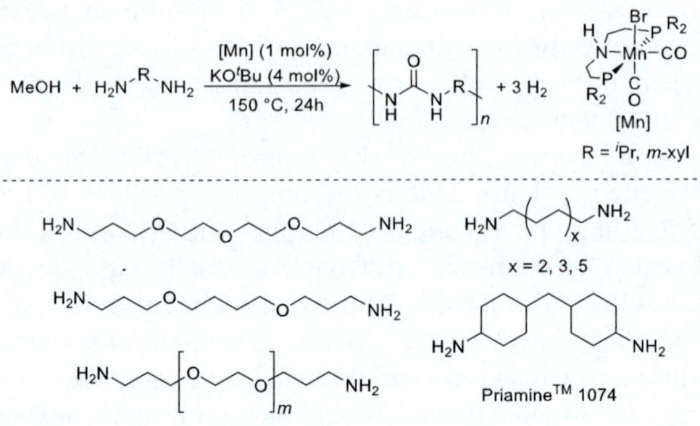

Scheme 22 Mn-catalysed preparation of polyurea from diamine and methanol [36, 127]

green and sustainable: (1) dehydrogenation of methanol to formaldehyde; (2) dehydrogenative coupling of formaldehyde and diamine to generate a formamide intermediate; (3) dehydrogenative coupling of formamide with another equivalent of diamine to provide a urea linkage in polyurea. This final step is proposed to proceed through dehydrogenation of formamide to isocyanate before reaction with amine, as opposed to going via an aminal-type intermediate [36, 127].

The initial reports (published at the same time) of the manganese catalysed dehydrogenative coupling of methanol with diamines for polyurea synthesis use very similar reaction conditions, requiring 1 mol% Mn-MACHO-Br or analogous precatalyst, base for precatalyst activation, 150 °C for 24 h in sealed vessels using solvents typical for these processes, i.e. Toluene, THF, Anisole, and obtaining high yields of polyurea product (68–99%) for a variety of aliphatic substrates (Scheme 22). Kumar also noted that aromatic diamines such as *p*-xylenediamine and *m*-xylenediamine were found to be unreactive, in line with the reduced nucleophilicity of these substrates.

Typically, the polyureas produced from methanol and diamine coupling have poor solubility. This adds a complication when it comes to accurately determining molecular weight of the polymeric products produced within these reactions [36]. Both Matrix Assisted Laser Desorption Ionisation (MALDI) mass spectrometry [36] and ^{1}H NMR spectroscopic end-group analysis [127] have been used to estimate molecular weight. However, observation of fragments with m/*z* > 5,000 by MALDI-MS is known to be difficult [135, 136] and solubility limits the accuracy of NMR spectroscopic end-group analysis. The molecular weights reported for polyurea samples produced from the dehydrogenative coupling of methanol and diamines are moderate: M_n 1,500–15,000 g mol^{-1}. Indeed, a recent report that includes the synthesis of a 100% water soluble polyurea from the Mn-catalysed dehydrogenative coupling of diamine with diformamide suggests a significant underestimation of molecular weight as determined by gel permeation chromatography (GPC) analysis vs MALDI-MS (×8) and end-group (×2) molecular weight determination techniques.

Of particular importance, Kumar and co-workers included priamine™ 1,074 (8,8′-(4-hexyl-3-octylcyclohexyane-1,2-diyl)*bis*(octylamine) in their substrate scope, obtaining polyurea in 80% yield. This diamine is commercialised by Croda and is fully renewable, being produced from vegetable oil [137], as such, it allows for the possibility of a 100% renewable polyurea product when coupled with methanol produced from sustainable sources [36].

This method of polyurea synthesis from sustainable alcohol dehydrogenation is a promising stepwise development in the preparation of these commercially important plastics. However, some work remains to be done, as for instance, these reactions do not tolerate aromatic amine substrates. This will be particularly important for the preparation of polyureas with mixed backbone functionality, e.g. both flexible and rigid backbone sections, to control their physical and mechanical properties [138, 139].

7 Polyethyleneimine Derivatives from the Coupling of Ethylene Glycol and Ethylenediamine

Branched polyethyleneimine (PEI) and polyethyleneimine ethoxylated (PEIE) have an annual global market estimated at $400 million USD [140] due to their wide range of applications, such as: textiles [141], cosmetics [142], adhesives [143] and water-treatment agents [144]. They have also been applied as the gold standard in CO_2 capture [145] and gene delivery [146], alongside uses in solar cells [147], tissue culture [146] and optoelectronic devices [148]. The current industrial preparation of PEI involves the acid-catalysed ring-opening polymerisation of aziridine (Scheme 23), that is an extremely toxic, carcinogenic and volatile liquid substrate [149]. This hazardous substrate is itself produced industrially one of two ways: (1) the Wenker

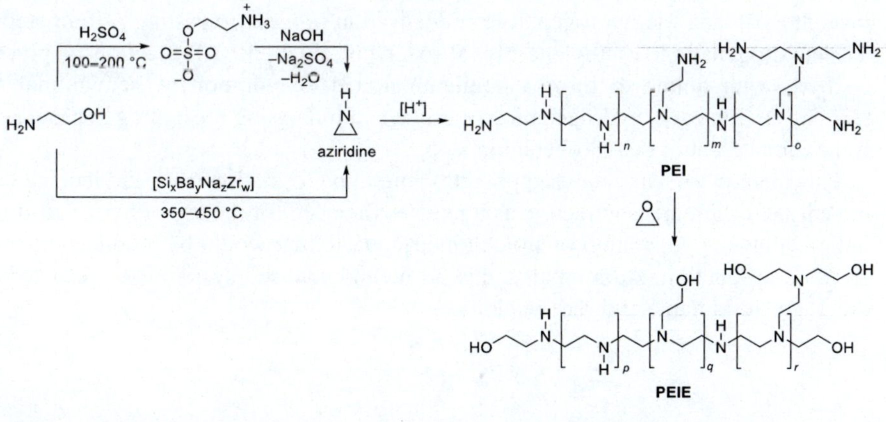

Scheme 23 Traditional routes to polyethyleneimine (PEI) and polyethyleneimine ethoxylated (PEIE)

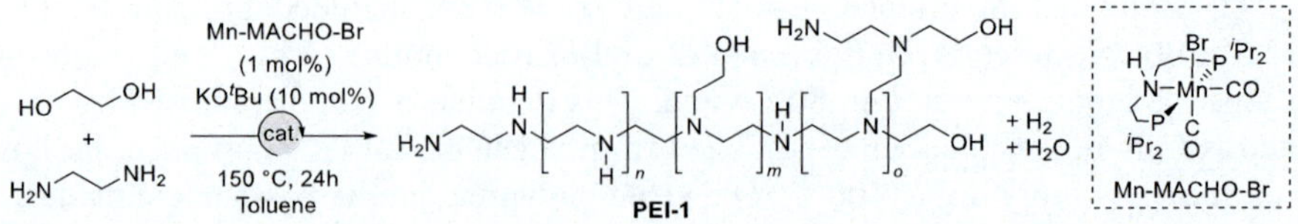

Scheme 24 Mn-MACHO catalysed preparation of polyethyleneimine derivative, **PEI-1** [154]

process (BASF) that involves the reaction of ethanolamine with sulphuric acid at 100–200 °C to give 2-aminoethyl hydrogen sulphate, that is subsequently reacted with aqueous sodium hydroxide to produce aziridine [149]; or (2) the Nippon Shobukai process that involves the dehydration of ethanolamine at high temperatures (350–450 °C) using a heterogeneous $Si_xBa_yNa_zZr_w$ catalyst [150]. PEIE is produced through ethoxylation of PEI with epoxide [151]. These processes either produce stoichiometric metal waste or require high energy conditions, and both have the disadvantages in the production of highly toxic aziridine. As such, there is a need for an alternative, less toxic and sustainable preparation of PEI and PEIE. There is some precedence in the form of registered technology from BASF for the coupling of amines and alcohols to generate PEI derivatives; however, these require precious metal catalysts, such as iridium or ruthenium, and a hydrogen atmosphere to ensure saturation within the product [152, 153].

A recent report from Kumar et al. demonstrates the synthesis of a derivative to PEI and PEIE, a partially ethoxylated polyethyleneimine (PEI-1, shown in Scheme 24) [154]. The authors exploit the (de)hydrogenative coupling capabilities of the well-known Mn-MACHO-Br precatalyst to successfully polymerise ethylene glycol and ethylenediamine to generate a branched polyamine structure in high molecular weight (>50,000 g mol^{-1}) with narrow polydispersity (Đ 1.1), releasing water as the only by-product.

Reaction monitoring, stoichiometric and computational experiments suggest a borrowing hydrogen mechanism is in place for the observed C-N bond formation through a cascade (de)hydrogenative condensation process consisting of four steps: (1) dehydrogenation of ethylene glycol to glycol aldehyde; (2) reaction of glycol aldehyde with amine to form a hemiaminal; (3) dehydration of hemiaminal to generate an imine; (4) hydrogenation of imine to amine using the hydrogen produced in the initial alcohol dehydrogenation step.

This method for Mn-catalysed polyethyleneimine formation from ethylene glycol and ethylene diamine is attractive as it negates the need for the use of toxic aziridine and prevents the generation of stoichiometric waste that would be produced during aziridine production. Additionally, this 3d metal-catalysed system used feedstocks which are sustainable and renewable.

8 Summary and Outlook

In conclusion, metal-ligand cooperative (MLC) complexes of the 3d transition metals allow facile bond activation of alcohols. The synergy between ligand and metal within MLC processes facilitates alcohol activation without the need of noble metals as the requirement for changes to metal oxidation state is removed. These earth-abundant metal-catalysed dehydrogenative alcohol transformations have been successfully applied to produce a wide range of alcohol up-conversion products such as esters, amides, acids, ureas, polyureas and polyethyleneimines. As can be seen above, manganese has been used more widely for various dehydrogenative transformations in comparison with iron, cobalt and nickel which have also been studied for dehydrogenative reactions. Key to this success lies in the use of the precursor $Mn(CO)_5Br$ that allows facile synthesis of stable manganese precatalysts containing bromide, and carbonyl ligands which are capable of exhibiting metal-ligand cooperation and have the desired electronic and steric effects. It is worth noting that activity of some of the base-metal catalysts are comparable or even higher in comparison to the precious metals such as ruthenium for some dehydrogenative reactions. However, some work does remain to be done to further expand the substrate scopes within some of these developments. There are a few dehydrogenative reactions that have been reported with precious metals such as ruthenium but not yet using a catalyst of base-metals, for example, (a) synthesis of primary amines from alcohols and ammonia, (b) synthesis of polyesters from the dehydrogenative coupling of diols, (c) synthesis of polyamides from the dehydrogenative coupling of diols and diamines. Additionally, the development of catalytic protocol to use other precursors such as $MnCl_2$ instead of $Mn(CO)_5Br$ will make the processes more cost-effective.

References

1. Sarp S, Gonzalez Hernandez S, Chen C, Sheehan SW (2021) Alcohol production from carbon dioxide: methanol as a fuel and chemical feedstock. Joule 5:59–76. https://doi.org/10.1016/j.joule.2020.11.005
2. Tse TJ, Wiens DJ, Reaney MJT (2021) Production of bioethanol – a review of factors affecting ethanol yield. Fermentation 7:1–18. https://doi.org/10.3390/fermentation7040268
3. Schubert T (2020) Production routes of advanced renewable C1 to C4 alcohols as biofuel components – a review. Biofuels, Bioprod Biorefining 14:845–878. https://doi.org/10.1002/bbb.2109
4. Kumar A, Daw P, Milstein D (2022) Homogeneous catalysis for sustainable energy: hydrogen and methanol economies, fuels from biomass, and related topics. Chem Rev 122:385–441. https://doi.org/10.1021/acs.chemrev.1c00412
5. Zhong Q, Liao J, Zhang Q, Qiu S, Meng Q, Wu X, Wang T (2022) Aqueous upgrading of ethanol to higher alcohol diesel blending and jet fuel precursors over Na-doped porous Ni@C nanocomposite. Fuel 324:124507. https://doi.org/10.1016/j.fuel.2022.124507

6. Te Molder TDJ, Kersten SRA, Lange JP, Ruiz MP (2021) Ethylene glycol from lignocellulosic biomass: impact of lignin on catalytic hydrogenolysis. Ind Eng Chem Res 60:7043–7049. https://doi.org/10.1021/acs.iecr.1c01063
7. Okolie JA (2022) Insights on production mechanism and industrial applications of renewable propylene glycol. iScience 25:104903. https://doi.org/10.1016/j.isci.2022.104903
8. De Bari I, Giuliano A, Petrone MT, Stoppiello G, Fatta V, Giardi C, Razza F, Novelli A (2020) From cardoon lignocellulosic biomass to bio-1,4 butanediol: an integrated biorefinery model. Processes 8:1–18. https://doi.org/10.3390/pr8121585
9. Nakagawa Y, Tomishige K (2012) Production of 1,5-pentanediol from biomass via furfural and tetrahydrofurfuryl alcohol. Catal Today 195:136–143. https://doi.org/10.1016/j.cattod.2012.04.048
10. He J, Burt SP, Ball M, Zhao D, Hermans I, Dumesic JA, Huber GW (2018) Synthesis of 1,6-hexanediol from cellulose derived tetrahydrofuran-dimethanol with Pt-WOx/TiO2 catalysts. ACS Catal 8:1427–1439. https://doi.org/10.1021/acscatal.7b03593
11. Zhao W, Wu W, Li H, Fang C, Yang T, Wang Z, He C, Yang S (2018) Quantitative synthesis of 2,5-bis(hydroxymethyl)furan from biomass-derived 5-hydroxymethylfurfural and sugars over reusable solid catalysts at low temperatures. Fuel 217:365–369. https://doi.org/10.1016/j.fuel.2017.12.069
12. Bonnin I, Mereau R, Tassaing T, de Oliveira VK (2020) One-pot synthesis of isosorbide from cellulose or lignocellulosic biomass: a challenge? Beilstein J Org Chem 16:1713–1721. https://doi.org/10.3762/BJOC.16.143
13. Tan HW, Abdul Aziz AR, Aroua MK (2013) Glycerol production and its applications as a raw material: A review. Renew Sustain Energy Rev 27:118–127. https://doi.org/10.1016/j.rser.2013.06.035
14. Langanke J, Wolf A, Hofmann J, Böhm K, Subhani MA, Müller TE, Leitner W, Gürtler C (2014) Carbon dioxide (CO2) as sustainable feedstock for polyurethane production. Green Chem 16:1865–1870. https://doi.org/10.1039/c3gc41788c
15. Econic Technologies – How It Works. https://econic-technologies.com/how-it-works/. Accessed 14 May 2023
16. Younus HA, Su W, Ahmad N, Chen S, Verpoort F (2015) Ruthenium Pincer Complexes: Synthesis and Catalytic Applications. Adv Synth Catal 357:283–330. https://doi.org/10.1002/adsc.201400777
17. Gunanathan C, Milstein D (2014) Bond activation and catalysis by ruthenium pincer complexes. Chem Rev 114:12024–12087. https://doi.org/10.1021/cr5002782
18. (2023) Esters Market Outlook 2022–2029. In: Futur. Mark. Insights. https://www.futuremarketinsights.com/reports/esters-market. Accessed 18 May 2023
19. Ambat I, Srivastava V, Sillanpää M (2018) Recent advancement in biodiesel production methodologies using various feedstock: a review. Renew Sustain Energy Rev 90:356–369. https://doi.org/10.1016/j.rser.2018.03.069
20. Aalto TR, Firman MC, Rigler NE (1953) p-hydroxybenzoic acid esters as preservatives. J Am Pharm Assoc 42:449–457. https://doi.org/10.1016/s0095-9561(16)33246-7
21. Tang X, Chen EYX (2019) Toward infinitely recyclable plastics derived from renewable cyclic esters. Chem 5:284–312. https://doi.org/10.1016/j.chempr.2018.10.011
22. Pereira CSM, Silva VMTM, Rodrigues AE (2011) Ethyl lactate as a solvent: properties, applications and production processes – a review. Green Chem 13:2658–2671. https://doi.org/10.1039/c1gc15523g
23. Aparicio S, Alcalde R, Dávila MJ, García B, Leal JM (2007) Properties and structure of aromatic ester solvents. J Phys Chem B 111:4417–4431. https://doi.org/10.1021/jp068560t
24. Reshna KR, Gopi S, Balakrishnan P (2023) Introduction to flavor and fragrance in food processing. American Chemical Society, Washington
25. Mitsui T (1997) Raw materials of cosmetics. In: Mitsui T (ed) New cosmetic science. Elsevier, pp 121–147

26. de Barros DPC, Azevedo AM, Cabral JMS, Fonseca LP (2012) Optimization of flavor esters synthesis by fusarium solani pisi cutinase. J Food Biochem 36:275–284. https://doi.org/10.1111/j.1745-4514.2010.00535.x
27. Kleoff M, Kiler P, Heretsch P (2022) Synthesis of odorants in flow and their applications in perfumery. Beilstein J Org Chem 18:754–768. https://doi.org/10.3762/bjoc.18.76
28. Siengalewicz P, Mulzer J, Rinner U (2014) Synthesis of esters and lactones. In: Comprehensive organic synthesis II2nd edn. Elsevier, Vienna, pp 355–410
29. Cherepakhin V, Williams TJ (2021) Direct oxidation of primary alcohols to carboxylic acids. Synthesis 53:1023–1034. https://doi.org/10.1055/s-0040-1706102
30. Smith PWG, Tatchell AR (1965) Aliphatic monocarboxyllic acids and their derivatives. In: Fundamental aliphatic chemistry. Elsevier, Oxford, pp 177–215
31. Zhang J, Leitus G, Ben-David Y, Milstein D (2005) Facile conversion of alcohols into esters and dihydrogen catalyzed by new ruthenium complexes. J Am Chem Soc 127:10840–10841. https://doi.org/10.1021/ja052862b
32. Blum Y, Shvo Y (1985) Catalytically reactive (n4-tetracyclone)(CO)2(H)2Ru and relative complexes in dehydrogenation of alcohols to esters. J Organomet Chem 282:C7–C10
33. Morton D, Cole-Hamilton DJ (1988) Molecular hydrogen complexes in catalysis: highly efficient hydrogen production from alcoholic substrates catalysed by ruthenium complexes. J Chem Soc Chem Commun 2:1154–1156. https://doi.org/10.1039/C39880001154
34. Chakraborty S, Lagaditis PO, Förster M, Bielinski EA, Hazari N, Holthausen MC, Jones WD, Schneider S (2014) Well-defined iron catalysts for the acceptorless reversible dehydrogenation-hydrogenation of alcohols and ketones. ACS Catal 4:3994–4003. https://doi.org/10.1021/cs5009656
35. Nguyen DH, Trivelli X, Capet F, Paul JF, Dumeignil F, Gauvin RM (2017) Manganese pincer complexes for the base-free, acceptorless dehydrogenative coupling of alcohols to esters: development, scope, and understanding. ACS Catal 7:2022–2032. https://doi.org/10.1021/acscatal.6b03554
36. Owen AE, Preiss A, McLuskie A, Gao C, Peters G, Bühl M, Kumar A (2022) Manganese-catalyzed dehydrogenative synthesis of urea derivatives and polyureas. ACS Catal 12:6923–6933. https://doi.org/10.1021/acscatal.2c00850
37. Armarego WLF, Chai CLL (2009) Purification of laboratory chemicals.6th edn. Elsevier, Oxford
38. Valerio O, Horvath T, Pond C, Misra M, Mohanty A (2015) Improved utilization of crude glycerol from biodiesel industries: synthesis and characterization of sustainable biobased polyesters. Ind Crops Prod 78:141–147. https://doi.org/10.1016/j.indcrop.2015.10.019
39. Elangovan S, Topf C, Fischer S, Jiao H, Spannenberg A, Baumann W, Ludwig R, Junge K, Beller M (2016) Selective catalytic hydrogenations of nitriles, ketones, and aldehydes by well-defined manganese pincer complexes. J Am Chem Soc 138:8809–8814. https://doi.org/10.1021/jacs.6b03709
40. Tondreau AM, Boncella JM (2016) 1,2-Addition of formic or oxalic acid to -N{CH2CH2(PiPr2)}2-supported Mn(I) dicarbonyl complexes and the manganese-mediated decomposition of formic acid. Organometallics 35:2049–2052. https://doi.org/10.1021/acs.organomet.6b00274
41. Peña-Lõpez M, Neumann H, Beller M (2015) Iron(II) pincer-catalyzed synthesis of lactones and lactams through a versatile dehydrogenative domino sequence. ChemCatChem 7:865–871. https://doi.org/10.1002/cctc.201402967
42. Kang B, Hong SH (2015) Hydrogen acceptor- and base-free N-formylation of nitriles and amines using methanol as C1 source. Adv Synth Catal 357:834–840. https://doi.org/10.1002/adsc.201400982
43. Shahane S, Fischmeister C, Bruneau C (2012) Acceptorless ruthenium catalyzed dehydrogenation of alcohols to ketones and esters. Catal Sci Technol 2:1425–1428. https://doi.org/10.1039/c2cy20066j

44. Tang Y, Meador RIL, Malinchak CT, Harrison EE, McCaskey KA, Hempel MC, Funk TW (2020) (Cyclopentadienone)iron-catalyzed transfer dehydrogenation of symmetrical and unsymmetrical diols to lactones. J Org Chem 85:1823–1834. https://doi.org/10.1021/acs.joc.9b01884
45. Indelicate JM, Pasini CE (1988) The acylating potential of γ-lactam antibacterials: base hydrolysis of bicyclic pyrazolidinones. J Med Chem 31:1227–1230. https://doi.org/10.1021/jm00401a026
46. Frau J, Coll M, Donoso J, Muñoz F, Vilanova B, García-Blanco F (1997) Alkaline and acidic hydrolysis of the β-lactam ring. Electron J Theor Chem 2:56–65. https://doi.org/10.1002/ejtc.34
47. Robinson SL, Christenson JK, Wackett LP (2019) Biosynthesis and chemical diversity of β-lactone natural products. Nat Prod Rep 36:458–475. https://doi.org/10.1039/c8np00052b
48. Pattabiraman VR, Bode JW (2011) Rethinking amide bond synthesis. Nature 480:471–479. https://doi.org/10.1038/nature10702
49. Lanigan RM, Starkov P, Sheppard TD (2013) Direct synthesis of amides from carboxylic acids and amines using B(OCH2CF3)3. J Org Chem 78:4512–4523. https://doi.org/10.1021/jo400509n
50. Bryan MC, Dunn PJ, Entwistle D, Gallou F, Koenig SG, Hayler JD, Hickey MR, Hughes S, Kopach ME, Moine G, Richardson P, Roschangar F, Steven A, Weiberth FJ (2018) Key green chemistry research areas from a pharmaceutical manufacturers' perspective revisited. Green Chem 20:5082–5103. https://doi.org/10.1039/c8gc01276h
51. Zultanski SL, Zhao J, Stahl SS (2016) Practical synthesis of amides via copper/ABNO-catalyzed aerobic oxidative coupling of alcohols and amines. J Am Chem Soc 138:6416–6419. https://doi.org/10.1021/jacs.6b03931
52. Chakraborty S, Gellrich U, Diskin-Posner Y, Leitus G, Avram L, Milstein D (2017) Manganese-catalyzed N-formylation of amines by methanol liberating H2: a catalytic and mechanistic study. Angew Chem Int Ed 56:4229–4233. https://doi.org/10.1002/anie.201700681
53. Gunanathan C, Ben-David Y, Milstein D (2007) Direct synthesis of amides from alcohols and amines with liberation of H2. Science 317:790–792
54. Nordstrøm LU, Vogt H, Madsen R (2008) Amide synthesis from alcohols and amines by the extrusion of dihydrogen. J Am Chem Soc 130:17672–17673
55. Ortega N, Richter C, Glorius F (2013) N-Formylation of amines by methanol activation. Org Lett 15:1776–1779. https://doi.org/10.1021/ol400639m
56. Kim K, Kang B, Hong SH (2015) N-Heterocyclic carbene-based well-defined ruthenium hydride complexes for direct amide synthesis from alcohols and amines under base-free conditions. Tetrahedron 71:4565–4569. https://doi.org/10.1016/j.tet.2015.02.016
57. Srimani D, Balaraman E, Hu P, Ben-David Y, Milstein D (2013) Formation of tertiary amides and dihydrogen by dehydrogenative coupling of primary alcohols with secondary amines catalyzed by ruthenium bipyridine-based pincer complexes. Adv Synth Catal 355:2525–2530. https://doi.org/10.1002/adsc.201300620
58. Xie X, Huynh HV (2015) Tunable dehydrogenative amidation versus amination using a single ruthenium-NHC catalyst. ACS Catal 5:4143–4151. https://doi.org/10.1021/acscatal.5b00588
59. Zhang Y, Chen C, Ghosh SC, Li Y, Hong SH (2010) Weil-defined N-heterocyclic carbene based ruthenium catalysts for direct amide synthesis from alcohols and amines. Organometallics 29:1374–1378. https://doi.org/10.1021/om901020h
60. Dam JH, Osztrovszky G, Nordstrøm LU, Madsen R (2010) Amide synthesis from alcohols and amines catalyzed by ruthenium N-heterocyclic carbene complexes. Chem A Eur J 16:6820–6827. https://doi.org/10.1002/chem.201000569
61. Prades A, Peris E, Albrecht M (2011) Oxidations and oxidative couplings catalyzed by triazolylidene ruthenium complexes. Organometallics 30:1162–1167. https://doi.org/10.1021/om101145y

62. Chen C, Zhang Y, Hong SH (2011) N-heterocyclic carbene based ruthenium-catalyzed direct amide synthesis from alcohols and secondary amines: Involvement of esters. J Org Chem 76: 10005–10010. https://doi.org/10.1021/jo201756z
63. Zweifel T, Naubron JV, Grützmacher H (2009) Catalyzed dehydrogenative coupling of primary alcohols with water, methanol, or amines. Angew Chem Int Ed 48:559–563. https://doi.org/10.1002/anie.200804757
64. Schleker PPM, Honeker R, Klankermayer J, Leitner W (2013) Catalytic dehydrogenative amide and ester formation with rhenium-triphos complexes. ChemCatChem 5:1762–1764. https://doi.org/10.1002/cctc.201200942
65. Andérez-Fernández M, Vogt LK, Fischer S, Zhou W, Jiao H, Garbe M, Elangovan S, Junge K, Junge H, Ludwig R, Beller M (2017) A stable manganese pincer catalyst for the selective dehydrogenation of methanol. Angew Chem Int Ed 56:559–562. https://doi.org/10.1002/anie.201610182
66. Lane EM, Uttley KB, Hazari N, Bernskoetter W (2017) Iron-catalyzed amide formation from the dehydrogenative coupling of alcohols and secondary amines. Organometallics 36:2020–2025. https://doi.org/10.1021/acs.organomet.7b00258
67. Lane EM, Hazari N, Bernskoetter WH (2018) Iron-catalyzed urea synthesis: dehydrogenative coupling of methanol and amines. Chem Sci 9:4003–4008. https://doi.org/10.1039/c8sc00775f
68. Espinosa-Jalapa NA, Kumar A, Leitus G, Diskin-Posner Y, Milstein D (2017) Synthesis of cyclic imides by acceptorless dehydrogenative coupling of diols and amines catalyzed by a manganese pincer complex. J Am Chem Soc 139:11722–11725. https://doi.org/10.1021/jacs.7b08341
69. Shao Z, Li Y, Liu C, Ai W, Luo SP, Liu Q (2020) Reversible interconversion between methanol-diamine and diamide for hydrogen storage based on manganese catalyzed (de)-hydrogenation. Nat Commun 11:1–7. https://doi.org/10.1038/s41467-020-14380-3
70. Kumar A, Janes T, Espinosa-Jalapa NA, Milstein D (2018) Selective hydrogenation of cyclic imides to diols and amines and its application in the development of a liquid organic hydrogen carrier. J Am Chem Soc 140:7453–7457. https://doi.org/10.1021/jacs.8b04581
71. Hu P, Fogler E, Diskin-Posner Y, Iron MA, Milstein D (2015) A novel liquid organic hydrogen carrier system based on catalytic peptide formation and hydrogenation. Nat Commun 6:1–7. https://doi.org/10.1038/ncomms7859
72. Hu P, Ben-David Y, Milstein D (2016) Rechargeable hydrogen storage system based on the dehydrogenative coupling of ethylenediamine with ethanol. Angew Chem Int Ed 55:1061–1064. https://doi.org/10.1002/anie.201505704
73. Kothandaraman J, Kar S, Sen R, Goeppert A, Olah GA, Prakash GKS (2017) Efficient reversible hydrogen carrier system based on amine reforming of methanol. J Am Chem Soc 139:2549–2552. https://doi.org/10.1021/jacs.6b11637
74. Jiang L, Fu H, Yang HK, Xu W, Wang J, Yang ST (2018) Butyric acid: applications and recent advances in its bioproduction. Biotechnol Adv 36:2101–2117. https://doi.org/10.1016/j.biotechadv.2018.09.005
75. Sjöblom M, Matsakas L, Christakopoulos P, Rova U (2016) Catalytic upgrading of butyric acid towards fine chemicals and biofuels. FEMS Microbiol Lett 363:1–7. https://doi.org/10.1093/femsle/fnw064
76. Brändle J, Domig KJ, Kneifel W (2016) Relevance and analysis of butyric acid producing clostridia in milk and cheese. Food Control 67:96–113. https://doi.org/10.1016/j.foodcont.2016.02.038
77. Dwidar M, Park JY, Mitchell RJ, Sang BI (2012) The future of butyric acid in industry. Sci World J 2012:471417. https://doi.org/10.1100/2012/471417
78. Vandák D, Zigová J, Šturdík E, Schlosser Š (1997) Evaluation of solvent and pH for extractive fermentation of butyric acid. Process Biochem 32:245–251. https://doi.org/10.1016/S0032-9592(96)00084-2

79. Fernández-Rubio C, Ordóñez C, Abad-González J, Garcia-Gallego A, Honrubia MP, Mallo JJ, Balaña-Fouce R (2009) Butyric acid-based feed additives help protect broiler chickens from Salmonella enteritidis infection. Poult Sci 88:943–948. https://doi.org/10.3382/ps.2008-00484
80. Kim J, Kim YM, Lebaka VR, Wee YJ (2022) Lactic acid for green chemical industry: recent advances in and future prospects for production technology, recovery, and applications. Fermentation 8:609. https://doi.org/10.3390/fermentation8110609
81. Kumar A, Singh J, Baskar C (2020) Bioactive natural products in drug discovery
82. Smith WP (1996) Epidermal and dermal effects of topical lactic acid. J Am Acad Dermatol 35: 388–391
83. Avinc O, Khoddami A (2009) Overview of poly(lactic acid) fibre. Fibre Chem 41:391–401
84. Datta R, Henry M (2006) Lactic acid: recent advances in products, processes and technologies – a review. J Chem Technol Biotechnol 81:1119–1129. https://doi.org/10.1002/jctb
85. Obst D, Wiese K-D (2006) Hydroformylation. In: Beller M (ed) Topics in organometallic chemistry: catalytic carbonylation reactions. Springer, Heidelberg, pp 1–35
86. Liu B, Wang Y, Huang N, Lan X, Xie Z, Chen JG, Wang T (2022) Review heterogeneous hydroformylation of alkenes by Rh-based catalysts. Chem 8:2630–2658. https://doi.org/10.1016/j.chempr.2022.07.020
87. Komesu A, De Oliveira JAR, de Martins LHS, Maciel MRW, Filho RM (2017) Lactic acid production to purification: a review. BioResources 12:4364–4383
88. Ojo AO, De Smidt O (2023) Lactic acid: a comprehensive review of production to purification. Processes 11:688
89. Brewster TP, Goldberg JM, Tran JC, Heinekey DM, Goldberg KI (2016) High catalytic efficiency combined with high selectivity for the aldehyde – water shift reaction using (paracymene)ruthenium precatalysts. ACS Catal 6:6302–6305. https://doi.org/10.1021/acscatal.6b02130
90. Santilli C, Makarov IS, Fristrup P, Madsen R (2016) Dehydrogenative synthesis of carboxylic acids from primary alcohols and hydroxide catalyzed by a ruthenium N-heterocyclic carbene complex. J Org Chem 81:9931–9938. https://doi.org/10.1021/acs.joc.6b02105
91. Choi JH, Heim LE, Ahrens M, Prechtl MHG (2014) Selective conversion of alcohols in water to carboxylic acids by in situ generated ruthenium trans dihydrido carbonyl PNP complexes. Dalton Trans 43:17248–17254. https://doi.org/10.1039/c4dt01634c
92. Malineni J, Keul H, Möller M (2015) A green and sustainable phosphine-free NHC-ruthenium catalyst for selective oxidation of alcohols to carboxylic acids in water. Dalton Trans 44: 17409–17414. https://doi.org/10.1039/c5dt01358e
93. Balaraman E, Khaskin E, Leitus G, Milstein D (2013) Catalytic transformation of alcohols to carboxylic acid salts and H 2 using water as the oxygen atom source. Nat Chem 5:122–125. https://doi.org/10.1038/nchem.1536
94. Hu P, Ben-David Y, Milstein D (2016) General synthesis of amino acid salts from amino alcohols and basic water liberating H2. J Am Chem Soc 138:6143–6146. https://doi.org/10.1021/jacs.6b03488
95. Zhang L, Nguyen DH, Raffa G, Trivelli X, Capet F, Desset S, Paul S, Dumeignil F, Gauvin RM (2016) Catalytic conversion of alcohols into carboxylic acid salts in water: scope, recycling, and mechanistic insights. ChemSusChem 9:1413–1423. https://doi.org/10.1002/cssc.201600243
96. Wang X, Wang C, Liu Y, Xiao J (2016) Acceptorless dehydrogenation and aerobic oxidation of alcohols with a reusable binuclear rhodium(II) catalyst in water. Green Chem 18:4605–4610. https://doi.org/10.1039/c6gc01272h
97. Gianetti TL, Annen SP, Santiso-Quinones G, Reiher M, Driess M, Grützmacher H (2016) Nitrous oxide as a hydrogen acceptor for the dehydrogenative coupling of alcohols. Angew Chem Int Ed 55:1854–1858. https://doi.org/10.1002/anie.201509288
98. Annen S, Zweifel T, Ricatto F, Grützmacher H (2010) Catalytic aerobic dehydrogenative coupling of primary alcohols and water to acids promoted by a Rhodium(I) amido

N-heterocyclic carbene complex. ChemCatChem 2:1286–1295. https://doi.org/10.1002/cctc.201000100
99. Cannizzaro von S (1853) On the alcohol corresponding to benzoic acid. Liebigs Ann der Chemie und Pharm 88:129–130
100. Brewster TP, Ou WC, Tran JC, Goldberg KI, Hanson SK, Cundari TR, Heinekey DM (2014) Iridium, rhodium, and ruthenium catalysts for the "Aldehyde – Water Shift" reaction. ACS Catal 4:3034–3038
101. Nguyen DH, Morin Y, Zhang L, Trivelli X, Capet F, Paul S, Desset S, Dumeignil F, Gauvin RM (2017) Oxidative transformations of biosourced alcohols catalyzed by earth-abundant transition metals. ChemCatChem 9:2652–2660. https://doi.org/10.1002/cctc.201700310
102. Sharninghausen LS, Mercado BQ, Crabtree RH, Hazari N (2015) Selective conversion of glycerol to lactic acid with iron pincer precatalysts. Chem Commun 51:16201–16204. https://doi.org/10.1039/c5cc06857f
103. Waiba S, Maji K, Maiti M, Maji B (2023) Sustainable synthesis of α-hydroxycarboxylic acids by manganese catalyzed acceptorless dehydrogenative coupling of ethylene glycol and primary alcohols. Angew Chem Int Ed 62:e202218329. https://doi.org/10.1002/anie.202218329
104. Ghosh AK, Brindisi M (2020) Urea derivatives in modern drug discovery and medicinal chemistry. J Med Chem 63:2751–2788. https://doi.org/10.1021/acs.jmedchem.9b01541
105. Gallou I, Eriksson M, Zeng X, Senanayake C, Farina V (2005) Practical synthesis of unsymmetrical ureas from isopropenyl carbamates. J Org Chem 70:6960–6963
106. Regan J, Breitfelder S, Cirillo P, Gilmore T, Graham AG, Hickey E, Klaus B, Madwed J, Moriak M, Moss N, Pargellis C, Pav S, Proto A, Swinamer A, Tong L, Torcellini C (2002) Pyrazole urea-based inhibitors of p38 MAP kinase: from lead compound to clinical candidate. J Med Chem 45:2994–3008. https://doi.org/10.1021/jm020057r
107. Melnikov NN (1971) Derivatives of urea and thiourea. In: Melnikov NN, Gunther FA, Gunther JD (eds) Chemistry of pesticides. Springer, New York, pp 225–239
108. Morais S, Correia M, Domingues V, Delerue-Matos C (2011) Pesticides – strategies for pesticides analysis. In: Stoytcheva M (ed) Urea pesticides. InTech, Rijeka, pp 241–262
109. Bremner JM, Krogmeier MJ (1988) Elimination of the adverse effects of urea fertilizer on seed germination, seedling growth, and early plant growth in soil. Proc Natl Acad Sci 85:4601–4604. https://doi.org/10.1073/pnas.85.13.4601
110. Bremner JM (1995) Recent research on problems in the use of urea as a nitrogen fertilizer. Fertil Res 42:321–329. https://doi.org/10.1111/j.1475-2743.1990.tb00804.x
111. Kaavessina M, Distantina S, Shohih EN (2021) A slow-release fertilizer of urea prepared via melt blending with degradable poly(Lactic acid): formulation and release mechanisms. Polymers 13:1856. https://doi.org/10.3390/polym13111856
112. Patel MJ, Tandel RC (2021) Synthesis of reactive dyes by the introduction of phenyl urea derivatives into the triazine ring and their application on different fibers. Mater Today Proc 46:6459–6464. https://doi.org/10.1016/j.matpr.2021.03.575
113. Christensen G (1980) Analysis of functional groups in amino resins. Prog Org Coatings 8:211–239. https://doi.org/10.1016/0300-9440(80)80017-8
114. Dehabadi VA, Buschmann HJ, Gutmann JS (2013) Durable press finishing of cotton fabrics: An overview. Text Res J 83:1974–1995. https://doi.org/10.1177/0040517513483857
115. Smith AR (1960) A review of the chemistry of thermosetting resins and related compounds and their application to textiles. Color Technol 77:416–434
116. Bigi F, Maggi R, Sartori G (2000) Selected syntheses of ureas through phosgene substitutes. Green Chem 2:140–148. https://doi.org/10.1039/b002127j
117. Mane M, Balaskar R, Gavade S, Pabrekar P, Mane D (2013) An efficient and greener protocol towards synthesis of unsymmetrical N,N′-biphenyl urea. Arab J Chem 6:423–427. https://doi.org/10.1016/j.arabjc.2011.01.030
118. Hardison LS, Wright E, Pizon AF (2014) Phosgene exposure: a case of accidental industrial exposure. J Med Toxicol 10:51–56. https://doi.org/10.1007/s13181-013-0319-6

119. Kotachi S, Tsuji Y, Kondo T, Watanabe Y (1990) Ruthenium catalysed N,N′-diarylurea synthesis from N-aryl substituted formamides and aminoarenes. J Chem Soc Chem Commun 2:549–550. https://doi.org/10.1039/C39900000549
120. Kondo T, Kotachi S, Tsuji Y, Watanabe Y, Mitsudo T (1997) Novel ruthenium-complex-catalyzed synthesis of ureas from formamides and amines. Organometallics 16:2562–2570. https://doi.org/10.1021/om970201k
121. Langsted CR, Paulson SW, Bomann BH, Suhail S, Aguirre JA, Saumer EJ, Baclasky AR, Salmon KH, Law AC, Farmer RJ, Furchtenicht CJ, Stankowski DS, Johnson ML, Corcoran LG, Dolan CC, Carney MJ, Robertson NJ (2022) Isocyanate-free synthesis of ureas and polyureas via ruthenium catalyzed dehydrogenation of amines and formamides. J Appl Polym Sci 139:52088. https://doi.org/10.1002/app.52088
122. Krishnakumar V, Chatterjee B, Gunanathan C (2017) Ruthenium-catalyzed urea synthesis by N-H activation of amines. Inorg Chem 56:7278–7284. https://doi.org/10.1021/acs.inorgchem.7b00962
123. Olah GA, Ohannesian L, Arvanaghi M (1987) Formylating agents. Chem Rev 87:671–686. https://doi.org/10.1021/cr00080a001
124. Allen CL, Williams JMJ (2011) Metal-catalysed approaches to amide bond formation. Chem Soc Rev 40:3405–3415. https://doi.org/10.1039/c0cs00196a
125. Roode-Gutzmer QI, Kaiser D, Bertau M (2019) Renewable methanol synthesis. Chem Bio Eng Rev 6:209–236. https://doi.org/10.1002/cben.201900012
126. Choi G, Hong SH (2018) Selective monomethylation of amines with methanol as the C 1 source. Angew Chem Int Ed 57:6166–6170. https://doi.org/10.1002/anie.201801524
127. Guo J, Tang J, Xi H, Zhao S-Y, Liu W (2023) Manganese catalyzed urea and polyurea synthesis using methanol as C1 source. Chinese Chem Lett 34:107731. https://doi.org/10.1016/j.cclet.2022.08.011
128. Kim SH, Hong SH (2016) Ruthenium-catalyzed urea synthesis using methanol as the C1 source. Org Lett 18:212–215. https://doi.org/10.1021/acs.orglett.5b03328
129. Luque-Urrutia JA, Solà M, Milstein D, Poater A (2019) Mechanism of the manganese-pincer-catalyzed acceptorless dehydrogenative coupling of nitriles and alcohols. J Am Chem Soc 141:2398–2403. https://doi.org/10.1021/jacs.8b11308
130. Santana JS, Cardoso ES, Triboni ER, Politi MJ (2021) Polyureas versatile polymers for new academic and technological applications. Polymers 13. https://doi.org/10.3390/polym13244393
131. Howarth GA (2003) Polyurethanes, polyurethane dispersions and polyureas: past, present and future. Surf Coatings Int B Coatings Trans 86:111–118. https://doi.org/10.1007/BF02699621
132. Rocas P, Cusco C, Rocas J, Albericio F (2018) On the importance of polyurethane and polyurea nanosystems for future drug delivery. Curr Drug Deliv 15:37–43. https://doi.org/10.2174/1567201814666171019102537
133. (2023) Polyurea coatings market. https://www.marketsandmarkets.com/Market-Reports/polyurea-coatings-market-152676861.html?gclid=Cj0KCQjw8rT8BRCbARIsALWiOvRwzrAtBwMcCRV3eXvOYKIdHfFk5OpjWw5rp5oVMIpe50yRVEX5LHcaAkN0EALw_wcB. Accessed 17 May 2023
134. Babad H, Zeiler AG (1973) The chemistry of phosgene. Chem Rev 73:75–91. https://doi.org/10.1021/cr60281a005
135. Byrd HCM, McEwen CN (2000) The limitations of MALDI-TOF mass spectrometry in the analysis of wide polydisperse polymers. Anal Chem 72:4568–4576. https://doi.org/10.1021/ac0002745
136. Kawasaki H, Takeda Y, Arakawa R (2007) Mass spectrometric analysis for high molecular weight synthetic polymers using ultrasonic degradation and the mechanism of degradation. Anal Chem 79:4182–4187. https://doi.org/10.1021/ac062304v
137. Croda PriamineTM 1074. https://www.crodasmartmaterials.com/en-gb/product-finder/product/448-priamine_1_1074. Accessed 18 May 2023

138. Choi T, Fragiadakis D, Roland CM, Runt J (2012) Microstructure and segmental dynamics of polyurea under uniaxial deformation. Macromolecules 45:3581–3589. https://doi.org/10.1021/ma300128d
139. Zhang R, Huang W, Lyu P, Yan S, Wang X, Ju J (2022) Polyurea for blast and impact protection: a review. Polymers 14:2670. https://doi.org/10.3390/polym14132670
140. (2023) Mordor intelligence: polyethyleneimine market size & share analysis – growth trends & forecasts (2023–2028). https://www.mordorintelligence.com/industry-reports/polyethyleneimine-market. Accessed 24 May 2023
141. Mohamed AL, Hassabo AG (2021) Cellulosic fabric treated with hyperbranched polyethyleneimine derivatives for improving antibacterial, dyeing, pH and thermo-responsive performance. Int J Biol Macromol 170:479–489. https://doi.org/10.1016/j.ijbiomac.2020.12.198
142. Yasin A, Ren Y, Li J, Sheng Y, Cao C, Zhang K (2022) Advances in hyaluronic acid for biomedical applications. Front Bioeng Biotechnol 10:1–12. https://doi.org/10.3389/fbioe.2022.910290
143. Shirmohammadli Y, Moradpour P, Abdulkhani A, Efhamisisi D, Pizzi A (2019) Water resistance improvement by polyethyleneimine of tannin-furfuryl alcohol adhesives. Int Wood Prod J 10:16–21. https://doi.org/10.1080/20426445.2019.1600814
144. Mubeen S, Asif A, Ahmad A, Khan M (2023) Recent advancements in sodium alginate-based nanogels. In: Sodium alginate-based nanomaterials for waste water treatment. Elsevier, Amsterdam, pp 225–233
145. Gupta S, Sharma S, Sharma P, Nadda AK, Raizada P, Singh P (2023) CO2-philic adsorbents: an overview. In: CO2-philic polymers, nanocomposites and chemical solvents. Elsevier, Amsterdam, pp 1–15
146. Chen Z, Lv Z, Sun Y, Chi Z, Qing G (2020) Recent advancements in polyethyleneimine-based materials and their biomedical, biotechnology, and biomaterial applications. J Mater Chem B 8:2951–2973. https://doi.org/10.1039/c9tb02271f
147. Collection E, Cells OS, Anefnaf I, Aazou S, Schmerber G, Refki S, Zimmermann N, Sekkat Z (2020) Polyethylenimine-ethoxylated interfacial layer for efficient electron collection in SnO2-based inverted organic solar cells. Crystals 10:731
148. Ohisa S, Kato T, Takahashi T, Suzuki M, Hayashi Y, Koganezawa T, McNeill CR, Chiba T, Pu YJ, Kido J (2018) Conjugated polyelectrolyte blend with polyethyleneimine ethoxylated for thickness-insensitive electron injection layers in organic light-emitting devices. ACS Appl Mater Interfaces 10:17318–17326. https://doi.org/10.1021/acsami.8b00752
149. Gleede T, Reisman L, Rieger E, Mbarushimana PC, Rupar PA, Wurm FR (2019) Aziridines and azetidines: building blocks for polyamines by anionic and cationic ring-opening polymerization. Polym Chem 10:3257–3283. https://doi.org/10.1039/C9PY00278B
150. Morimoto Y, Shimasaki Y, Tsuneki H, Yamamoto K (1994) Process for producing aziridines. EP0305543B1
151. Batchelor SN, Tucker I, Petkov JT, Penfold J, Thomas RK (2014) Sodium dodecyl sulfate-ethoxylated polyethylenimine adsorption at the air-water interface: how the nature of ethoxylation affects the pattern of adsorption. Langmuir 30:9761–9769. https://doi.org/10.1021/la502848a
152. Strautmann J, Schaub T, Hüffer S, Paciello R (2014) Increasing the molar mass of polyalkylenepolyamines by homogeneously catalysed alcohol amination. US 2014/0309460A1
153. Ebert S, Schaub T, Strautmann J, Hüffer S (2014) Alkoxylated polyalkylenepolyamines. US 2014/0288265A1
154. Brodie CN, Owen AE, Kolb JS, Buehl M, Kumar A (2023) Synthesis of polyethyleneimines from the manganese catalyzed coupling of ethylene glycol and ethylenediamine. Angew Chem Int Ed. https://doi.org/10.1002/anie.202306655

Top Organomet Chem (2024) 73: 257–306
https://doi.org/10.1007/3418_2023_111

Published online: 14 November 2023

A Mechanistic Analysis of Dehydrogenation Reactions with First-Row Transition Metal Complexes

Priyanka Chakraborty, Subhankar Pradhan, and Basker Sundararaju

Contents

Abstract Over the past decade, the dominance of first-row transition metals in catalytic dehydrogenation reactions has outpaced traditional precious metal-catalyzed processes. This progress has been made possible by the implementation of bifunctional catalysts, where both the metal and ligand actively contribute to the reaction's outcome. Insights into the behavior of first-row transition metals in catalytic dehydrogenations are being continually upgraded through a combination

P. Chakraborty, S. Pradhan, and B. Sundararaju (✉)
Department of Chemistry, Indian Institute of Technology Kanpur, Kanpur, Uttar Pradesh, India
e-mail: basker@iitk.ac.in

of experimental and computational studies, thus, greatly advancing our understanding of the underlying reaction mechanisms. In this chapter, we aim to offer a comprehensive perspective on the evolution of rational catalyst design with 3d metals in dehydrogenation reactions, specifically focusing on their preferred mechanistic pathways, thus laying the conceptual groundwork for future developments.

Keywords Chromium · Cobalt · Dehydrogenation · Iron · Manganese · Mechanism · Nickel · Zinc

1 Introduction

1.1 3d Metal-Based Dehydrogenation Reactions: A General Overview

The chemistry of first-row transition metal compounds is often intriguing as it may involve variable oxidation states, one-electron changes, paramagnetism, weak metal-carbon bonds, and radical intermediates, which in turn pose unique opportunities and challenges for the rational design of well-defined catalytic systems [1]. Mechanistic understanding of the catalytic cycle is crucial in finding novel reactivity aided by conceptual catalyst design for future developments. It is essential to consider here that the intrinsic properties and the coordination geometry of the 3d transition metal can significantly influence the catalytic processes [2]. In general, transition metals in different groups have a rich diversity in oxidation states, electronic structures, and atomic radii [3]. These differences have an important impact on the geometry, reaction mechanism and catalytic activity [4].

Over the years, transition metal-catalyzed dehydrogenation reactions have captivated researchers with their continuous evolution and intrigue. Traditionally, it was believed a “classical” metal hydride formation through β-hydride elimination was the only potent catalytic intermediate in alcohol dehydrogenation. But, especially with first-row transition metals, this is not always the case. Sometimes, it is not favorable, and even when it happens, these first-row metal hydrides can be tricky due to their low stability. This is because 3d metal complexes have weaker M-H bonds than their 4d and 5d counterparts [5], adding complexity. Hence, the catalytic pathway may often involve other intermediates in equilibria facilitated by non-innocent ligands. The ancillary ligands in any 3d-metal catalyst have much more to offer than just provide an electronic environment [6–9]. Traditional organometallic catalysts operate via single-site mechanisms in which all key transformations occur exclusively at the metal center and the ligands act as mere spectators. However, with the advent of redox active ligands [8] or structurally responsive ligand systems [9], the reactivity is no longer restricted to the metal center alone, and the concept of metal-ligand cooperativity (MLC) came into light. While the mechanistic pathways of 3d-metal complexes in (de)hydrogenation catalysis is not entirely mapped out, unlike their noble metal counterparts, reasonable progress has been made. This perspective analyses successes in this area from a mechanistic point of view.

1.2 Scope of This Chapter

The primary purpose of this chapter is to highlight the mechanistic landscape of non-precious transition metal complexes with respect to dehydrogenation reactions and to showcase the opportunities and challenges that lie therein. The impact of the substrate-catalyst interaction mode, metal-ligand cooperativity and their consequences will be considered. With focus on underlying concepts, this chapter is essentially categorized into dehydrogenation pathways for each metal, followed by some representative prominent examples of a broad diversity of ligand types and promoted reaction pathways. A selection of examples will be drawn from stoichiometric reactions that are relevant to the catalytic pathway and the connection between ligand dynamics and reaction intermediates will be looked into.

1.3 Substrate Interaction with Catalyst in Dehydrogenation Reactions

In non-precious metal-catalyzed dehydrogenation, alcohols are extensively investigated, while substrates like formic acid or heterocycles have received more limited attention. The activation of the substrate by the metal complex can involve two possible pathways: outer-sphere or inner-sphere. In the outer-sphere pathway, for instance with alcohols, the proton and C-H atom transfer to the catalyst occurs simultaneously via a four-membered transition state. Alternatively, a stepwise process can occur, involving separate proton and hydride transfers, or proton delivery to an external base (Scheme 1). In the inner-sphere mechanism, it is a stepwise process where a ligand initially withdraws a proton to form a metal alkoxide intermediate, which then transforms into a metal hydride. Variants with external base assistance also exist. Thus, a myriad of different mechanisms of substrate interaction is available in this area of catalysis [10]. A dehydrogenation catalyst's activity (in other words, TOF) hinges on two crucial structures: the resting state (TOF-determining intermediate, TDI) and the transition state that governs the turnover frequency (TDTS) [11]. There have been multiple research studies in this realm throughout the past decade that challenge the traditional notion of concerted reactions and investigate the actual stepwise nature of key transition states in a much more complicated mechanistic pool [12].

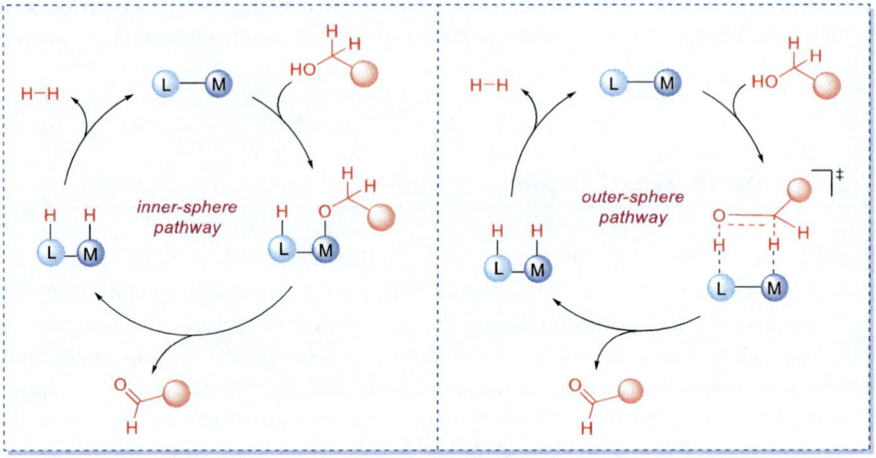

Scheme 1 Mode of alcohol activation either by an inner-sphere or an outer-sphere pathway

1.4 Cooperative Effects in 3d Transition Metals

The bifunctional pathway is possible for catalysts that possess a Lewis acidic metal center and a Brønsted basic cooperative site in a ligand. (De)hydrogenation reactions by metal-ligand cooperativity (MLC) typically involve polar substrates (e.g., alcohols, formic acid) that utilize the combination of proton and hydride transfers [13, 14]. Most explored MLC systems are based on N-donor ligands or other donor atoms (carbon, oxygen, sulfur, and boron). This characteristic cooperativity plays a vital role in the enhancement of the performance of first-row transition metal catalysts, where both the metal and ligand work synergistically in the bond activation process [15]. Multifunctional ligands of such kinds can be broadly divided into three categories based on their mode of action: (a) cooperativity via C-H/N-H (de)-protonation (b) hemilability via coordination/decoordination, and (c) redox activity (Scheme 2). First-row transition metals are also capable of existing in multiple spin-states and favorable electronic structures, which can effectively aid catalysis. Thus, not only this is a more attractive option, but also a promising ground for novel reactivities and selectivities [16].

Coined by Jeffrey and Rauchfuss in 1979 [17], the term hemilability in a ligand refers to a hybrid polydentate ligand where one donor atom can reversibly dissociate in solution to open up a coordination site, while maintaining at least one group firmly bound to the metal center to stabilize reactive catalytic intermediates. The prime advantage in this strategy is that the labile donor group can re-associate rapidly to the metal center when needed as compared to dissociation of a monodentate ligand. Hemilability of a bidentate chelate leaves the ligand vulnerable to further dissociation, but in case of a pincer the hemi-dissociated state is still firmly held by the chelate effect and is thus expected to be much more robust [18].

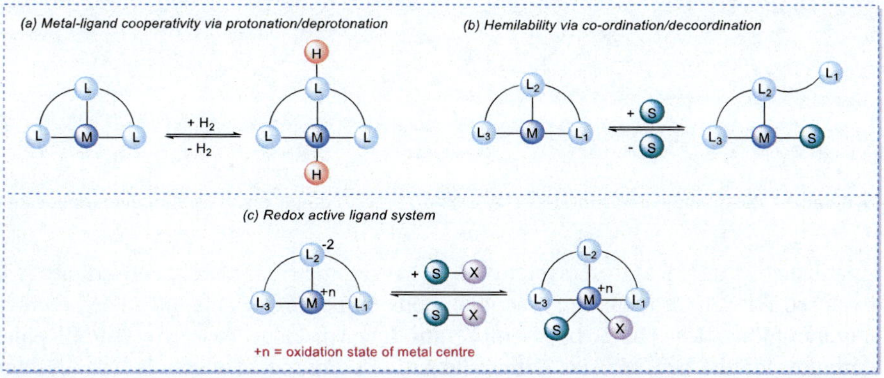

Scheme 2 Modes of cooperative effects in transition metal catalysis involving multifunctional ligands

Redox non-innocent ligand systems are characterized by their ability to facilitate distinct electronic events directly involved in catalytic cycles or stoichiometric transformations [19, 20]. These redox-active ligands play an active role in catalytic reactions, as they can either donate or accept single electrons within their framework, ensuring that the metal center remains in a constant oxidation state throughout the entire process.

2 Case Studies on the Mechanism of Manganese-Catalyzed Dehydrogenation Reactions

As the third most abundant transition metal, manganese (Mn) has proved its mettle to be one of the best contenders for replacement of noble metals for catalyst development. Incidentally, Mn-pincer complexes in dehydrogenation reactions that are a rage (intensively investigated) now, were barely explored up until 2016. Active manganese complexes reported prior to that are mainly based on salox/salen ligands or redox-active bis(imino)pyridine (PDI) ligands. Such ligands act as electron reservoirs, enabling the metal center to maintain common oxidation states and undergo two-electron redox processes to equate catalytic performance with noble-metal catalysis [16, 21, 22].

2.1 Bifunctional Manganese(I) Complexes

An in-depth literature survey shows that Mn(I) complexes dominate the contributions in the field of dehydrogenation reactions. Among them, alcohol dehydrogenation is one of the largest-volume applications till date (2016–2023). While many

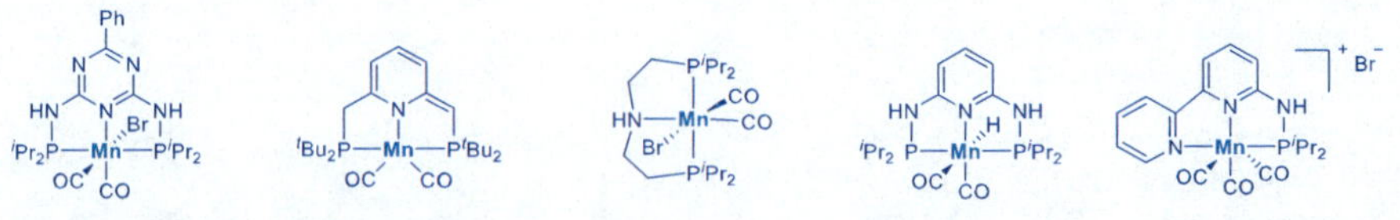

Scheme 3 Phosphine-based Mn(I)- (PNP or NNP) pincer complexes in alcohol dehydrogenations

established Mn(I) catalyst systems with a symmetric ligand scaffold are well explored for dehydrogenation reactions, nonsymmetric variants have been comparatively studied less. In 2016, Kempe introduced triazine base Mn(I)-PNP pincer complex for the acceptorless dehydrogenative coupling of alcohols and amidines [23]. Simultaneously, the groups of Milstein and Beller also reported Mn(I) pincer complexes for alcohol dehydrogenation [24, 25]. Since then, phosphine-based Mn (I)- (PNP or NNP) pincer complexes have gone through rapid development as an emerging field in alcohol dehydrogenation as summarized in (Scheme 3). In the reaction pathway these Mn(I) complexes initially leads to the formation of stable alkoxide-Mn species which is the resting state of the catalytic cycle. A couple of pathways for converting alkoxide-Mn intermediates into catalytically active Mn-hydride species have been proposed. The most widely accepted among them are (a) concerted proton/hydride transfer (outer sphere) and (b) stepwise activation (inner-sphere β-H elimination mechanisms) assisted by an internal base.

In the concerted pathway, the key species for Mn(I) pincer complexes is the generation of a Mn(I)-hydride via a deprotonated PNP ligand in the presence of base. Upon comparison of Kempe's **[Mn]-1** [23] and Kirchner's **[Mn]-4** [26], both complexes feature almost the same ligand framework with the important difference that the former contains a halide ligand, and the latter in its hydride form. This detail makes the two complexes follow different pathway for alcohol dehydrogenation. In the first case, complex **[Mn]-1a** is generated by salt elimination which in the presence of an alcohol undergoes concerted activation to form alkoxy complex **[Mn]-1b** (Scheme 4a). The alkoxy complex is then proposed to undergo a β-hydride elimination liberating the aldehyde and the hydride complex **[Mn]-1c**. Proton shuttle liberates H_2 and thereby regenerating **[Mn]-1a** [27–33]. In the latter case, the precatalyst **[Mn]-4** undergoes a proton transfer to the hydride forming a dihydrogen complex **[Mn]-4a** [26]. H_2 liberation generates the active catalytic species **[Mn]-4b** which subsequently dehydrogenates the incoming alcohol in a concerted approach (Scheme 4b). It is noteworthy to mention here that in a separate report by Kempe, complex **[Mn]-1** showcases a cation (K or Na) dependent switchable reactivity between borrowing hydrogen or hydrogen-autotransfer reaction under identical catalytic conditions [28]. Mechanistic study by an in-depth DFT calculations showed direct impact of the potassium ion on lowering energy of the involved transition states thereby providing stability to promote borrowing hydrogen reactivity [34].

PNP-Mn(I) complex **[Mn]-3** as mentioned earlier was first introduced by Beller for dehydrogenation [15]. The precatalyst **[Mn]-3** undergoes an initial base mediated

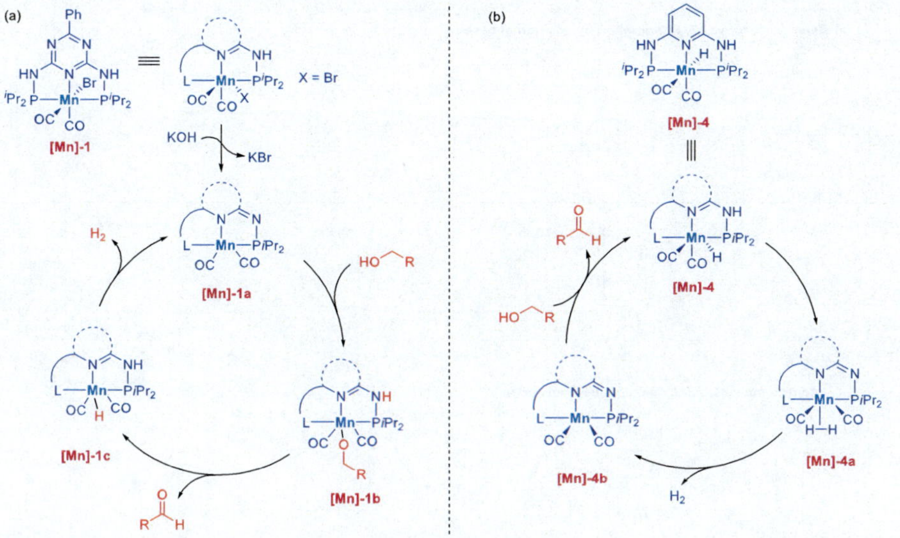

Scheme 4 Comparative catalytic pathway of alcohol dehydrogenation between phosphine-based Mn(I)-PNP pincer complexes (**a**) [Mn]-1 and (**b**) [Mn]-4

debromination to form a Mn-amido complex **[Mn]-3a** which is the key active catalytic intermediate in the mechanistic pathway (Scheme 5a) [15, 35–39]. This amido complex activates an external alcohol to give **[Mn]-3b.** At this stage, owing to the lack of a vacant site in the coordination sphere, Beller proposed the formation of the Mn-hydride **[Mn]-3c** to proceed by internal base assisted β-H abstraction via an amidate transition state **[Mn]-3b-TS** followed by protonation. Liberation of H_2 regenerates **[Mn]-3a**. This amido pathway, as originally proposed, has later been proved by other groups on multiple occasions [40–42]. However, there is still no direct evidence for the involvement β-H transfer to an internal base. All the reports in the subsequent years, have suggested the formation of **[Mn]-3c** directly from **[Mn]-3b** [40, 43–46].

In 2017, Gauvin used DFT calculations to uncover the true pathway of alcohol dehydrogenation in a Mn-amido-mediated mechanism involving **[Mn]-3** [47]. As proposed, the alkoxy species formed after alcohol activation with amido complex **[Mn]-3a** is energetically stable (considered resting state). In this context, Gauvin found that hydrogen release from the amino-hydride intermediate was found to be more energetically determining. Instead, a pathway involving an outer-sphere dehydrogenation of the alcohol into aldehyde via **[Mn]-3b'** was found to be less energetically demanding (Scheme 5b). Thus, the Mn-amido pathway for alcohol dehydrogenation, particularly in case of **[Mn]-3**, does not involve the formation of an "Mn-alkoxy intermediate." This behavior was attributed to trans effect of the strongly π-accepting carbonyl groups.

In 2017, Milstein introduced a Mn(I)-PNP **[Mn]-6** complex featuring a benzylic backbone structure that enables activation of the benzylic C–H bond within the

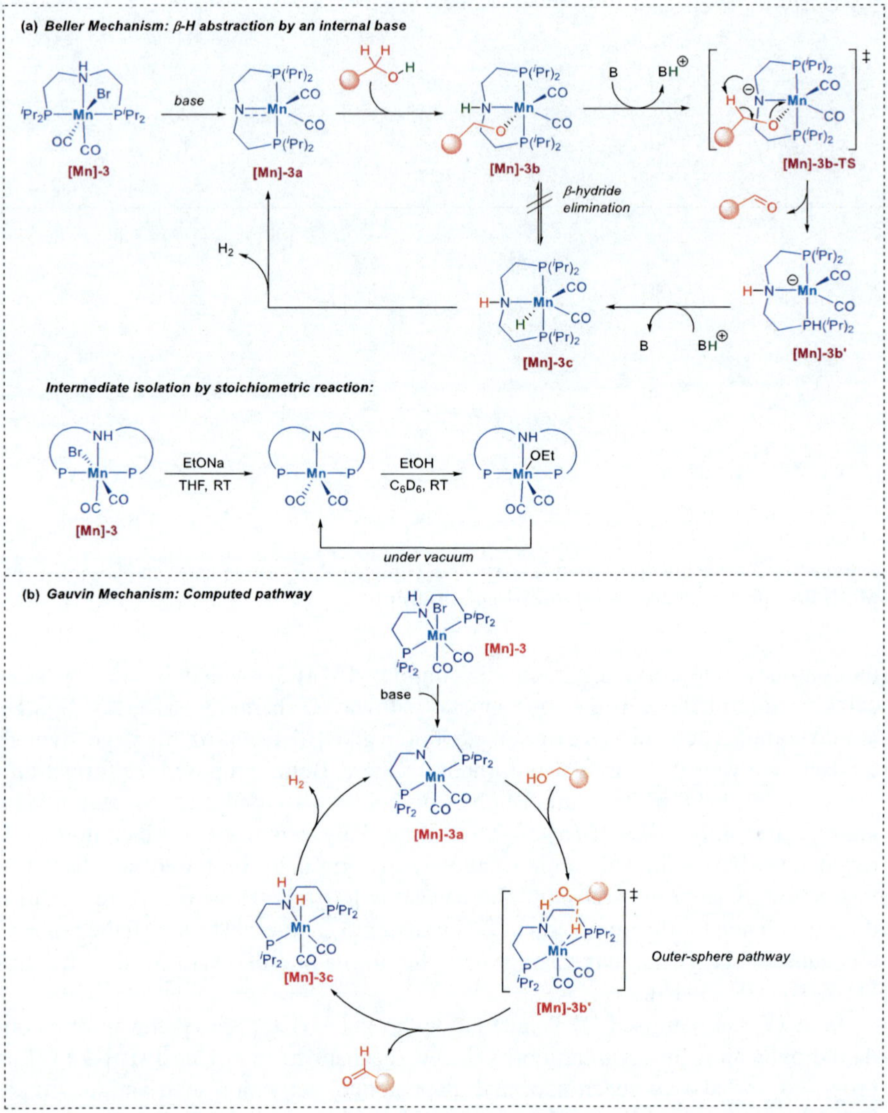

Scheme 5 (**a**) Mechanism of internal base assisted β-H abstraction pathway for phosphine-based Mn(I)-PNP pincer complex [Mn]-3. (**b**) Computed mechanistic pathway of [Mn]-3 for alcohol dehydrogenation

ligand framework. **[Mn]-6** dehydrogenates methanol in a metal-amido pathway but via an inner sphere mechanism. It was found that **[Mn]-6** favors the formation of a thermodynamically stable C-metalated complex **[Mn]-6′** attained via intramolecular C–H bond activation (Scheme 6) [41]. Under stoichiometric reaction conditions, the catalytic intermediates **[Mn]-6** and **[Mn]-6'** were isolated and structurally

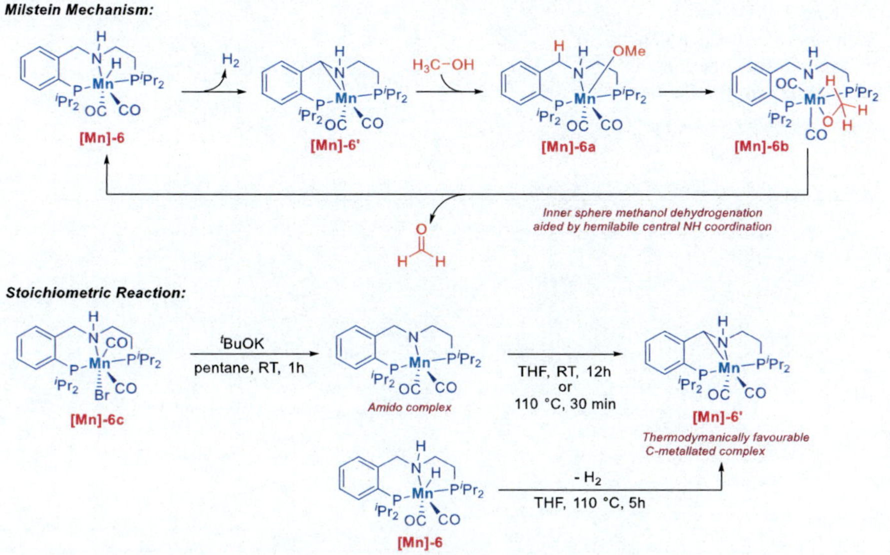

Scheme 6 Mechanism of Mn(I)-PNP benzylic pincer complex [Mn]-6 for alcohol dehydrogenations

characterized High level DFT calculations reveal that a polar molecule (water or an alcohol) has prominent role in facilitating the proton transfer for the interconversion between the kinetic complex **[Mn]-6** and its thermodynamic isomer **[Mn]-6′** [48].

In the same year, Milstein introduced Mn(I)-PNN complex **[Mn]-7** wherein he proposed an inner sphere dehydrogenation pathway [49, 50]. Given the ligand structure, it can be expected to exhibit MLC via aromatization/dearomatization as known in literature [51, 52]. Nonetheless, according to stoichiometric NMR experiments, the deprotonation took place on the NH site rather than at CH_2 position of the ligand arm, to form an amido complex **[Mn]-7a**, which after alcohol activation gave the intermediate **[Mn]-7b**, undergoing β-hydride elimination of the bound alkoxy group likely via amine "arm" opening intermediate **[Mn]-7c** (Scheme 7). Thereby, making this pathway different from those of PNP systems which do not bear a potential hemilabile pincer amine arm, wherein an outersphere mechanism has been proposed to be operative (e.g., Mn(I)-PNP complex [Mn]-3). It is noteworthy to mention here that in complex **[Mn]-6**, despite having a PNP ligand system, the alcohol dehydrogenation is proposed by an inner sphere mechanism aided by hemilability of central NH coordination site (intermediate **[Mn]-6b**, Scheme Mn-6) [41].

In another interesting variant of Mn(I) complexes, Milstein demonstrated the synthesis of acridine-based Mn-PNP complex **[Mn]-8** [53]. Reaction with an excess of sodium borohydride leads to formation of a novel azaborametallacyclic complex **[Mn]-8a** which was found to be active for dehydrogenation reactions (Scheme 8a). To isolate the possible active organometallic amido intermediate, without the boron-

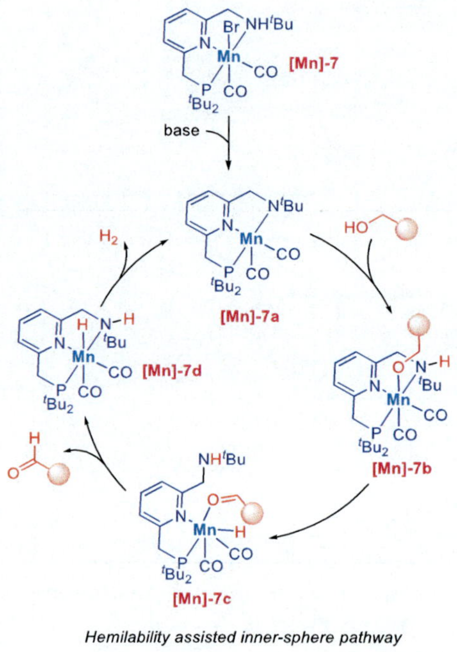

Scheme 7 Mechanistic pathway of a Mn(I)-PNN complex [Mn]-7 bearing a hemilabile arm for alcohol dehydrogenation

bridged moiety, stoichiometric experiments conducted in the presence of NH_3 gas as external ligand, yielded valuable insight. This experiment indicated towards the fact external alcohol can be the driving force for generation of a five-coordinated amido active species **[Mn]-8c** from either **[Mn]-8a** or **[Mn]-8b**. Owing to the less basic nature of the amido nitrogen of the ligand, alkoxide-assisted alcohol dehydrogenation via a six-membered transition state and dihydrogen liberation mechanism is proposed (Scheme 8b).

Speaking of phosphine free Mn(I)-(NNN or NNS) catalytic systems (Scheme 9), initially Maji described a Mn complex derived from hydrazone type NNN pincer ligand [54, 55]. Deprotonation in presence of base generates the active catalyst **[Mn]-9** which undergoes alcohol activation in an inner-sphere concerted pathway (Scheme 10a). The prime advantage is that such nitrogen donor-based ligands [56, 57] are bench stable, and in situ catalyst formation provides opportunity rapid optimization without any loss in activity. Srimani introduced NNS pincer Mn (I) complex **[Mn]-11** which participates in dehydrogenation involving an amido pathway (via intermediate **[Mn]-11a**) which is discussed in detail in the previous examples [58–61]. DFT calculations carried out by Srivastava and Srimani revealed that the amide complex **[Mn]-11a** dehydrogenates alcohol by forming a hydrogen bonded alcohol adduct **[Mn]-11b** followed by an outer sphere hydride and proton abstraction (Scheme 10b) [62].

The dehydrogenation of formic acid is highly significant due to its abundant hydrogen content and its ability to provide hydrogen storage with zero carbon emissions. By catalytically decomposing formic acid, an ideal mixture of carbon

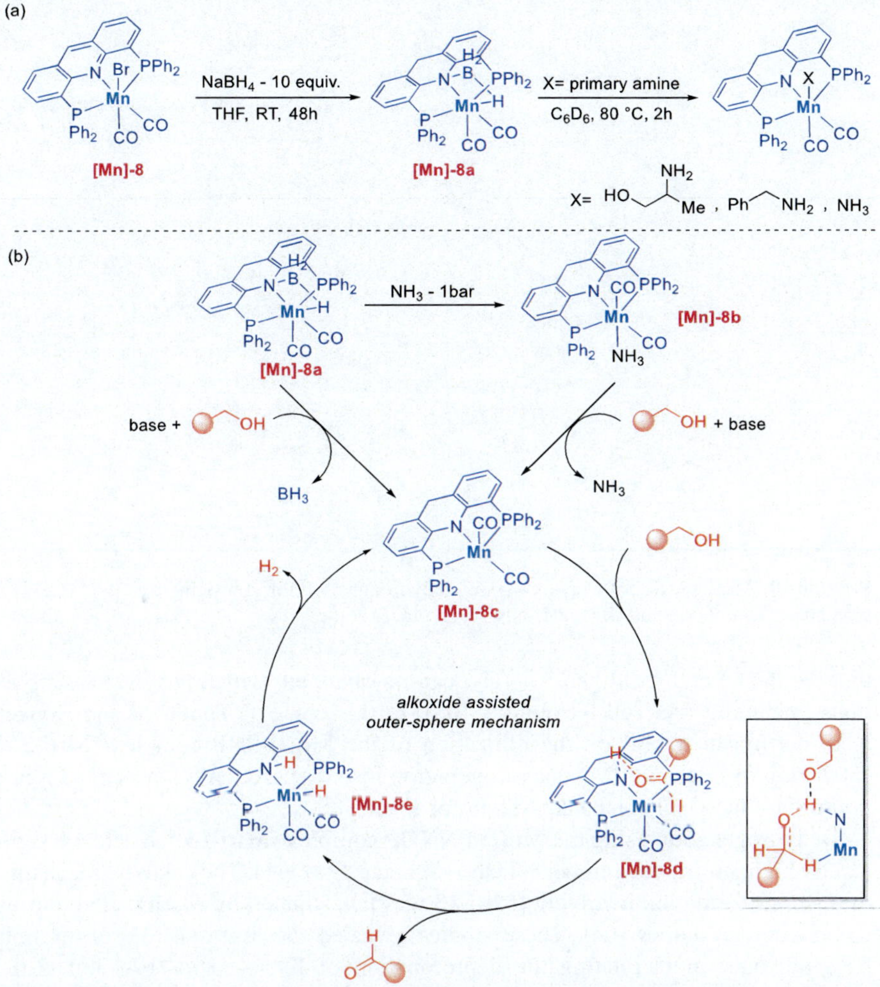

Scheme 8 (**a**) Formation of a novel azaborametallacyclic complex [Mn]-8a and its reactivity (**b**) Proposed mechanistic pathway of acridine-based Mn(I)-PNP complex [Mn]-8 for alcohol dehydrogenation

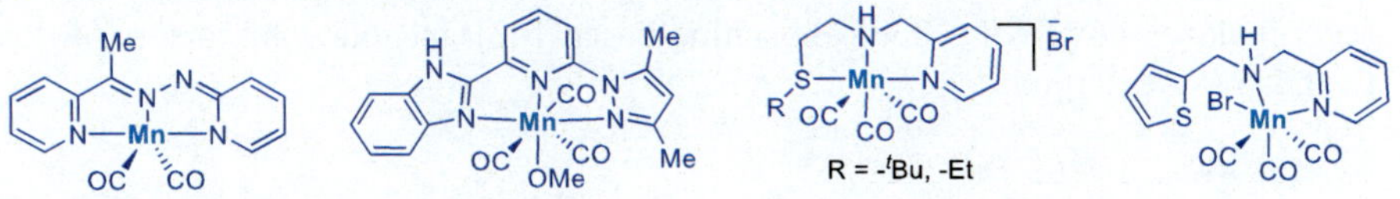

Maji 2018, [Mn]-9 *Yu 2018*, [Mn]-10 *Srimani 2018*, [Mn]-11 *Maji 2020*, [Mn]-12

Scheme 9 Phosphine free Mn(I)- (NNN or NNS) pincer complexes in alcohol dehydrogenations

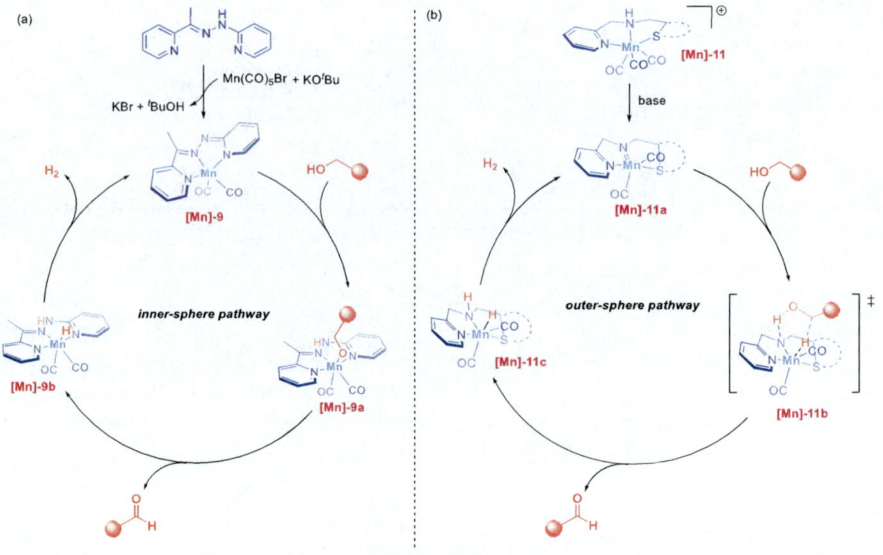

Scheme 10 Mechanistic pathway of alcohol dehydrogenation for phosphine-free (**a**) Mn(I)-NNN type pincer complexes and (**b**) Mn(I)-NNS type pincer complexes

dioxide (CO_2) and hydrogen gas (H_2) can be obtained, which has diverse applications, including fuel cell technology [63]. In this context, Tondreau and Boncella were the first to report on the utilization of the Mn(I)-PNP complex **[Mn]-3** for catalyzing formic acid (FA) dehydrogenation [64]. Subsequently, the same research group introduced an enhanced system for FA.

Dehydrogenation using the Mn(I)-PNNOP complex **[Mn]-13** can achieve significantly higher turnover numbers (TON) (Scheme 11a) [65]. The catalytic mechanism of this transformation involving **[Mn]-13** was investigated by Marino et al. through computational studies [66]. These studies revealed the essential role of an amine base not only in facilitating the deprotonation of formic acid (FA) but also in enabling the liberation of H_2, acting as a proton-donor species (Scheme 11b). Computational analysis indicated that the reaction proceeds through the formation of a formate complex **[Mn]-13a**, which subsequently releases CO_2 and forms the Mn (I)-hydride intermediate **[Mn]-13c**. The liberation of H_2 then takes place via an amine-assisted pathway. In the following years, additional reports on Mn (I) complexes employing both phosphine-based [67] and phosphine-free [68, 69] ligands have emerged.

2.2 Manganese(II/III) Complexes

In the context of a continuous quest for methods that can apply non-precious metals and simpler ligands, Kaur introduced the use of 2,2′:6′,2″-terpyridine (Terpy)

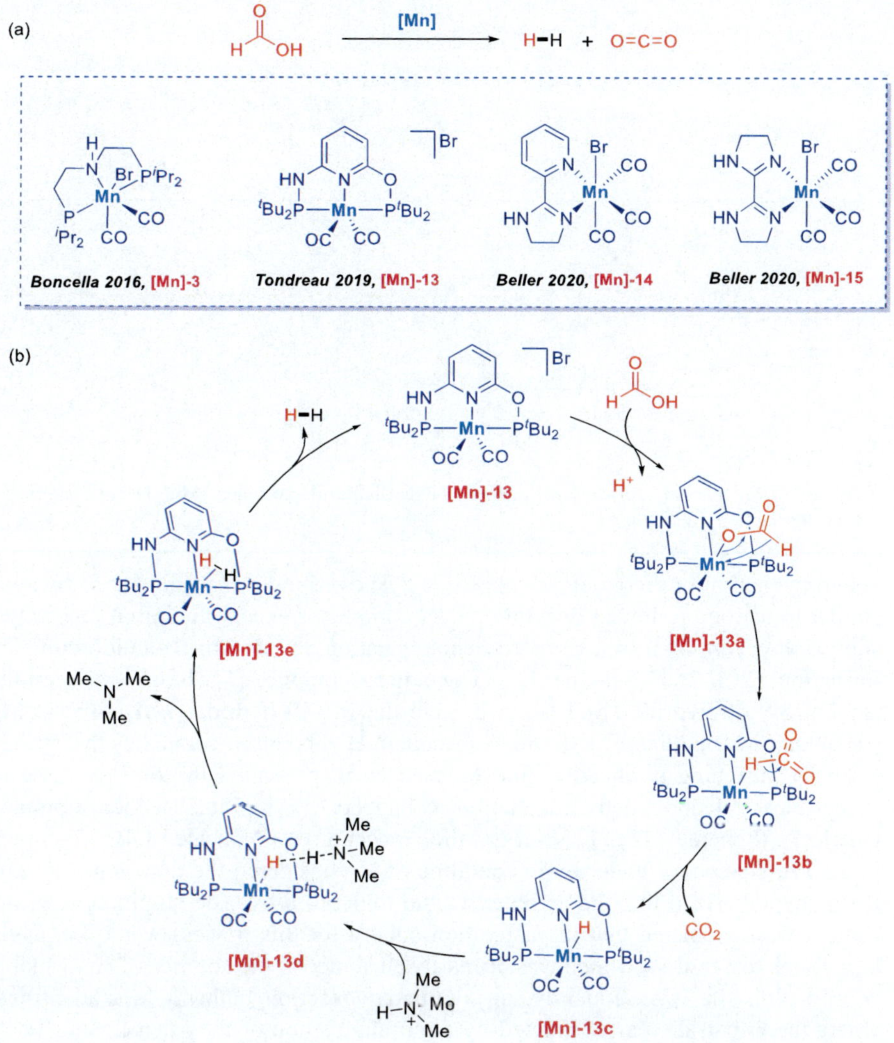

Scheme 11 (**a**) Mn(I) complexes employed for dehydrogenation of formic acid. (**b**) Proposed mechanistic pathway of Mn(I)-PNNOP complex [Mn]-13 for FA dehydrogenation

ligands with Mn(II) salts in dehydrogenative functionalization of alcohols [70]. The Mn-Terpy system was speculated to form a dinuclear complex, which can control the necessary oxidation state change during dehydrogenation. Independently, Banerjee [71] and Balaraman [72] have also developed and disclosed inexpensive manganese-catalytic systems, free of phosphine ligands, which effectively facilitate the dehydrogenation of alcohols. However, no experimental or theoretical investigations have been conducted to gain a deeper understanding of the underlying mechanism at play. Later, Balaraman et al. reported a NNN-Mn(II) pincer complex **[Mn]-16** for

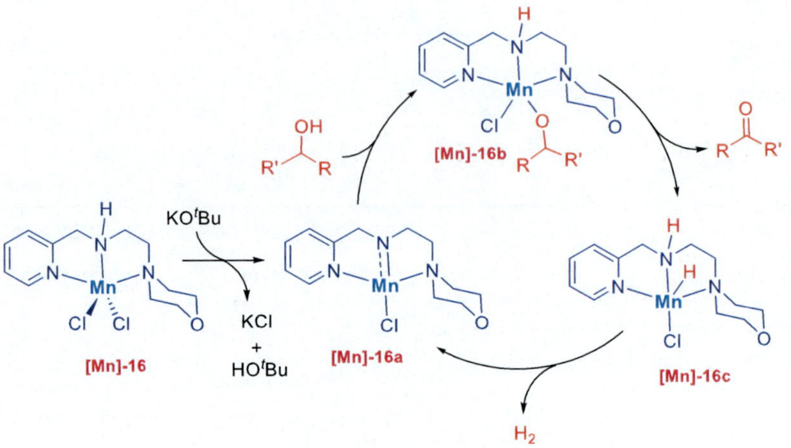

Scheme 12 Mechanistic pathway of a Mn(II)-NNN dichloride complex [Mn]-16 for dehydrogenation of secondary alcohols

dehydrogenation of secondary alcohols [57]. The proposed mechanistic pathway is similar to the one followed by Mn(I)-NNN complexes except the initial precatalyst activation occurs by a base assisted dehalogenation and an Mn(II)-amido complex formation **[Mn]-16a** (Scheme 12). The amido complex **[Mn]-16a** activates the alcohol by a β-hydride elimination to give the Mn(II)-hydride **[Mn]-16c** via Mn (II) alkoxy intermediate **[Mn]-16b**. Subsequent H_2 liberation completes the cycle.

In an intriguing publication, the Madsen et al. presented the first example of acceptorless alcohol dehydrogenation catalyzed by a Mn(III)-salen complex **[Mn]-17** (Scheme 13) [73]. For the mechanism, Mn(salen)OBn complex **[Mn]-17a** generation under basic conditions was considered the first step. Initially, it was hypothesized that the complex would undergo β-hydride elimination. However, it was calculated that the activation energy for this process was excessively high (37.9 kcal mol^{-1}), which was primarily attributed to the absence of an available coordination site. Consequently, an alternative reaction pathway was identified, where the β-hydride was abstracted by the imine carbon of the salen ligand (intermediate **[Mn]-17c**). Thus, the key active catalytic species **[Mn]-17d**, where one imine is hydrogenated to the amine and the other is reduced to an amide ligand formed. Subsequently, **[Mn]-17d** dehydrogenates another unit of alcohol by an outer-sphere hydrogen transfer pathway leading to the formation of a Mn(III) hydride **[Mn]-17i**. Liberation of hydrogen gas from the Mn(III) hydride intermediate completes the cycle. The same group has also reported the use of Mn(III) porphyrin chloride complexes in dehydrogenation of both primary and secondary alcohols [74].

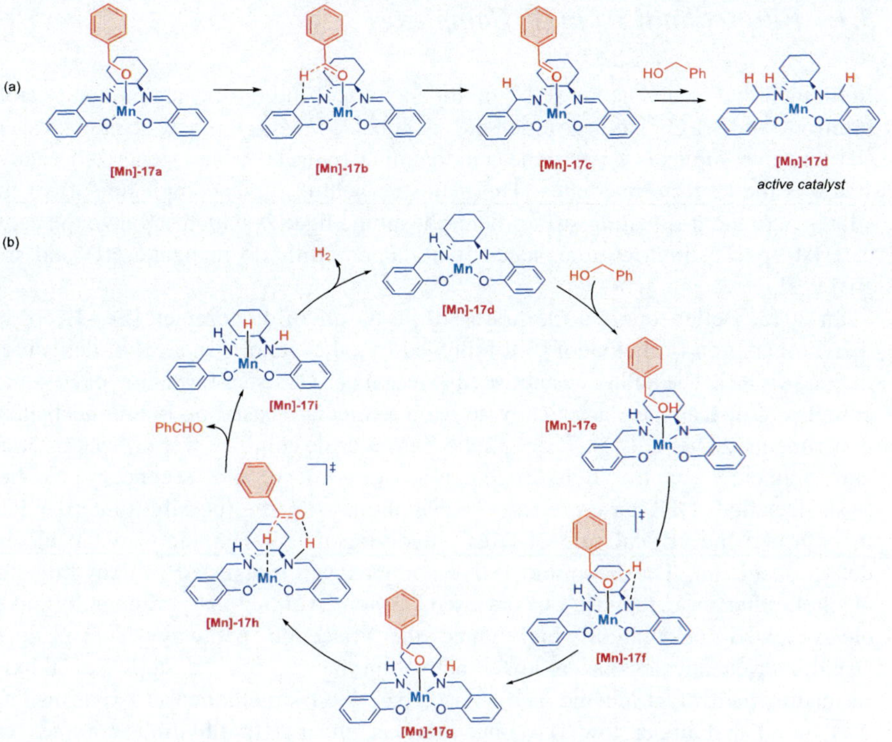

Scheme 13 Proposed mechanistic pathway of a Mn(III)-salen complex [Mn]-17 for dehydrogenation of alcohols

3 Case Studies on the Mechanism of Iron-Catalyzed Dehydrogenation Reactions

The abundance of iron (Fe) in earth's crust and it being an integral part of metabolic processes including non-toxic disposal pathways in biological systems has prompted a renaissance in the use of Fe-catalysts [75]. Notably, iron is hailed as the most versatile catalytic metal next to palladium in homogeneous catalysis as it finds uses in diverse areas of organic synthesis [76]. Nonetheless, its reactivity and selectivity currently exhibit comparable performance to state-of-the-art catalysts based on precious metals [77]. Iron-catalyzed synthetic transformations have demonstrated remarkable efficiency, making them a prominent subject of focus in contemporary scientific research. Contrary to their widespread applicability in other fields such as catalytic hydrogenation, the development of homogeneous Fe-catalytic systems for the dehydrogenation process has been comparatively limited [78].

3.1 Bifunctional Iron(II) Complexes

In bifunctional catalysis, the [P,N] multidentate scaffold has been continually at the forefront of MLC. Isoelectronic and isolobal to Mn(I) pincer complexes, Fe (II) pincer complexes have same amido/amine hydride in situ generated catalyst states as the key intermediates. The mainstay behind the accomplishments of this strategy are the σ-donating strong-field phosphine ligands which stabilize the active catalyst species in low spin states [particularly with d^6 manganese(I) and iron (II)] [79].

In 2013, Beller reported the first PNP Fe-borohydride complex **[Fe]-1a** for the dehydrogenation of methanol [80] followed by other reports in alcohol dehydrogenation [81, 82]. The authors established external base assisted formation of Fe-amido complex **[Fe]-1** as the catalytically active intermediate based on initial mechanistic experiments (Scheme 14a). The same Fe-borohydride complex was utilized by Jones and Schneider in the dehydrogenation of primary and secondary alcohols [83]. Detailed DFT-based energy calculations for the possible intermediates reconfirmed the critical role of MLC involving the N–H moiety in the alcohol dehydrogenation. The Fe-amido active complex was suggested to dehydrogenate alcohol either in a stepwise or concerted fashion. Although experimentally inconclusive, DFT calculations highlighted the concerted pathway to form a Fe (II) dihydride species has a lower activation barrier over a stepwise alkoxide formation pathway (Scheme 14b). Also, based on stoichiometric experiments it was found that under low H_2 concentrations, the Fe(II) dihydride complex can undergo decomposition to Fe(0) species via reductive elimination of H_2. Bernskoetter et al. subsequently reported an improved procedure for the direct use of catalytically active PNP Fe-amido complex **[Fe]-1** for methanol reformation [84].

In a separate study, Kirchner presented the application of a comparable Fe-complex containing his triazine-based PNP pincer ligand framework (Scheme 14c) [85, 86]. The precatalyst **[Fe]-2a** dehydrogenates alcohol in the presence of base and forms the active hydride complex **[Fe]-2** through an alkoxy intermediate **[Fe]-2b**. The active hydride complex **[Fe]-2** is regenerated by H_2 liberation via a dihydrogen complex **[Fe]-2d** initiated by another molecule of incoming alcohol.

Redox-noninnocent azoaromatic pincers featuring NNN donor atoms represent an intriguing category of Fe(II)-catalytic systems for alcohol dehydrogenations [87]. The iron(II) catalyst **[Fe]-3** was shown to undergo one-electron reduction on the ligand framework to form **[Fe]-3a** as the catalytically active species. Mechanistically the reaction proceeds by coordination of the deprotonated alcohol to the active catalyst **[Fe]-3a** followed by hydrogen atom abstraction by the azo framework from the α-carbon of the coordinated alcohol (Scheme 15). The generated O-coordinated ketyl radical anion **[Fe]-3c** is proposed to undergo rapid one-electron oxidation to afford the desired carbonyl compound. Intermediate **[Fe]-3d** activates another incoming molecule of alcohol forming the anionic Fe (I) complex **[Fe]-3e**. The Fe(I) intermediate is oxidized back to Fe(II) by aerobic oxidation leaving H_2O_2 as the by-product. On the basis of kinetic and spectroscopic

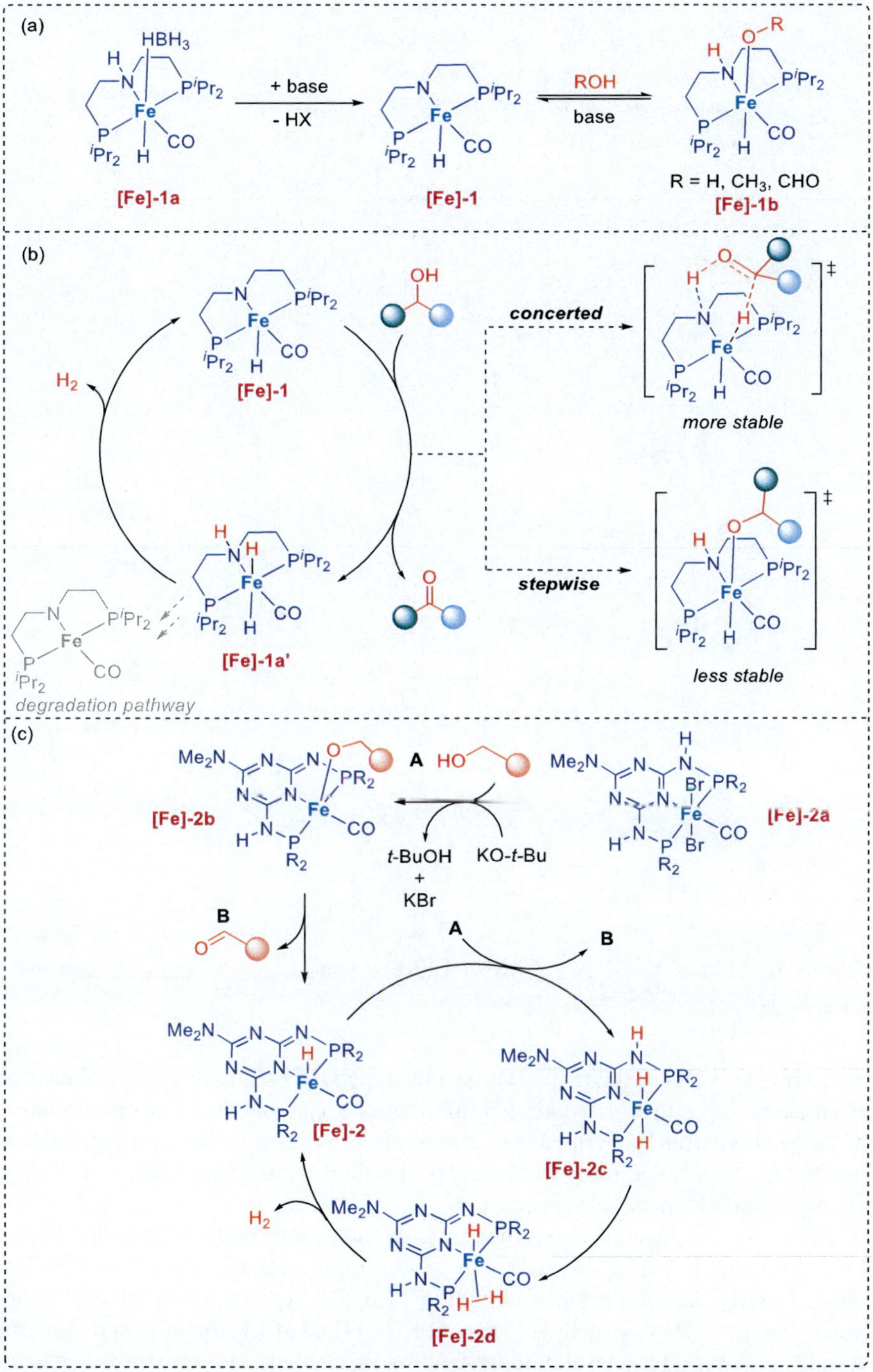

Scheme 14 (**a**) Fe-amido complex [Fe]-1 as the in situ catalytically active intermediate (**b**) General mechanistic pathway for PNP Fe(II)-pincer complexes for alcohol dehydrogenation. (**c**) Pathway for triazine-based Fe-PNP complex [Fe]-2a in alcohol dehydrogenation

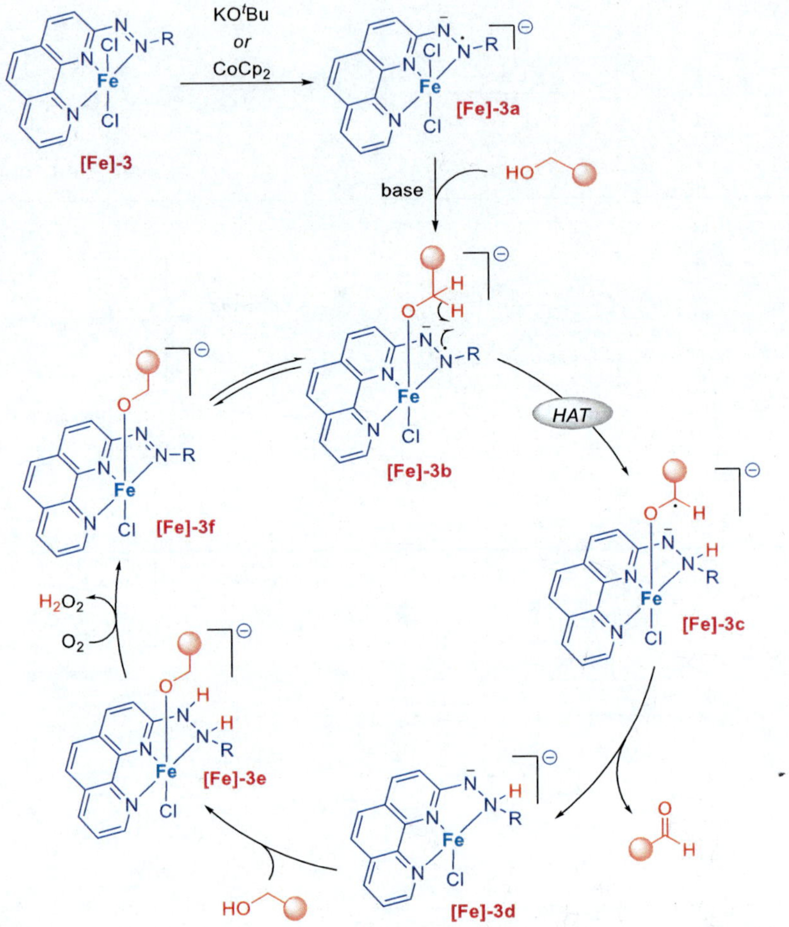

Scheme 15 Mechanistic pathway for NNN Fe(II)-azoaromatic pincers for dehydrogenation of alcohols via aerobic oxidation

data, this 2e−/2H+ oxidation reaction via stepwise HAT process is satisfactorily established. Nevertheless, additional information regarding the electronic structures of the putative iron(I) intermediates is necessary to obtain a conclusive understanding of the mechanistic framework at play. Further investigations and clarifications are needed to elucidate this aspect.

Beller et al. made a nice contribution to the field of formic acid dehydrogenation by developing an iron-catalyzed process with high catalytic efficiency [88]. They utilized Fe(II) salts in combination with PP_3 (tris[(2 diphenylphosphino)ethyl]phosphine) ligands. This opened new avenues for efficient hydrogen generation from formic acid and advanced the understanding of catalytic systems for this reaction [89]. Based on high pressure in situ NMR studies and DFT insights, the authors

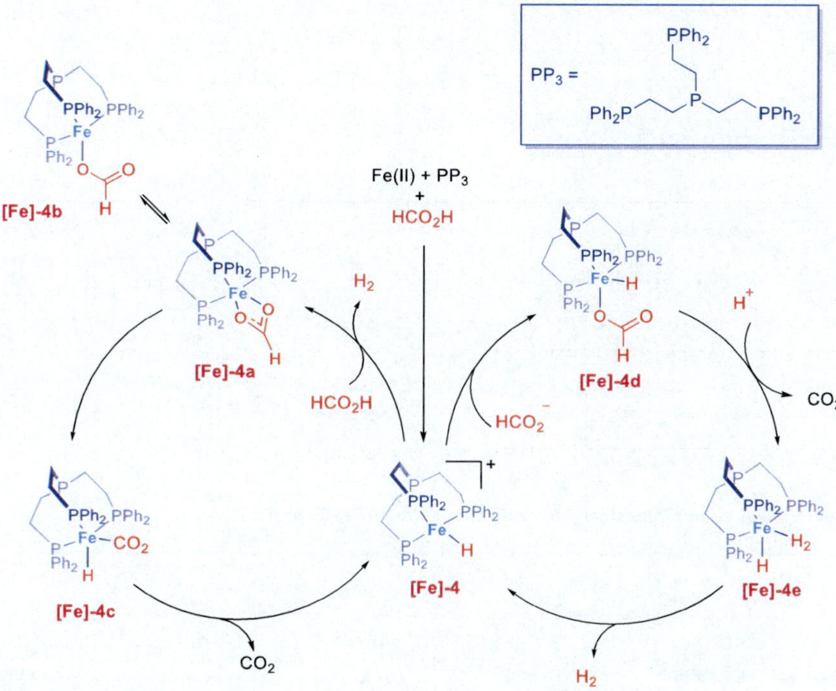

Scheme 16 Mechanistic pathway for Fe(II) and PP_3-catalyzed dehydrogenation of FA

proposed two possible catalytic routes as shown in (Scheme 16) both of which involves $[FeH(PP_3)]^+$ **[Fe]-4** as the catalytically active species. In first case, initial coordination of FA leads to a formate species **[Fe]-4a** along with the concerted liberation of H_2. Further β-hydride elimination followed by CO_2 elimination regenerates the active species **[Fe]-4**. Alternatively, initial coordination of FA can lead to hydrido formate species **[Fe]-4d** which was computationally found to be much more energetically stable than formate complex **[Fe]-4a**. Another β-hydride elimination accompanied by protonation, releases CO_2 to form species **[Fe]-4e**. Subsequent elimination of H_2 completes the cycle.

Use of Iron pincer catalysts for selective FA decomposition was first demonstrated by Milstein et al. by employing a *trans* dihydride Fe(II)-PNP complex **[Fe]-5** [90]. Following this, Kirchner and Gonsalvi also described the use of Fe(II)-PNP pincer hydrido carbonyl complex **[Fe]-6** based on the 2,6-diaminopyridine scaffold for decomposition of formic acid under similar catalytic conditions [91]. Mechanistically, both these reactions rely on the formation of η^1-O hydrido formate intermediate for efficient activity. In the first case, since the precatalyst **[Fe]-5** is a dihydride complex, there is a need for generation of vacant side for dehydrogenation to occur. Therefore, the authors proposed that an initial protonation assisted by formic acid occurs leading to the formation of a σ-H_2 complex **[Fe]-5a**. Subsequent H_2 liberation creates a vacant site for the coordination of formate to form **[Fe]-5c** (Scheme 17a). It

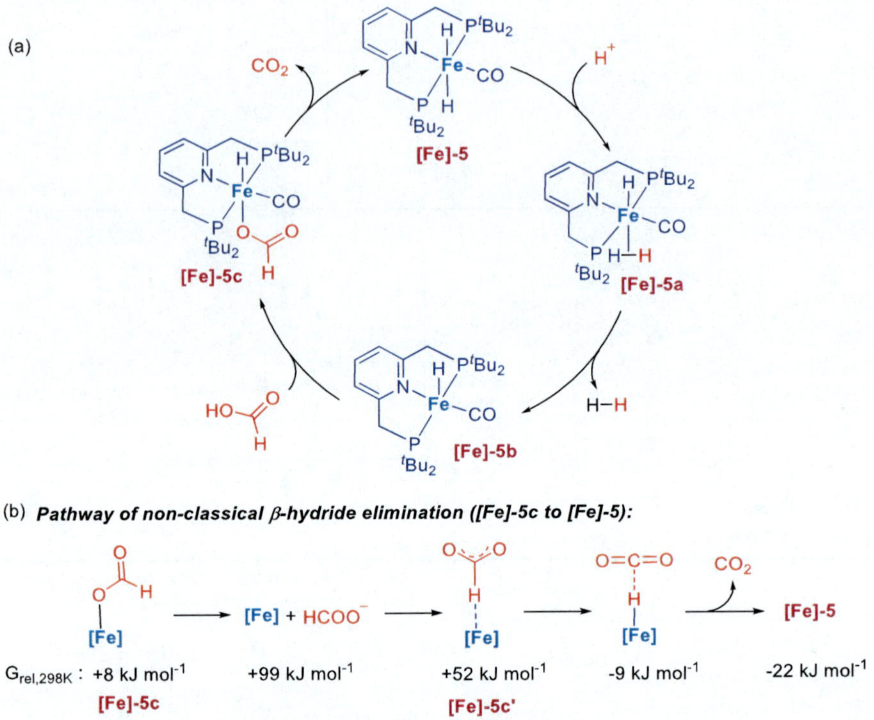

Scheme 17 Mechanistic pathway for (**a**) PNP-Fe(II)-*trans* dihydrido complex for formic acid decomposition. (**b**) non-classical β-hydride elimination from [Fe]-5

is noteworthy to highlight here that **[Fe]-5c** is a coordinatively saturated species and a traditional β-hydride elimination is unlikely unless there is another hemilabile ligand involved. Based on computational calculations, the authors proposed a non-classical β-hydride elimination wherein regeneration of the dihydride complex **[Fe]-5** from η^1-O hydrido formate species **[Fe]-5c** involves a η^1-H hydrido formate intermediate **[Fe]-5c'** (Scheme 17b). Although there is no isolated structural evidence, this η^1-H hydrido formate intermediate for facile CO_2 elimination is also proposed by Kirchner and Gonsalvi [91].

Here the precatalyst **[Fe]-6** activation occurs by bromide abstraction giving a coordinatively unsaturated cationic species **[Fe]-6**a (Scheme 18a). Similar to Milstein's proposed mechanism, formate coordination followed by η^1-O to η^1-H coordination switch gives dihydride intermediate by CO_2 elimination. Finally, a protonaion and subsequent release of H_2 completes the cycle. In a seminal report, Hazari and Schneider showed that the use of catalytic amounts of Lewis acid as the co-catalyst can dramatically increase the turn over number (TON) of the formic acid decomposition when an PNP type Fe(II) pincer complex **[Fe]-7** is used [92]. With the support of stoichiometric experiments, it was reasoned that Lewis acid assists in

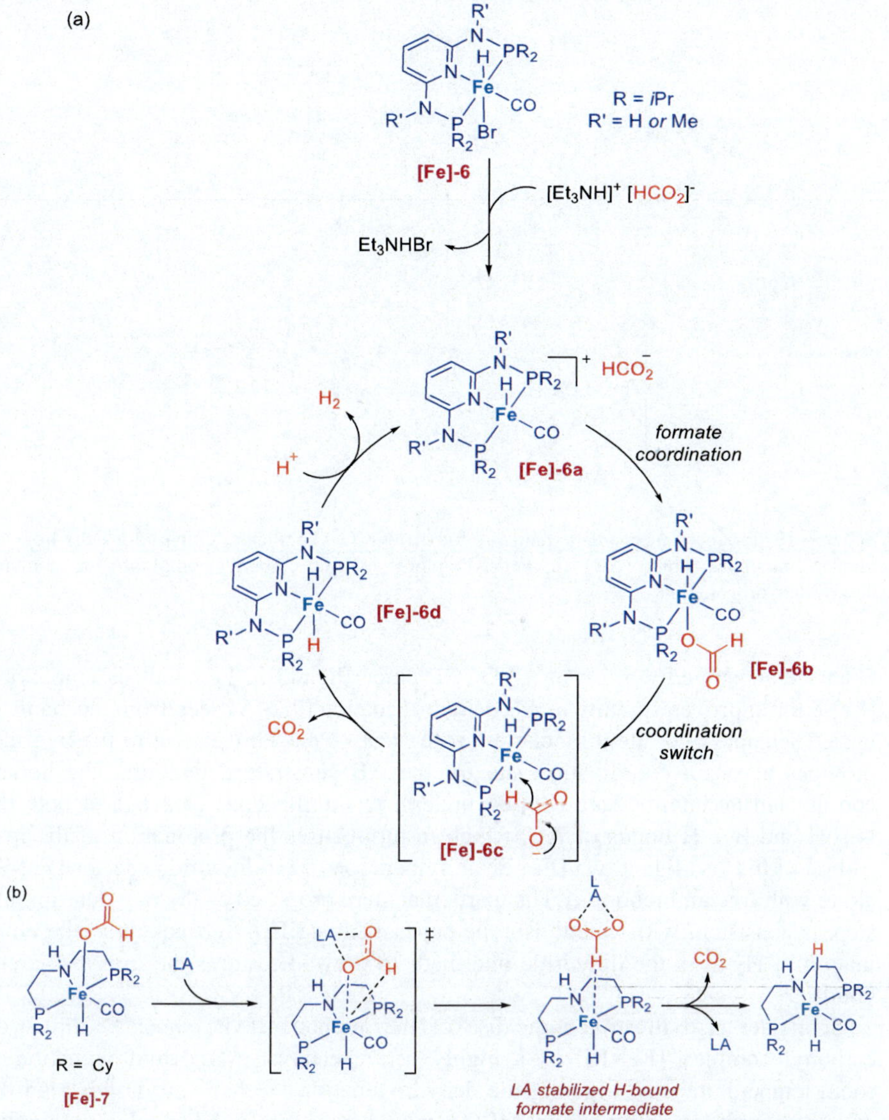

Scheme 18 Mechanistic pathway for (**a**) Fe(PNP) pincer hydrido carbonyl complexes for formic acid decomposition. (**b**) Lewis acid as co-catalyst along with Fe(PNP) pincer complexes

the stabilization of the H-bound formate intermediate prior to decarboxylation (Scheme 18b).

Dehydrogenation of nitrogen-containing compounds has garnered significant attention as well in the recent years. Dehydrogenation of ammonia-borane (AB) is of interest owing to its hydrogen storage capability. In a mechanistic study report,

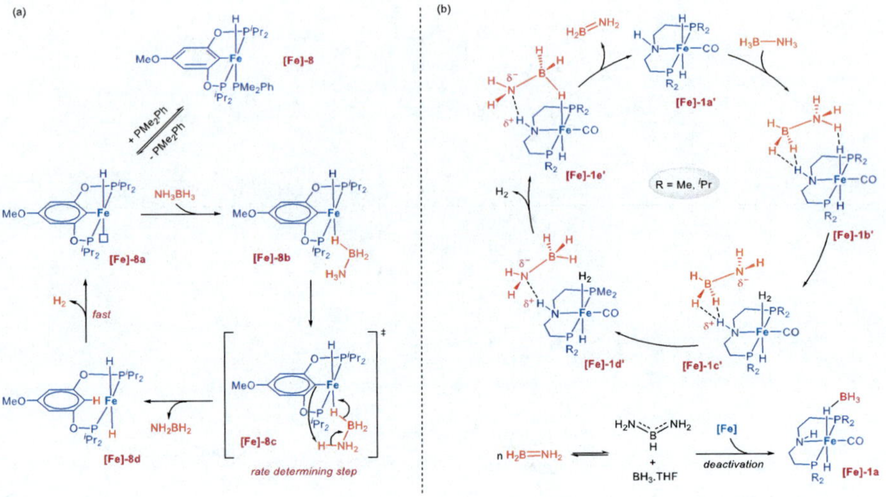

Scheme 19 Proposed mechanistic pathway for (**a**) POCOP-Fe pincer complex for dehydrogenation of ammonia boarane (AB). (**b**) Fe(PNP) pincer hydrido carbonyl complexes for ammonia borane dehydrogenation

Guan demonstrated the use of a POCOP-pincer ligand in Fe catalysis employing **[Fe]-8** for improved stability in AB dehydrogenation [93]. As seen from mechanistic cycle (Scheme 19a), the dissociation of the *trans*-Pme$_2$Ph (relative to the hydride) provides a vacant coordination site for the AB substrate activation. The borane coordinated Fe-intermediate **[Fe]-8b** undergoes simultaneous cleavage of both the N – H and B – H bonds of AB, which in turn causes the protonation of the ipso carbon of the aryl ligand via **[Fe]-8c** and liberation of Fe-dihydride species **[Fe]-8d** along with free aminoborane. This particular step, proposed as the rate-determining step, is consistent with kinetic isotope experiments (KIE). Subsequently, the elimination of H_2 from the dihydride intermediate **[Fe]-8d** restores the iron ipso-arene bond.

Schneider also illustrated the use of the aliphatic PNP pincer Fe-dihydrido carbonyl complex **[Fe]-1a'** as a highly active catalyst AB dehydrogenation at room temperature. Investigating the dehydrogenation mechanism via DFT led the authors to propose a bifunctional MLC catalytic cycle in which the dehydrogenation of AB proceeds via an outer sphere stepwise 1,2 hydride addition to across the Fe-N bond of the catalyst (Scheme 19b) [94]. It is essential to note here that ligand cooperativity plays a crucial role in elimination of H_2 as well as in stabilizing intermediates through hydrogen bridging of the pincer ligand. The aliphatic PNP pincer Fe-hydrido carbonyl complex is also utilized for dehydrogenation of tetrahydroquinoline, wherein it follows a similar mechanistic pathway with dihydrido Fe-complex as the key intermediate [95].

3.2 Bifunctional Iron(0) Complexes

Following the groundbreaking studies on Shvo's highly effective catalytic system [96], the utilization of cyclopentadieneone ligand systems has undergone significant development into various reaction platforms. Considering its abundance and non-toxic nature, iron emerged as the primary candidate for incorporating cyclopentadienone ligands. The Fe(0) complex **[Fe]-9**, initially synthesized by Knölker [97], has found versatile applications in diverse reactions. Guan demonstrated the initial Fe-based dehydrogenation of alcohols, which was accomplished in the presence of acetone acting as both the solvent and hydrogen acceptor (Scheme 20a). Thereafter, various other Knölker type Fe(0) complexes has been reported and extensively studied for alcohol dehydrogenations, particularly in borrowing hydrogen reactions [98–105]. In all these cases, the proposed mechanism for alcohol activation primarily centers around an outer-sphere dehydrogenation pathway, as illustrated in Scheme 20b.

Upon thermal or photolytic treatment of **[Fe]-9**, mono decarbonylation takes place, leading to the formation of a vacant site **[Fe]-9a**. The presence of a base (particularly its cationic part) facilitates an outer-sphere concerted alkoxy dehydrogenation process by lowering the activation energy of relevant transition state **[Fe]-9e**. This leads to the liberation of corresponding carbonyl product along with the formation of an active Fe-H catalyst **[Fe]-9f** with an η^5-cyclopentadienyl ligand.

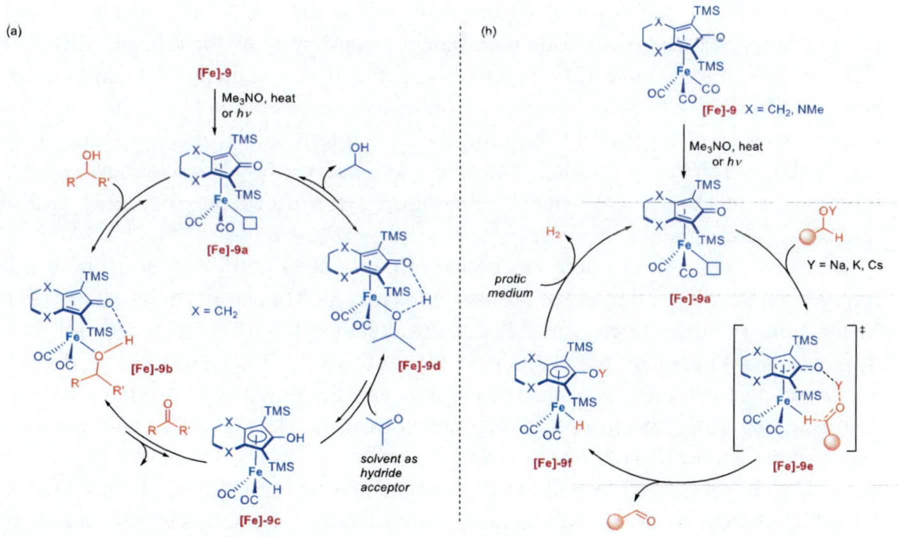

Scheme 20 Mechanistic pathway of Knölker type Fe(0) complexes bearing cyclopenadieneone ligands for (**a**) dehydrogenation of alcohols in presence of hydrogen acceptor (**b**) acceptorless dehydrogenation of alcohols

4 Case Studies on the Mechanism of Cobalt-Catalyzed Dehydrogenation Reactions

Cobalt is another base metal that enjoyed persistent attention ever since many developed techniques by the noble metals were reinvestigated with base metals to achieve the same or the unique reactivity offered by the 3d metals [106]. This has resulted in notable advancements and led to the discovery of uncharted mechanistic pathways in the realm of catalysis. The reason being that cobalt complexes can exhibit variety of oxidation states (+3, +2, +1, 0, and −1) and can proceed through both one- and two-electron redox processes [107]. This ability to undergo multiple oxidation and spin states is a unique characteristic of cobalt complexes and contributes to their exceptional catalytic activity and selectivity.

Owing to such unique reactivity, researchers across the world continue to investigate the potential of cobalt-catalyzed reactions, seeking new applications and uncovering novel reaction mechanisms. In particular, among many other reactions, cobalt-catalyzed alcohol dehydrogenations and related transformations have also emerged as a stirring area of research [108].

4.1 Bifunctional Cobalt(II/III) Complexes

In literature, a significant portion of the catalytically active complexes involved in dehydrogenation reactions are Co(II)-pincer complexes [109]. While many of these complexes feature a phosphine ligand framework, there have also been instances where phosphine-free NNN tridentate pincer ligand systems have been utilized. In 2013, in a seminal work Hanson presented the first cobalt(II)-PNP alkyl pincer complex (**[Co]-1**) for the dehydrogenation of primary alcohols [110]. In a stoichiometric experiment targeted at trapping reactive intermediate, the formation of the cobalt(III) acetylphenyl hydride complex was observed [111] (Scheme 21a). Furthermore, a reaction with 1-phenylethanol-d*8* reconfirmed the source of hydride (Scheme 21b).

Contrary to an outer-sphere mechanism that requires participation of the central nitrogen on the pincer ligand (as discussed in cases of Mn and Fe; Scheme Mn-5 and Scheme Fe-1b respectively), in this case of cobalt catalytic system, alcohol dehydrogenation pathway did not proceed by the traditional MLC pathway. The $PNHP_{Cy}$ backbone behaved only as a spectator ligand and the innocence of NH moiety was confirmed by other spectroscopic techniques as well. Thus, **[Co]-1a** as the catalyst resting state was validated. The reaction proceeds as follows: Reductive elimination from **[Co]-1a** produces a cobalt(I) intermediate and acetophenone, which allows for ligand exchange at the cobalt center. Subsequently, 1-phenylethanol enters the cobalt(I) coordination sphere, leading to the oxidative addition of the O − H bond and the formation of a cobalt(III) alkoxide complex **[Co]-1c**. This complex then undergoes β-hydride elimination, resulting in the generation of a cobalt(III)

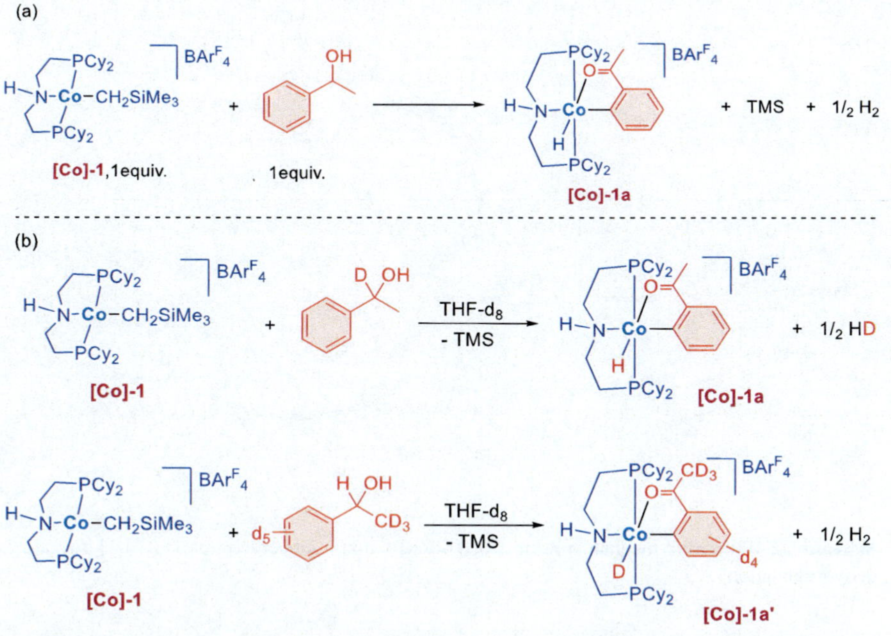

Scheme 21 (**a**) Formation of cobalt (III) acetylphenyl hydride complex in the dehydrogenation of secondary alcohols. (**b**) Experimental evidence for the source of the hydride

dihydride complex **[Co]-1d**. The catalytic cycle is completed by the loss of hydrogen and thc coordination of either acetophenone or 1-phenylethanol (Scheme 22). The detailed energetics is studied by Yang in a separate report which confers with these observations [112]. It is worth noting that the overall reaction was found to be reversible, as experimental evidence had demonstrated. Such a two-electron cobalt (I)/(III) redox cycle has previously been reported by Caulton et al., wherein the unusual unsaturated, paramagnetic and a high-spin triplet configuration cobalt(I) (PNP′) complex showed similar cobalt(I)/(III) redox cycle upon hydrogen addition [113]. Later, **[Co-1]** was also explored for various other applications involving alcohol dehydrogenations [114–116] and heterocycle dehydrogenation [117].

Kempe [118, 119], Kirchner [120], and Milstein [121, 122] have each reported on the application of Co(II)-PNP complexes for the dehydrogenation of primary alcohols (Scheme 23a). Notably, these ligand frameworks have been previously employed with both Mn and Fe (refer to Sects. 1.1 and 2.1). What is intriguing is despite having bifunctional ligand backbone, Kempe and Kirchner's cobalt catalytic systems require the use of stoichiometric amounts of base to activate the precatalyst, whereas Milstein's Co(II)-PNP complex does not. Although it would have been insightful to explore the behavior of these complexes through stoichiometric reactions, a limitation of these reports is that none of them offer a plausible catalytic cycle or any stoichiometric experimental justification for the observed reactivity.

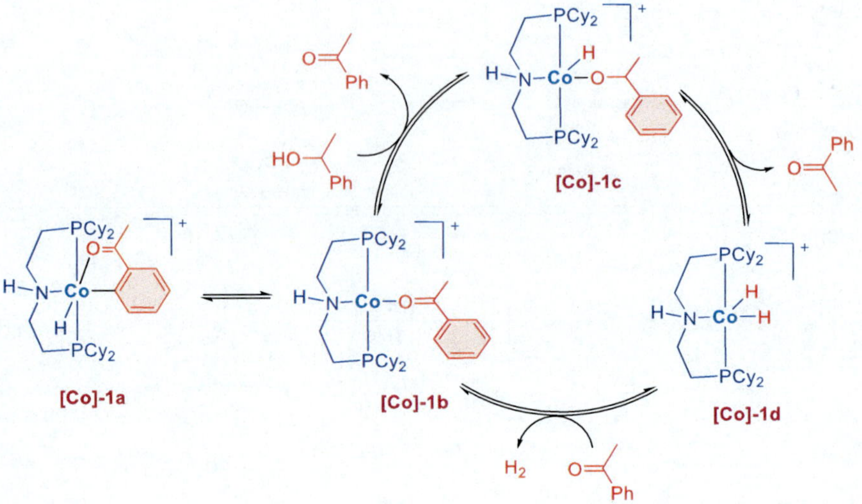

Scheme 22 Proposed mechanism for Co(II)-(PNP) alkyl pincer complex [Co-1] for alcohol dehydrogenation

In 2019, Madsen revealed a rigid Co(II)-PNP catalytic system ([Co-6]) that was formed in situ through the combination of cobalt salts and a specific ligand [123]. Catalytic amounts of base and reductants were enough to drive the reaction. Contrary to the previously proposed Co(I)-Co(III) catalytic cycle where the alcohol first undergoes oxidative addition to the Co(I) complex to form a Co(III) alkoxide, the alcohol dehydrogenation was proposed to proceed through an outer-sphere hydrogen transfer mechanism instead (Scheme 23b). The initially formed Co (II) complex is reduced in the presence of Zn to an unsaturated Co(I) intermediate. Outer sphere alcohol activation generates a monohydride intermediate assisted by the involvement of the NH moiety in the ligand. Following this, hydrogen gas is eliminated to regenerate the active catalytic Co(I) species.

Ding demonstrated the use of a $^{iPr}PPPN^{H}Py^{Me}$ tetradentate tripodal ligand featuring MLC by introducing a N – H linker in the dehydrogenation of both secondary and primary alcohols [124–127]. As a part of initial study to understand reaction mechanism, the authors synthesized the derivative complex **[Co]-8** bearing a dearomatized pyridine arm which showed comparable activity with the precatalyst **[Co]-7** and KO^{t}Bu as base (Scheme 24a). The findings suggested the involvement of species **[Co]-8** in the generation of the actual catalyst via dehydrohalogenation by the base. Intriguingly, a N – Me substituted complex **[Co]-9** exhibited similar activity, indicating that the N – H linker does not play a crucial role in the metal-ligand cooperation (MLC). Despite **[Co]-7** being paramagnetic, the detection of a diamagnetic hydride signal suggests the possible generation of a Co(I) or Co(III) hydride [128] (Scheme 24b), which aligns with the mechanistic investigation conducted by Hanson [111]. However, given that the N – H linker does not involve in metal-ligand cooperativity, the mode of alcohol activation remains elusive.

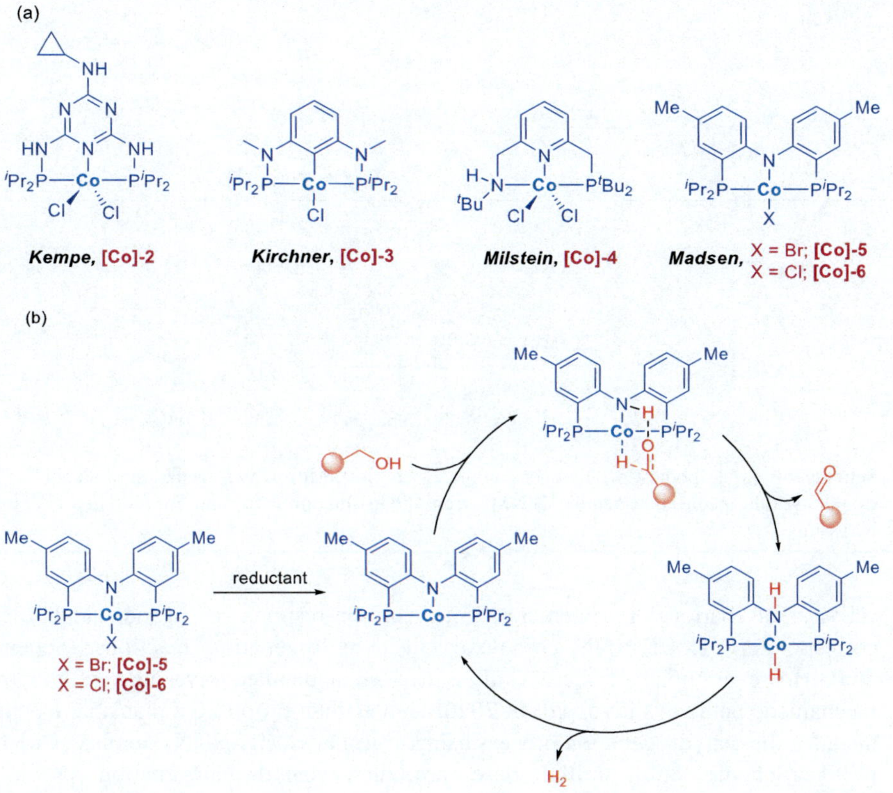

Scheme 23 (**a**) Phosphine-based Co(II) pincer complexes employed for dehydrogenation of alcohols. (**b**) Proposed mechanism for Co(II)-(PNP) pincer complex [Co-5] for alcohol dehydrogenation

Commercially available Co(II) salts and a tetradentate phosphine ligand P $(CH_2CH_2PPh_2)_3(PP_3)$ along with a base, is also a highly efficient catalytic system in alcohol dehydrogenation, especially methanol. Liu utilized this catalytic system for dehydrogenations applied in methylation reactions [129, 130]. Kundu studied this catalytic system in detail leading to useful insights into the behavior of the Co (II)/ PP_3 system [131, 132] (Scheme 25). According to the proposed mechanism, the initial step involves the base-assisted formation of a Co(I)-alkoxy intermediate. Again, DFT calculations indicated that β-hydride elimination necessitates the dissociation of one phosphine arm to create a vacant site. After the β-hydride elimination, a Co(I)-H intermediate is formed which proceeds further in the catalytic cycle. The existence of the Co(I)-H species is experimentally supported and validated.

The effectiveness of phosphine-free Co(II)-NNN pincer complexes has been comparable to complexes containing PNP ligands. In 2018, Kundu [133, 134] and Balaraman [135] independently reported on the use of Co(II)-NNN complexes for alcohol dehydrogenation. Subsequently, Li [136] and Gupta [137] disclosed the

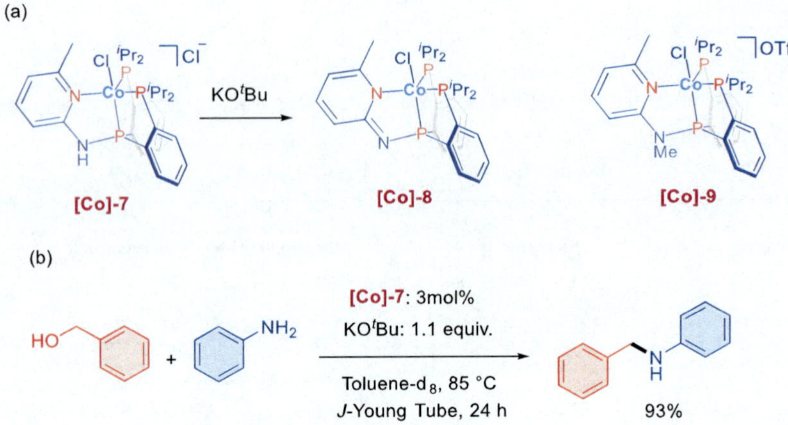

Scheme 24 (**a**) Tripodal Co(II) pincer complexes employed for dehydrogenation of alcohols. (**b**) Cobalt hydride species detected by ^{1}H NMR from the in-situ amine forming reaction in a J. Young NMR tube

activity of various nitrogen-containing proton-responsive ligands and their corresponding Co(II)-NNN complexes in dehydrogenation reactions (Scheme 26a). However, none of these studies invoked a detailed investigation into the mechanistic pathways involved. In 2020, Gunanathan proposed a plausible mechanism for the dehydrogenation process using a similar Co(II)-NNN complex **[Co]-15** [138] (Scheme 26b). Initial base promoted dehydrohalogenation provides coordinatively unsaturated intermediate **[Co]-15a**. This intermediate upon subsequent interaction with alcohol, results in O – H bond activation via MLC, and generates alkoxy-complex intermediate **[Co]-15b**. Following this, β-hydride elimination of the alkoxy ligand in leads to formation of a monohydrido cobalt intermediate **[Co]-15c** and liberation of aldehyde. Intermediate **[Co]-15c** is proposed to be in equilibrium with the nonclassical hydrogen-coordinated **[Co]-15d**, which can regenerate **[Co]-15a** by liberation of molecular hydrogen.

Now, in the context of alcohol dehydrogenation, Co(III) complexes have been relatively less explored in the literature. Bala introduced an Im-NHC-based cationic Co(III) complex that exhibits transfer hydrogenation (TH) activity [139]. A major advantageous feature of Im-NHC ligands is their ability in fine-tuning the steric and electronic properties of bound metal centers utilized in homogeneous catalysis. In the presence of catalytic amounts of KOH as a base, the aforementioned Co(III) complex demonstrated dehydrogenation of isopropanol. Based on insights from liquid chromatography-mass spectrometry (LC – MS) data, the authors proposed a mechanistic pathway as follows: the hemilabile imine N-donor allows for the coordination of isopropanol, resulting in the formation of a metal alkoxy complex **[Co]-16a**. Simultaneous formation of a metal hydride via β-hydride elimination and external carbonyl insertion releases acetone along with intermediate **[Co]-16c** (Scheme 27). Importantly, the presence of two hemilabile donor ligands in this

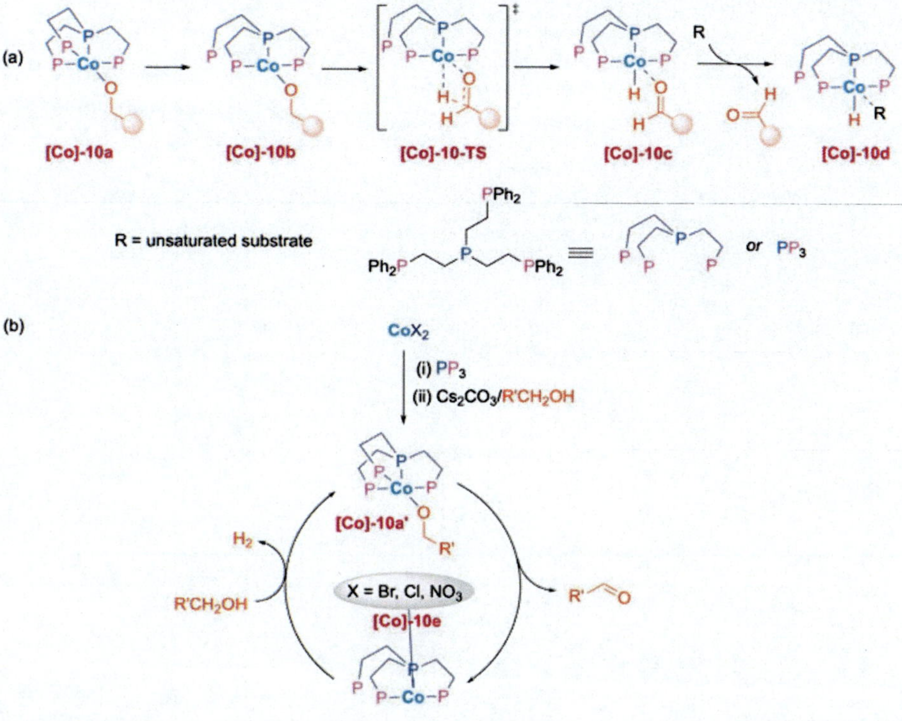

Scheme 25 Proposed reaction pathway for a Co(II)/ PP_3 catalytic system for dehydrogenation of primary alcohols

saturated Co(III) system is crucial for creating vacant site enabling the formation of the alkoxy complex and subsequent hydride generation. The cycle is completed through subsequent hydride transfer and alkoxide exchange. Fang and Yang also demonstrated the use of phosphine-free cobalt-NHC pincer complex in TH reactions [140].

In a seminal report, Sundararaju reported on the dehydrogenation of secondary alcohols using Cp*Co(III) complexes in the presence of acetone as solvent [141]. These high valent coordinatively saturated Cp*Co(III) complexes have found diverse applications, especially in borrowing hydrogen reactions [142–144]. Owing to the lack of a vacant site at the metal center (which is crucial, as established already in the aforementioned example), stoichiometric KO^{t}Bu as base assists in the dehalogenation via the formation of KI. For elucidating the mechanistic pathway involved in high-valent Cp*Co(III)-catalyzed dehydrogenations, detailed and comprehensive energetic calculations using DFT on a well-defined Cp*Co(N,O) I catalytic system by Poli and Sundararaju provided deeper insights [145]. Preliminary DFT calculations suggested the possibility of a catalytic resting state involving the formation of a formally charged Co(I) species **[Co]-17c** (Scheme 28a). According to the proposed mechanism, from the Co(III)-alkoxy intermediate (stable

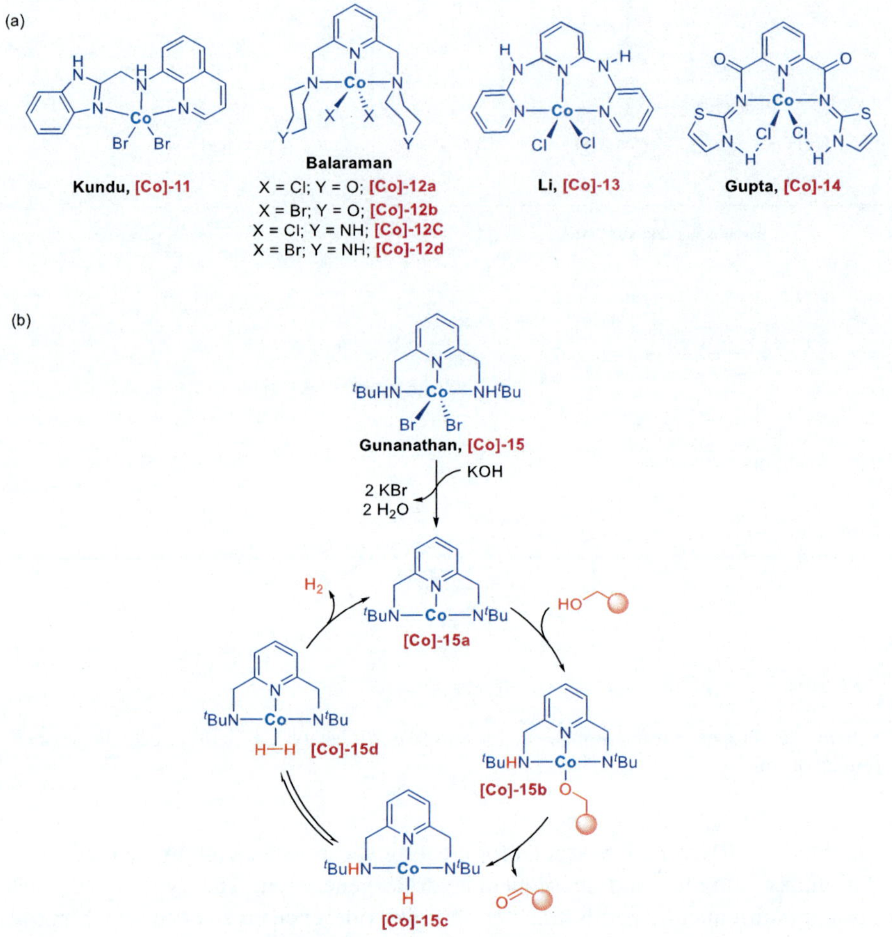

Scheme 26 (**a**) Phosphine-free Co(II)-NNN pincer complexes employed for alcohol dehydrogenation. (**b**) Proposed mechanism for Co(II)-NNN complex [Co]-15 in dehydrogenation of primary alcohol

in the spin triplet state) **[Co]-17a**, the oxygen atom of the ligand arm abstracts the β-hydride as a proton while simultaneously transferring 2 electrons to the metal center. The metastable Cp*Co(III) alkoxy complex **[Co]-17a-ii** was spectroscopically characterized by utilizing alcohols containing electron-withdrawing CF_3 groups (Scheme 28b).

This stabilization was subsequently verified in the established DFT pathway, which reconfirmed that electron-withdrawing groups imparted stability to the intermediates with energies up to −27.3 kcal/mol. Although direct evidence for the electron-rich Co(I) resting state species is lacking, it was successfully trapped by the subsequent addition of CO, resulting in the formation of an unusual spin-triplet

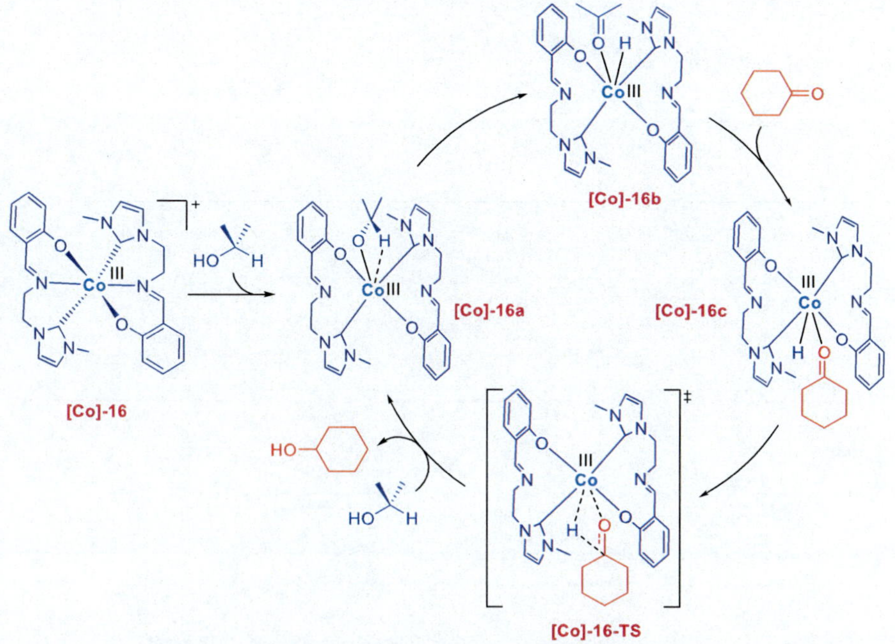

Scheme 27 Proposed mechanism of a high valent NHC-based Co(III) complex for dehydrogenation of isopropanol

20-electron adduct **[Co]-17c-i** that was characterized conclusively. Overall, the combined findings from DFT calculations and experimental investigations provide a robust confirmation of each other's outcomes, shedding light on the previously unexplored reductive alcohol activation pathway in metal-catalyzed dehydrogenations. Also, this pathway, as supported by the findings, further reinforces the validity of a Co(III)/Co(I) redox mechanism similar to the one proposed by Hanson [111].

In a recent study, Tang reported the use of a high-valent Cp*Co(N,N)I catalytic system **[Co]-18** for acceptorless dehydrogenation of alcohols, specifically targeting coupling reactions to produce diverse heterocycles [146]. The complex incorporates an NH group in the pyrazole unit, which serves as the site for metal-ligand cooperation (MLC). This was confirmed by the isolation of a neutral complex **[Co]-18a** after treatment with a base. Interestingly, when a catalyst with a methyl-substituted pyrazole was employed, it exhibited poor reactivity, indicating the absence of MLC. In contrast to the earlier proposed reductive activation pathway in Scheme Co-8, the authors propose an outersphere metal-ligand synergistic dehydrogenation mechanism similar to the one proposed by Madsen [123] (Scheme 29). According to this proposal, the authors suggest the formation of a Cp*Co(III)-hydride via an intermediate of the type **[Co]-18c**, wherein dehydrogenation is proposed without the generation of a coordinatively saturated alkoxy complex. However, no direct evidence is provided for the generation of a cobalt-hydride species. The diversity of

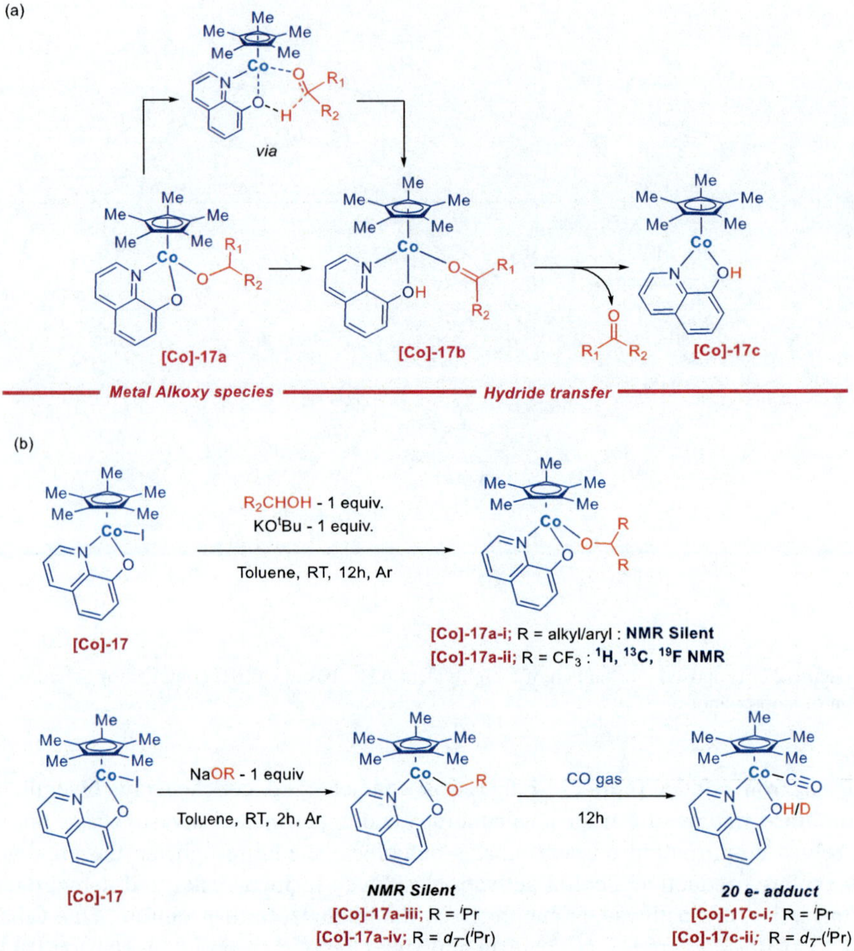

Scheme 28 (**a**) A new reductive activation computed pathway for dehydrogenation of secondary alcohols using high valent Cp*Co(III) catalysts. (**b**) Trapping of the Co(III)-alkoxy and Co (I) resting state by electron withdrawing groups/stabilizing agents

reaction pathways observed in cobalt-catalyzed dehydrogenation reactions, as discussed here, underscores the intricate nature of these mechanisms. This complexity emphasizes the range of possibilities for cobalt and probably other first-row transition-metal complexes.

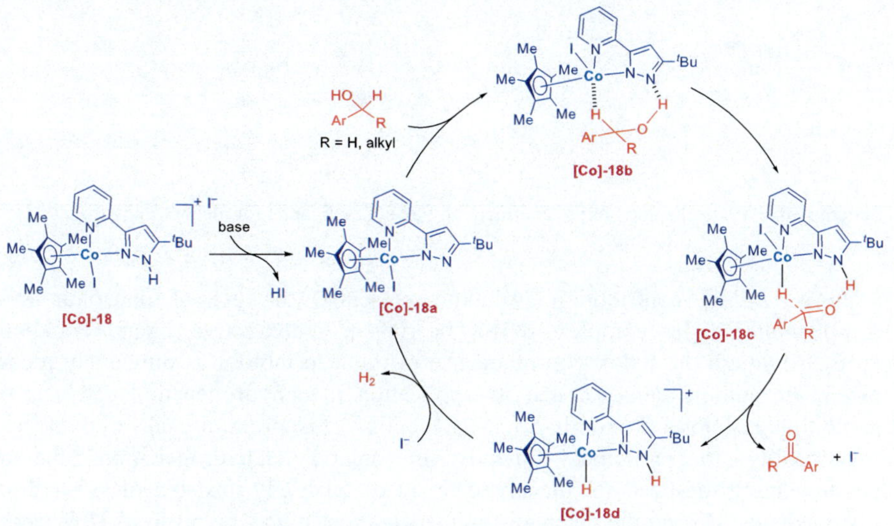

Scheme 29 Proposed mechanism for acceptorless dehydrogenative coupling of alcohols using a Cp*Co(N,N)I catalyst with a proton responsive unit

5 Case Studies on the Mechanism of Nickel-Catalyzed Dehydrogenation Reactions

5.1 Bifunctional Nickel(II) Complexes

The prominence of 3d-metal homogeneous dehydrogenation catalysis surged after 2013; however, the utilization of nickel in such reactions commenced at a much later stage [147]. Nickel is known for its distinctive redox properties, which make it an excellent candidate for catalysis and provide a range of rewarding and unique reactivity. Nickel can readily switch between different oxidation states, such as Ni(0), Ni(I), Ni(II), and Ni(III), enabling it to participate in various catalytic processes. Despite being the second most abundant transition metal present in the earth's crust and having other promising attributes, the true potential of nickel is yet to be fully realized. In a significant development, the research groups of Banerjee [148–150], Bo Tang [151], Maji [152], and Baidya [153] independently put forth reports on the application of Ni(II) salts, combined with bidentate or tridentate ligands, for the dehydrogenation of alcohols. This marked the beginning of nickel catalysis in this domain of dehydrogenation reactions (Scheme 30). In the following years, two other commercially available Ni(II) systems were also demonstrated to be applicable [154, 155]. To date, no experimental or theoretical investigations have been undertaken to shed light on the mechanism of dehydrogenation by these systems, thereby limiting a comprehensive understanding on the catalytic cycle.

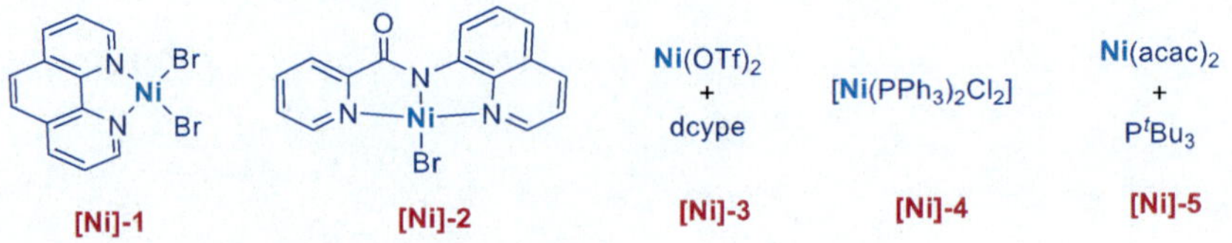

Scheme 30 Ni(II) salts and simple commercially available ligands for alcohol dehydrogenation

In a seminal contribution, Adhikari presented the use of phosphine-free, azo-phenolic Ni(II) complex **[Ni]-6** bearing a redox-active ligand backbone [156]. Although the redox non-innocence of this azo motif was previously recognized, the authors demonstrated its application in dehydrogenation catalysis by exploiting its close proximity to metal center. This proximity allowed the azo functionality to participate directly in alcohol dehydrogenation, thereby circumventing the need for the involvement of relatively unstable nickel hydride intermediates. During the mechanistic investigation, it was found from DFT calculations that the binding of an alkoxide to an already tetra coordinated nickel complex is not feasible due to the very high affinity for square planar geometry. Rather, one of the phenolate arms of the ligand in **[Ni]-6** was found to act as a hemilabile ligand upon protonation, facilitating substrate(alcohol) binding (Scheme 31). Simultaneous monoreduction of the azo ligand in presence of base generates intermediate **[Ni]-6a**. At the right alignment of β-hydrogen with the azo backbone, hydrogen atom transfer (HAT) occurs to generate ketyl radical intermediate **[Ni]-6b** which was trapped as a TEMPO adduct. Unstability of **[Ni]-6b** leads to one electron transfer to the azo backbone leading to **[Ni]-6c**. Subsequently, protonation of the anionic azo motif by the phenolic proton occurs which generates **[Ni]-6d** and along with the release of ketone. Another incoming alcohol unit completes the catalytic cycle. Later, the same group utilized this nickel catalytic system for dehydrogenation of alcohols applicable to diverse reactions [157–160]. In 2023, Erande and Metre reported a modified catecholaldimine-based azo-phenolic Ni(II) complex, that participates in alcohol dehydrogenation in a similar pathway [161].

Over time, stable and catalytically active Ni(II)-(NNO, NNN, PPP) pincer complexes have emerged, showcasing their potential in alcohol dehydrogenation. As this field is still in its early stages of development, these complexes hold the promise of diverse applications, potentially even in unprecedented new reactions. In 2020, Ramesh first reported the synthesis and utility of Ni(II)-NNO complex **[Ni]-7** bearing hydrazone ligands with N^N^O donor atoms in dehydrogenation of primary alcohols [162]. Following this, Liang Lu also reported another acetate bound Ni(II)-NNO pincer complex **[Ni]-8** for the same [163]. As shown in Scheme 32a, both **[Ni]-7** and **[Ni]-8** complexes share a common feature: each possesses a labile spectator ligand that creates an available site for coordination, to facilitate the formation of a Ni(II)-alkoxy species. This alkoxy species is proposed to undergo a β-H elimination to generating a Ni − H intermediate and releases the aldehyde (Scheme 32b). Another molecule of alcohol completes the catalytic cycle. The

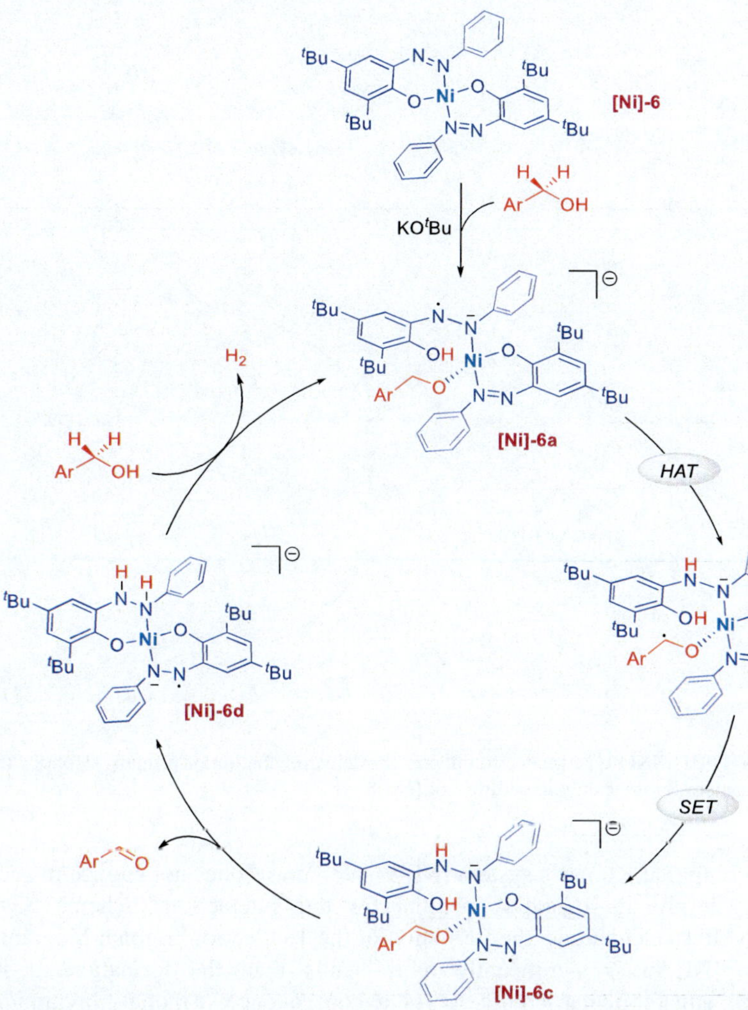

Scheme 31 Proposed mechanism for redox-active azo-phenolic Ni(II)-catalyzed alcohol dehydrogenation

authors also state that despite multiple attempts, isolation of the Ni-hydride species from the reaction mixture remained unsuccessful due to its unstable nature [162]. Ramesh et al. also investigated the application of Ni(II)-N^S bis-chelated complexes in the dehydrogenation of primary alcohols, following a similar mechanistic pathway [164].

Srivastava and Kumar [165, 166], followed by Balaraman [167] introduced penta-coordinated Ni(II)-NNN pincer type complexes for use in the dehydrogenation of primary alcohols. Notably, both complexes bear ligand framework capable of participating in metal-ligand cooperativity via aromatization/dearomatization. It is

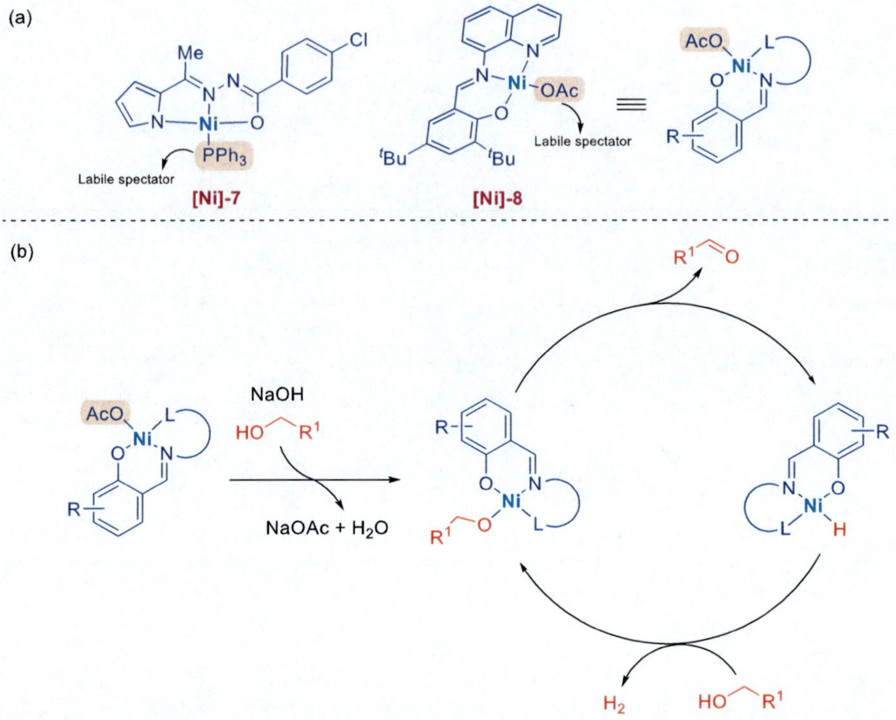

Scheme 32 (**a**) Ni(II)-NNO type pincer complexes for dehydrogenation of primary alcohols. (**b**) Proposed mechanism for the catalytic pathway of [Ni]-8

worth mentioning that Kumar's system **[Ni]-9**, in contrast, does not engage in such cooperativity despite its ligand's potential for this interaction (Scheme 33a). According to DFT calculations, the formation of the 18-electron trigonal-bipyramidal complex **[Ni]-9a** is significantly more stable than the formation of its corresponding square planar complex. As a consequence, trigonal-bipyramidal complex **[Ni]-9a** is the catalytic active species that participates in reaction mechanism. It is to note here that the formation of the Ni(II) alkoxy species **[Ni]-9a** occurs through salt metathesis due to the absence of a vacant coordination site, rather than by metal-ligand cooperation. A β-hydride elimination leads to Ni-hydride species **[Ni]-9c** and liberates the product. The regeneration of the catalytic active species is facilitated by another incoming alcohol and subsequent liberation of H_2. Whereas in Balaraman's system **[Ni]-10**, the Ni-hydride **[Ni]-10c** (generated after a β-hydride elimination from the alkoxy intermediate) undergoes metal-ligand cooperativity to release H_2 and form intermediate **[Ni]-10d** which continues further in the catalytic cycle (Scheme 33b).

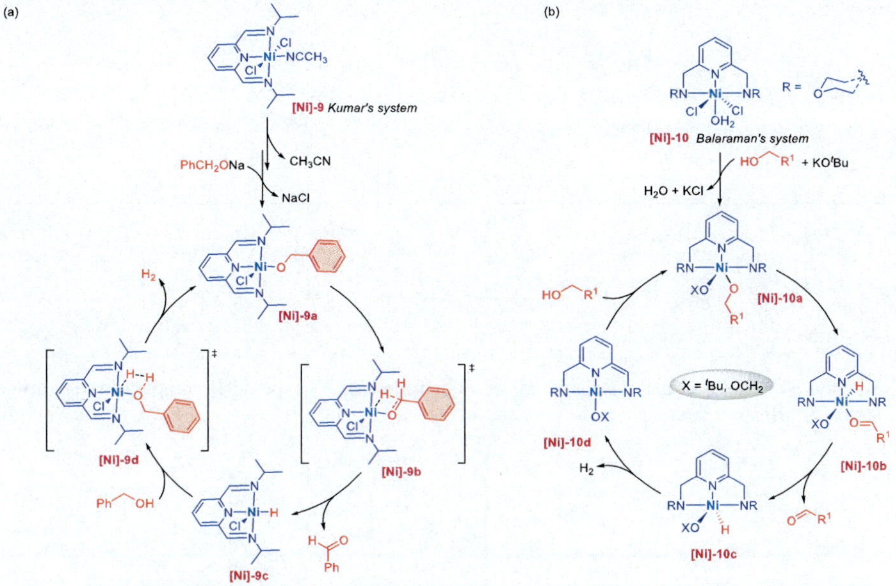

Scheme 33 Proposed mechanistic pathway of alcohol dehydrogenation for (**a**) Kumar's Ni(II)-NNN pincer complex [Ni]-9 (**b**) Balaraman's Ni(II)-NNN pincer complex [Ni]-10

5.2 *Low Valent Nickel(0/I) Complexes*

The utilization of Ni(0) and Ni(I) complexes for catalyzed conversions is relatively uncommon and challenging due to their scarcity and inherent instability. These species, featuring low oxidation states of nickel, often possess high reactivity and tend to undergo rapid transformations or decompose under typical reaction conditions. Consequently, their practical application in catalysis requires careful design and precise control to harness their potential for promoting specific reactions effectively. The use of such complexes in catalyzed dehydrogenation is very rare, with two reports till date. In 2019, García published a study detailing the utilization of an in situ-generated Ni(0) system, in the presence of $Ni(COD)_2$ and a bisphosphine catalyst, for transfer hydrogenation employing ethanol as the hydrogen source [168]. The authors postulate the development of a crucial Ni(II)-dihydride species during the dehydrogenation process, which serves as the pivotal intermediate in the transfer hydrogenation reaction. Very recently, Zhang et al. reported a novel pincer nickel(I) complex **[Ni]-11** based on the 1,2,4-triazole backbone ligand system for dehydrogenation of primary alcohols [169]. Based on initial preliminary control studies, an inner sphere β-hydride elimination pathway via intermediate **[Ni]-11b** is proposed (Scheme 34).

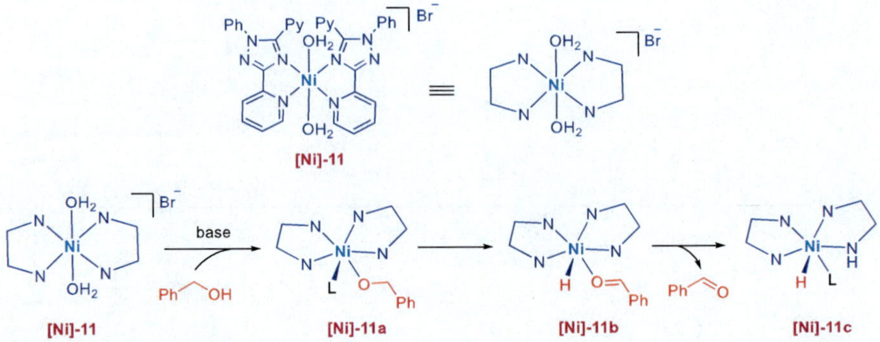

Scheme 34 Proposed reaction pathway of a quadridentate N_4 type Ni(I)-complex in dehydrogenation of primary alcohols

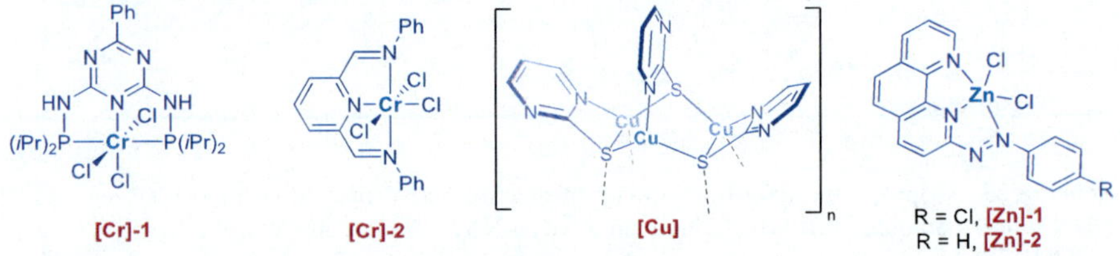

Scheme 35 Chromium, copper and zinc complexes utilized for alcohol dehydrogenation

6 Miscellaneous Case Studies

In the past 2–3 years, there has been a growing interest among research groups to explore the untapped potential of 3d metals such as chromium, copper, and zinc in dehydrogenation reactions (Scheme 35). This pursuit of lesser-known metals leads to exciting new opportunities and advancements in the fields of coordination chemistry and catalysis. In 2020, Kempe [170] and Kumar [171] demonstrated the compatibility of their respective ligand systems with chromium in catalyzing alcohol dehydrogenation. However, due to the nascent stage of this research, the true mechanistic behavior of these reactions stands elusive. The proposed mechanism based on initial control experiments is analogous to other 3d-metals with these ligand systems. Among other reports, Ping-lang reported a copper(I) N-heterocyclic thiolate complex [172] and Mannath reported a ligand free zinc hydrate complex in the presence of base to be active for dehydrogenation of primary alcohols [173]. In 2023, Paul put forth a well-defined Zn(II)-NNN pincer catalyst **[Zn]-1** featuring an azo backbone applied in an elegant approach for pyridine synthesis by alcohol dehydrogenation [174]. In the proposed mechanism, the active intermediate is **[Zn]-1b** generated in a similar way as discussed in Scheme 31. The one electron reduced azo-chromophore abstracts the β-hydrogen atom of the Zn-alkoxy

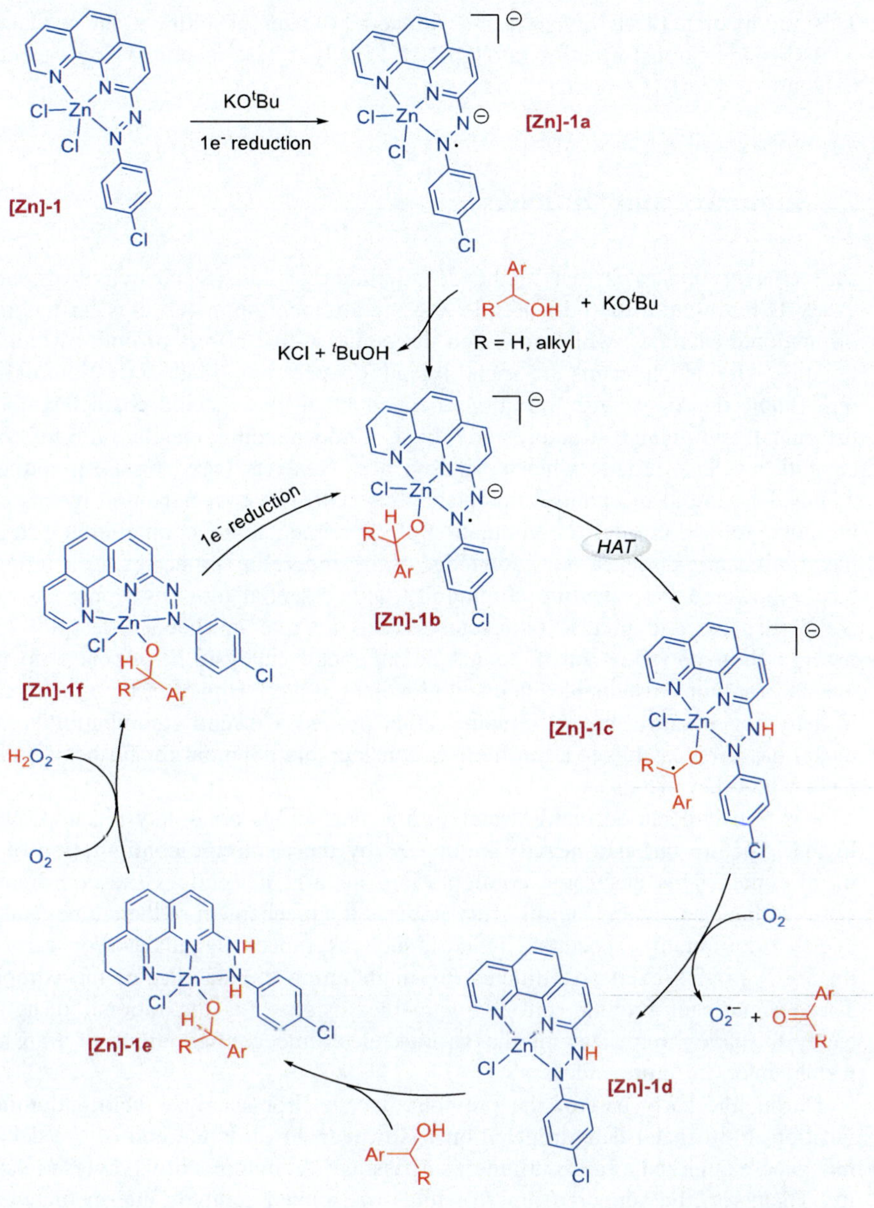

Scheme 36 Proposed mechanism of the Zn(II)-NNN pincer complex-catalyzed alcohol dehydrogenation

intermediate forming a ketyl radical intermediate **[Zn]-1c** (Scheme 36). Subsequent rapid one-electron oxidation under air liberates the desired carbonyl product, along with the formation of intermediate **[Zn]-1d**. For the regeneration of catalyst, activation of another incoming alcohol occurs leading to a dihydrazo intermediate **[Zn]-1e**.

This intermediate **[Zn]-1e** then reacts with aerial oxygen, resulting in the production of H_2O_2 and its transformation into **[Zn]-1f**. Finally, a base-mediated single-electron reduction of **[Zn]-1f** completes the cycle.

7 Summary and Outlooks

This chapter illustrates that in the development of base metal dehydrogenation catalysts throughout the past decade, the predominant approach thus far has been bifunctional catalysis, which relies on the collaborative efforts of both metals and ligands. The mechanisms of metal-ligand cooperative bond activation exhibit remarkable diversity, with variations dependent on the specific characteristics of the metal and ligand structures involved. Understanding the accessibility of a catalytic cycle is crucial when designing new catalysts based on earth-abundant metals. The initial mechanistic understanding with first-row transition metals continuously refined through a combination of experimental and computational studies has significantly enriched our knowledge of the underlying processes. This progress has empowered the scientific community with essential tools to propose a more comprehensive and precise representation of the true mechanisms at play. The incorporation of redox-active ligands in 3d metal catalysis has opened up new avenues for more sustainable chemical processes as well. However, their application in dehydrogenation catalysis remains rather limited. Currently, compatibility with nickel has been established, but there is considerable potential for further development with other metals.

It is also understood that a "catalytic mechanism" is not solely dictated by the ligand structure but also heavily influenced by the electronic configuration of the metal center. This electronic configuration, in turn, determines the coordination status of the catalyst and significantly impacts the mechanism of the entire catalytic cycle. Additionally, spectator ligands, such as pincer ligands and/or carbonyl ligands, can also exert an influence by maintaining the stability of the complex. These significant findings provide valuable insights for the rational design of catalysts, taking into account the various electronic configurations of transition metal center, in future endeavors.

During the early part of the previous decade, iron catalysts held a dominant position in 3d metal dehydrogenations. However, towards the end of the decade, manganese emerged as a frontrunner, surpassing iron in terms of its diverse reactivity. Thanks to the support from structures of isolated catalytic intermediates, the possible mechanistic pathways for iron and manganese are now fairly well-established. Conversely, cobalt's open-shell electronic states present challenges, hindering the development of a comprehensive mechanistic understanding similar to iron and manganese. In recent years, the use of other remaining 3d-metals has started to gain momentum, offering ample opportunities for future explorations in this domain. Thus, the first-row transition metals, with their distinctive redox properties and readily switchable range of oxidation states, although extensively investigated, still hold significant potential to unlock new applications in dehydrogenation catalysis, making them an exciting area for further research and discovery.

References

1. Bullock RM (2010) Catalysis without precious metals. Wiley-VCH, Weinheim
2. Falicov LM, Somorjai GA (1985) Correlation between catalytic activity and bonding and coordination number of atoms and molecules on transition metal surfaces: theory and experimental evidence. Proc Natl Acad Sci 82:2207–2211
3. McCleverty J (1999) Chemistry of the first-row transition metals. OUP Oxford
4. Houa C, Lib Y, Keb Z (2020) Transition metal center effect on the mechanism of homogenous hydrogenation and dehydrogenation. Inorg Chim Acta 511:119808
5. Guan H, Chakraborty S (2010) First-row transition metal catalyzed reduction of carbonyl functionalities: a mechanistic perspective. Dalton Trans 39:7427–7436
6. Filonenko GA, van Putten R, Hensen EJM, Pidko EA (2018) Catalytic (De)hydrogenation promoted by non-precious metals –Co, Fe and Mn: recent advances in an emerging field. Chem Soc Rev 47:1459–1483
7. Werkmeister S, Neumann J, Junge K, Beller M (2015) Pincer type complexes for catalytic (De)hydrogenation and transfer (De)hydrogenation reactions: recent progress. Chem A Eur J 21:12226–12250
8. Luca OR, Crabtree RH (2013) Redox-active ligands in catalysis. Chem Soc Rev 42:1440–1459
9. Blacquiere JM (2021) Structurally-responsive ligands for high-performance catalysts. ACS Catal 11:5416–5437
10. Poli R (2023) A new classification for the ever-expanding mechanistic landscape of catalyzed hydrogenations, dehydrogenations and transfer hydrogenations. Adv Organomet Chem 79: 87–133
11. Kozuch S, Shaik S (2011) How to conceptualize catalytic cycles? The energetic span model. Acc Chem Res 44:101–110
12. Dub PA, Gordon JC (2017) Metal–ligand bifunctional catalysis: the "Accepted" mechanism, the issue of concertedness, and the function of the ligand in catalytic cycles involving hydrogen atoms. ACS Catal 7:6635–6655
13. Colebatch AL, Stevens MA (2022) Cooperative approaches in catalytic hydrogenation and dehydrogenation. Chem Soc Rev 51:1881–1898
14. Li H, Hall MB (2015) Computational mechanistic studies on reactions of transition metal complexes with noninnocent pincer ligands: aromatization-dearomatization or not. ACS Catal 5:1895–1913
15. Khusnutdinova JR, David M (2015) Metal–ligand cooperation. Angew Chem Int Ed 54: 12236–12273
16. Alig L, Fritz M, Schneider S (2019) First-row transition metal (De)hydrogenation catalysis based on functional pincer ligands. Chem Rev 119:2681–2751
17. Jeffrey JC, Rauchfuss TB (1979) Metal complexes of hemilabile ligands. Reactivity and structure of dichlorobis(o-(diphenylphosphino)anisole)ruthenium(II). Inorg Chem 18:2658–2666
18. Peris E, Crabtree RH (2018) Key factors in pincer ligand design. Chem Soc Rev 47:1959–1968
19. Suarez ALO, Lyaskovskyy V, Reek JNH et al (2013) Complexes with nitrogen-centered radical ligands: classification, spectroscopic features, reactivity, and catalytic applications. Angew Chem Int Ed 52:12510–12529

20. Winter A, Newkome GR, Schubert US (2011) Catalytic applications of terpyridines and their transition metal complexes. ChemCatChem 3:1384–1406
21. Chirik PJ (2015) Iron- and cobalt-catalyzed alkene hydrogenation: catalysis with both redox-active and strong field ligands. Acc Chem Res 48:1687–1695
22. Wang Y, Wang M, Li Y, Liu Q (2021) Homogeneous manganese-catalyzed hydrogenation and dehydrogenation reactions. Chem 7:1180–1223
23. Deibl N, Kempe R (2017) Manganese-catalyzed multicomponent synthesis of pyrimidines from alcohols and amidines. Angew Chem Int Ed 56:1663–1666
24. PeÇa-Llpez M, Piehl P, Elangovan S, Neumann H, Beller M (2016) Manganese-catalyzed hydrogen-autotransfer C-C bond formation: α-alkylation of ketones with primary alcohols. Angew Chem Int Ed 55:14967–14971
25. Mukherjee A, Nerush A, Leitus G, Shimon LJW, Ben-David Y, Espinosa-Jalapa NA, Milstein D (2016) Manganese-catalyzed environmentally benign dehydrogenative coupling of alcohols and amines to form aldimines and H_2: a catalytic and mechanistic study. J Am Chem Soc 138: 4298–4301
26. Mastalir M, Glatz M, Gorgas N, Stçger B, Pittenauer E, Allmaier G, Veiros LF, Kirchner K (2016) Divergent coupling of alcohols and amines catalyzed by isoelectronic hydride Mn (I) and Fe(II) PNP pincer complexes. Chem A Eur J 22:12316–12320
27. Kallmeier F, Dudziec B, Irrgang T, Kempe R (2017) Manganese-catalyzed sustainable synthesis of pyrroles from alcohols and amino alcohols. Angew Chem Int Ed 56:7261–7265
28. Fertig R, Irrgang T, Freitag F, Zander J, Kempe R (2018) Manganese-catalyzed and base-switchable synthesis of amines or imines via borrowing hydrogen or dehydrogenative condensation. ACS Catal 8:8525–8530
29. Zhang G, Irrgang T, Dietel T, Kallmeier F, Kempe R (2018) Manganese-catalyzed dehydrogenative alkylation or α-olefination of alkyl-substituted N-heteroarenes with alcohols. Angew Chem Int Ed 57:9131–9135
30. Pandia BK, Chidambaram G (2021) Manganese(I) catalyzed α-alkenylation of amides using alcohols with liberation of hydrogen and water. J Org Chem 86:9994–10005
31. Gawali SS, Pandia BK, Pal S, Gunanathan C (2019) Alcohols, manganese(I)-catalyzed cross-coupling of ketones and secondary alcohols with primary. ACS Omega 4:10741–10754
32. Homberg L, Roller A, Hultzsch KC (2019) A highly active PN^3 manganese pincer complex performing N-alkylation of amines under mild conditions. Org Lett 21:3142–3147
33. Waiba S, Barman MK, Maji B (2019) Manganese-catalyzed acceptorless dehydrogenative coupling of alcohols with sulfones: a tool to access highly substituted vinyl sulfones. J Org Chem 84:973–982
34. Lupp D, Huang KW (2020) The importance of metal−ligand cooperativity in the phosphorus−nitrogen PN3P platform: a computational study on Mn-catalyzed pyrrole synthesis. Organometallics 39:18–24
35. Rana J, Gupta V, Balaraman E (2019) Manganese-catalyzed direct C–C coupling of α-C–H bonds of amides and esters with alcohols via hydrogen autotransfer. Dalton Trans 48:7094–7099
36. Kaithal A, Gracia LL, Camp C, Quadrelli EA (2019) Direct synthesis of cycloalkanes from diols and secondary alcohols or ketones using a homogeneous manganese catalyst. J Am Chem Soc 141:17487–17492
37. Reed-Berendt BG, Morrill LC (2019) Manganese-catalyzed N-alkylation of sulfonamides using alcohols. J Org Chem 84:3715–3724
38. Yadav V, Landge VG, Subaramanian M, Balaraman E (2020) Manganese-catalyzed α-olefination of nitriles with secondary alcohols. ACS Catal 10:947–954
39. Jana A, Kumar A, Maji B (2021) Manganese catalyzed C-alkylation of methyl N-heteroarenes with primary alcohols. Chem Commun 57:3026–3029
40. Fu S, Shao Z, Wang Y, Liu Q (2017) Manganese-catalyzed upgrading of ethanol into 1-butanol. J Am Chem Soc 139:11941–11948

41. Chakraborty S, Gellrich U, Diskin-Posner Y, Leitus G, Avram L, Milstein D (2017) Manganese-catalyzed N-formylation of amines by methanol liberating H_2: a catalytic and mechanistic study. Angew Chem Int Ed 56:4229–4233
42. Chakraborty S, Daw P, Ben-David Y, Milstein D (2018) Manganese-catalyzed α-alkylation of ketones, esters, and amides using alcohols. ACS Catal 8:10300–10305
43. Wang Y, Shao Z, Zhang K, Liu Q (2018) Manganese-catalyzed dual-deoxygenative coupling of primary alcohols with 2-arylethanols. Angew Chem Int Ed 57:15143–15147
44. Waiba S, Maiti M, Maji B (2022) Manganese-catalyzed reformation of vicinal glycols to α-hydroxy carboxylic acids with the liberation of hydrogen gas. ACS Catal 12:3995–4001
45. Borghs JC, Zubar V, Azofra LM, Sklyaruk J, Rueping M (2020) Manganese-catalyzed regioselective dehydrogenative C-versus N-alkylation enabled by a solvent switch: experiment and computation. Org Lett 22:4222–4227
46. Shao Z, Wang Y, Liu Y, Wang Q, Fub X, Liu Q (2018) A general and efficient Mn-catalyzed acceptodehydrogenative coupling of alcohols with hydroxides into carboxylates. Org Chem Front 5:1248–1256
47. Nguyen DC, Trivelli X, Capet F, Paul JF, Dumeignil F, Gauvin RM (2017) Manganese pincer complexes for the base-free, acceptorlesss dehydrogenative coupling of alcohols to esters: development, scope, and understanding. ACS Catal 7:2022–2032
48. Luque-Urrutia JA, Solà M, Milstein D, Poater A (2019) Mechanism of the manganese-pincer-catalyzed acceptorless dehydrogenative coupling of nitriles and alcohols. J Am Chem Soc 41: 12398–12403
49. Espinosa-Jalapa NA, Kumar A, Leitus G, Diskin-Posner Y, Milstein D (2017) Synthesis of cyclic imides by acceptorless dehydrogenative coupling of diols and amines catalyzed by a manganese pincer complex. J Am Chem Soc 139:11722–11725
50. Kumar A, Espinosa-Jalapa NA, Leitus G, Diskin-Posner Y, Avram L, Milstein D (2017) Direct synthesis of amides by dehydrogenative coupling of amines with either alcohols or esters: manganese pincer complex as catalyst. Angew Chem Int Ed 56:14992–14996
51. Das UK, Ben-David Y, Leitus G, Diskin-Posner Y, Milstein D (2019) Dehydrogenative cross-coupling of primary alcohols to form cross-esters catalyzed by a manganese pincer complex. ACS Catal 9:479–484
52. Das UK, Ben David Y, Diskin-Posner Y, Milstein D (2018) N-substituted hydrazones by manganese-catalyzed coupling of alcohols with hydrazine: borrowing hydrogen and acceptorless dehydrogenation in one system. Angew Chem Int Ed 57:2179–2182
53. Daw P, Kumar A, Espinosa-Jalapa NA, Diskin-Posner Y, Ben-David Y, Milstein D (2018) Synthesis of pyrazines and quinoxalines via acceptorless dehydrogenative coupling routes catalyzed by manganese pincer complexes. ACS Catal 8:7734–7741
54. Barman MK, Waiba S, Maji B (2018) Manganese-catalyzed direct olefination of methyl-substituted heteroarenes with primary alcohols. Angew Chem Int Ed 57:9126–9130
55. Barman MK, Jana A, Maji B (2018) Phosphine-free NNN-manganese complex catalyzed a-alkylation of ketones with primary alcohols and Friedländer quinoline synthesis. Adv Synth Catal 360:3233–3238
56. Liu T, Wang L, Wu K, Yu Z (2018) Manganese-catalyzed β-alkylation of secondary alcohols with primary alcohols under phosphine-free conditions. ACS Catal 8:7201–7207
57. Rana J, Nagarasu P, Subaramanian M, Mondal A, Madhu V, Balaraman E (2021) Manganese-catalyzed C(α)-alkylation of oxindoles with secondary alcohols via borrowing hydrogen. Organometallics 40:627–634
58. Das K, Mondal A, Srimani D (2018) Phosphine free Mn-complex catalysed dehydrogenative C–C and C–heteroatom bond formation: a sustainable approach to synthesize quinoxaline, pyrazine, benzothiazole and quinoline derivatives. Chem Commun 54:10582–10585
59. Das K, Mondal A, Srimani D (2018) Selective synthesis of 2-substituted and 1,2-disubstituted benzimidazoles directly from aromatic diamines and alcohols catalyzed by molecularly defined nonphosphine manganese(I) complexes. J Org Chem 83:9553–9560

60. Mondal A, Sharma R, Pal D, Srimani D (2021) Manganese catalyzed switchable C-alkylation/alkenylation of fluorenes and indene with alcohols. Chem Commun 57:10363–10366
61. Waiba S, Jana SK, Jati A, Jana A, Maji B (2020) Manganese complex-catalysed α-alkylation of ketones with secondary alcohols enables the synthesis of β-branched carbonyl compounds. Chem Commun 56:8376–8379
62. Das K, Mondal A, Pal D, Shrivastava HK, Srimani D (2019) Phosphine-free well-defined Mn (I) complex-catalyzed synthesis of amine, imine, and 2,3-Dihydro-1H-perimidine via hydrogen autotransfer or acceptorless dehydrogenative coupling of amine and alcohol. Organometallics 38:1815–1825
63. Onishi N, Kanega R, Kawanami H, Himeda Y (2022) Recent progress in homogeneous catalytic dehydrogenation of formic acid. Molecules 27:455
64. Tondreau AM, Boncella JM (2016) 1,2-addition of formic or oxalic acid to –N $\{CH_2CH_2(P^iPr_2)\}_2$-supported Mn(I) dicarbonyl complexes and the manganese-mediated decomposition of formic acid. Organometallics 35:2049–2052
65. Anderson NH, Boncella J, Tondreau AM (2016) Manganese-mediated formic acid dehydrogenation. Chem A Eur J 25:10557–10560
66. Marino T, Mario Prejanò M (2021) Dehydrogenation of formic acid to CO_2 and H_2 by manganese(I)–complex: theoretical insights for green and sustainable route. Catalysts 11:141
67. Indranil Dutta I, Alobaid NA, Menicucci FL, Chakraborty P, Guan C, Han D, Huang KW (2023) Dehydrogenation of formic acid mediated by a phosphoruse nitrogen PN^3P-manganese pincer complex: catalytic performance and mechanistic insights. Int J Hydrogen Energy 48: 26559–26567
68. Léval A, Agapova A, Steinlechner C, Alberico E, Junge H, Beller M (2020) Hydrogen production from formic acid catalyzed by a phosphine free manganese complex: investigation and mechanistic insights. Green Chem 22:913–920
69. Léval A, Junge H, Beller M (2020) Manganese(I) κ2-NN complex-catalyzed formic acid dehydrogenation. Cat Sci Technol 10:3931–3937
70. Lim H, Chohan P, Moustafa D, Sweet C, Calalpa B, Kaur P (2018) New manganese-terpyridine-based catalytic system for the dehydrogenative coupling of alcohols and amines for the synthesis of aldimines. ChemistrySelect 3:9443–9447
71. Kabadwal LM, Das J, Banerjee D (2018) Mn(II)-catalysed alkylation of methylene ketones with alcohols: direct access to functionalised branched products. Chem Commun 54:14069–14072
72. Landge VG, Mondal A, Kumar V, Nandakumar A, Balaraman E (2018) Manganese catalyzed N-alkylation of anilines with alcohols: ligand enabled selectivity. Org Biomol Chem 16:8175–8180
73. Samuelsen SV, Santilli C, Ahlquist MSG, Madsen R (2019) Development and mechanistic investigation of the manganese(III) salen-catalyzed dehydrogenation of alcohols. Chem Sci 10:1150–1157
74. Azizi K, Akrami S, Madsen R (2019) Manganese(III) porphyrin-catalyzed dehydrogenation of alcohols to form imines, tertiary amines and quinolines. Chem A Eur J 25:6439–6446
75. Gorgas N, Kirchner K (2018) Isoelectronic manganese and iron hydrogenation/dehydrogenation catalysts: similarities and divergences. Acc Chem Res 51:1558–1569
76. Bauer I, Knolker HJ (2015) Iron catalysis in organic synthesis. Chem Rev 115:3170–3387
77. Alois F (2016) Iron catalysis in organic synthesis: a critical assessment of what it takes to make this base metal a multitasking champion. ACS Cent Sci 2:778–789
78. Zell T, Milstein D (2015) Hydrogenation and dehydrogenation iron pincer catalysts capable of metal–ligand cooperation by aromatization/dearomatization. Acc Chem Res 48:1979–1994
79. Elsby MR, Baker RT (2020) Strategies and mechanisms of metal–ligand cooperativity in first-row transition metal complex catalysts. Chem Soc Rev 49:8933–8987
80. Alberico E, Sponholz P, Cordes C, Nielsen M, Drexler HJ, Baumann W, Junge H, Beller M (2013) Selective hydrogen production from methanol with a defined iron pincer catalyst under mild conditions. Angew Chem Int Ed 52:14162–14166

81. Peña-López M, Neumann H, Beller M (2015) Iron (II) pincer-catalyzed synthesis of lactones and lactams through a versatile dehydrogenative domino sequence. ChemCatChem 7:865–871
82. Budweg S, Wei Z, Jiao H, Junge K, Beller M (2019) Iron–PNP-pincer-catalyzed transfer dehydrogenation of secondary alcohols. ChemSusChem 12:2988–2993
83. Chakraborty S, Lagaditis PO, Förster M, Bielinski EA, Hazari N, Holthausen MC, Jones WD, Schneider S (2014) Well-defined iron catalysts for the acceptorless reversible dehydrogenation-hydrogenation of alcohols and ketones. ACS Catal 4:3994–4003
84. Bielinski EA, Förster M, Zhang Y, Bernskoetter WH, Hazari N, Holthausen MC (2015) Base-free methanol dehydrogenation using a pincer-supported iron compound and Lewis acid co-catalyst. ACS Catal 5:2404–2415
85. Mastalir M, Stçger B, Pittenauer E, Puchberger M, Allmaier G, Veiros LF, Kirchner K (2016) Air stable iron(II) PNP pincer complexes as efficient catalysts for the selective alkylation of amines with alcohols. Adv Synth Catal 358:3824–3831
86. Mastalir M, Glatz M, Gorgas N, Stçger B, Pittenauer E, Puchberger M, Allmaier G, Veiros LF, Kirchner K (2016) Divergent coupling of alcohols and amines catalyzed by isoelectronic hydride Mn(I) and Fe(II) PNP pincer complexes. Chem A Eur J 22:12316–12320
87. Sinha S, Das S, Sikari R, Parua S, Brandaõ P, Demeshko S, Meyer F, Paul ND (2017) Redox noninnocent azo-aromatic pincers and their iron complexes. Isolation, characterization, and catalytic alcohol oxidation. Inorg Chem 56:14084–14100
88. Boddien A, Mellmann D, Felix Gärtner F, Jackstell R, Junge H, Dyson PJ, Laurenczy G, Ludwig R, Beller M (2011) Efficient dehydrogenation of formic acid using an iron catalyst. Science 333:1733–1736
89. Montandon-Clerc M, Dalebrook AF, Laurenczy G (2016) Quantitative aqueous phase formic acid dehydrogenation using iron(II) based catalysts. J Catal 343:62–67
90. Zell T, Butschke B, Ben-David Y, Milstein D (2013) Efficient hydrogen liberation from formic acid catalyzed by a well-defined iron pincer complex under mild conditions. Chem A Eur J 19: 8068–8072
91. Mellone I, Gorgas N, Bertini F, Peruzzini M, Kirchner K, Gonsalvi L (2016) Selective formic acid dehydrogenation catalyzed by Fe-PNP pincer complexes based on the 2,6-diaminopyridine scaffold. Organometallics 35:3344–3349
92. Bielinski EA, Lagaditis PO, Zhang Y, Mercado BQ, Würtele C, Bernskoetter WH, Hazari N, Schneider S (2014) Lewis acid-assisted formic acid dehydrogenation using a pincer-supported iron catalyst. J Am Chem Soc 136:10234–10237
93. Bhattacharya P, Krause JA, Guan H (2014) Mechanistic studies of ammonia borane dehydrogenation catalyzed by iron pincer complexes. J Am Chem Soc 136:11153–11161
94. Arne Glüer A, Moritz Förster M, Celinski VR, Günne JS, Holthausen MC, Schneider S (2015) Highly active iron catalyst for ammonia borane dehydrocoupling at room temperature. ACS Catal 5:7214–7217
95. Chakraborty S, Brennessel WW, Jones WD (2014) A molecular iron catalyst for the acceptorless dehydrogenation and hydrogenation of N-heterocycles. J Am Chem Soc 136: 8564–8567
96. Shvo Y, Czarkie D, Rahamim Y, Chodosh DF (1986) A new group of ruthenium complexes: structure and catalysis. J Am Chem Soc 108:7400–7402
97. Knölker HJ, Baum E, Goesmann H, Klauss R (1999) Demetalation of tricarbonyl (cyclopentadienone)iron complexes initiated by a ligand exchange reaction with NaOH – X-ray analysis of a complex with nearly square-planar coordinated sodium. Angew Chem 38: 2064–2066
98. Quintard A, Constantieux T, Rodriguez J (2013) An iron/amine-catalyzed cascade process for the enantioselective functionalization of allylic alcohols. Angew Chem Int Ed 52:12883–12887
99. Yan T, Feringa BL, Barta K (2016) Benzylamines via iron-catalyzed direct amination of benzyl alcohols. ACS Catal 6:381–388

100. Elangovan S, Sortais JB, Beller M, Darcel C (2015) Iron-catalyzed a-alkylation of ketones with alcohols. Angew Chem Int Ed 54:14483–14486
101. Rawlings AJ, Diorazio LJ, Wills M (2015) C−N bond formation between alcohols and amines using an iron cyclopentadienone catalyst. Org Lett 17:1086–1089
102. Yan T, Feringa BL, Barta K (2014) Iron catalysed direct alkylation of amines with alcohols. Nat Commun 5:5602
103. Brown TJ, Cumbes M, Diorazio LJ, Clarkson GJ, Wills M (2017) Use of (Cyclopentadienone)iron tricarbonyl complexes for C−N bond formation reactions between amines and alcohols. J Org Chem 82:10489–10503
104. Pan HJ, Ng TW, Zhao Y (2015) Iron-catalyzed amination of alcohols assisted by Lewis acid. Chem Commun 51:11907–11910
105. Vayer M, Morcillo SP, Dupont J, Gandon V, Bour C (2018) Iron-catalyzed reductive ethylation of imines with ethanol. Angew Chem Int Ed 57:3228–3232
106. Liu W, Sahoo B, Junge K, Beller M (2018) Cobalt complexes as an emerging class of catalysts for homogeneous hydrogenations. Acc Chem Res 51:1858–1869
107. Smith AL, Hardcastle KI, Soper JD (2010) Redox-active ligand-mediated oxidative addition and reductive elimination at square planar cobalt(III): multielectron reactions for cross-coupling. J Am Chem Soc 132:14358–14360
108. Borthakur I, Sau A, Kundu S (2022) Cobalt-catalyzed dehydrogenative functionalization of alcohols: progress and future prospect. Coord Chem Rev 451:214257
109. Junge K, Papa V, Beller M (2019) Cobalt–Pincer complexes in catalysis. Chem A Eur J 25: 122–143
110. Zhang G, Hanson SK (2013) Cobalt-catalyzed acceptorless alcohol dehydrogenation: synthesis of imines from alcohols and amines. Org Lett 15:650–653
111. Zhang G, Vasudevan KV, Scott BL, Hanson SK (2013) Understanding the mechanisms of cobalt-catalyzed hydrogenation and dehydrogenation reactions. J Am Chem Soc 135:8668–8681
112. Jing Y, Chen X, Yang X (2015) Computational mechanistic study of the hydrogenation and dehydrogenation reactions catalyzed by cobalt pincer complexes. Organometallics 34:5716–5722
113. Ingleson M, Fan H, Pink M, Tomaszewski J, Caulton KG (2006) Three-coordinate Co (I) provides access to unsaturated dihydrido-Co(III) and seven-coordinate Co(V). J Am Chem Soc 128:1804–1805
114. Zhang G, Yin Z, Zheng S (2016) Cobalt-catalyzed N-alkylation of amines with alcohols. Org Lett 18:300–303
115. Zhang G, Wu J, Zeng H, Zhang S, Yin Z, Zheng S (2017) Cobalt-catalyzed α-alkylation of ketones with primary alcohols. Org Lett 19:1080–1083
116. Zhang G, Yin Z, Tan J (2016) Cobalt(II)-catalysed transfer hydrogenation of olefins. RSC Adv 6:22419–22423
117. Xu R, Chakraborty S, Yuan H, Jones WD (2015) Acceptorless, reversible dehydrogenation and hydrogenation of N-heterocycles with a cobalt pincer catalyst. ACS Catal 5:6350–6354
118. Rosler S, Ertl M, Irrgang T, Kempe R (2015) Cobalt-catalyzed alkylation of aromatic amines by alcohols. Angew Chem Int Ed 54:15046–15050
119. Freitag F, Irrgang T, Kempe R (2017) Cobalt-catalyzed alkylation of secondary alcohols with primary alcohols via borrowing hydrogen/hydrogen autotransfer. Chem A Eur J 23:12110–12113
120. Mastalir M, Tomsu G, Pittenauer E, Allmaier G, Kirchner K (2016) Co(II) PCP pincer complexes as catalysts for the alkylation of aromatic amines with primary alcohols. Org Lett 18:3462–3465
121. Daw P, Chakraborty S, Garg JA, Ben-David Y, Milstein D (2016) Direct synthesis of pyrroles by dehydrogenative coupling of diols and amines catalyzed by cobalt pincer complexes. Angew Chem Int Ed 55:14373–14377

122. Daw P, Ben-David Y, Milstein D (2017) Direct synthesis of benzimidazoles by dehydrogenative coupling of aromatic diamines and alcohols catalyzed by cobalt. ACS Catal 7:7456–7460
123. Fabrizio Bottaro F, Madsen R (2017) In situ generated cobalt catalyst for the dehydrogenative coupling of alcohols and amines into imines. Chem A Eur J 11:2707–2712
124. Xu S, Alhthlol LM, Paudel K, Reinheimer E, Tyer DL, Taylor DK, Smith AM, Holzmann J, Lozano E, Ding K (2018) Tripodal N,P mixed-donor ligands and their cobalt complexes: efficient catalysts for acceptorless dehydrogenation of secondary alcohols. Inorg Chem 57: 2394–2397
125. Paudel K, Pandey B, Xu S, Taylor DK, Tyer DL, Torres CL, Sky Gallagher S, Lin Kong L, Ding K (2018) Cobalt-catalyzed acceptorless dehydrogenative coupling of primary alcohols to esters. Org Lett 20:4478–4481
126. Pandey B, Xu S, Ding K (2019) Selective ketone formations via cobalt-catalyzed β-alkylation of secondary alcohols with primary alcohols. Org Lett 21:7420–7423
127. Pandey B, Xu S, Ding K (2020) α-Alkylation of nitriles with primary alcohols by a well-defined molecular cobalt catalyst. J Org Chem 85:14980–14988
128. Paudel K, Xu S, Hietsoi O, Pandey B, Onuh C, Ding K (2021) Switchable imine and amine synthesis catalyzed by a well-defined cobalt complex. Organometallics 40:418–426
129. Liu Z, Yang Z, Yu X, Zhang H, Yu B, Zhao Y, Liu Z (2017) Methylation of C(sp3)−H/C (sp2)−H bonds with methanol catalyzed by cobalt system. Org Lett 19:5228–5231
130. Liu Z, Yang Z, Yu X, Zhang H, Yu B, Zhao Y, Liu Z (2017) Efficient cobalt-catalyzed methylation of amines using methanol. Adv Synth Catal 359:4278–4283
131. Paul B, Maji M, Kundu S (2019) Atom-economical and tandem conversion of nitriles to N-methylated amides using methanol and water. ACS Catal 9:10469–10476
132. Samim SA, Roy BC, Nayak S, Kundu S (2020) Cobalt-catalyzed tandem transformation of 2-aminobenzonitriles to quinazolinones using hydration and dehydrogenative coupling strategy. J Org Chem 85:11359–11367
133. Shee S, Ganguli K, Jana K, Kundu S (2018) Cobalt complex catalyzed atom-economical synthesis of quinoxaline, quinoline and 2-alkylaminoquinoline derivatives. Chem Commun 54:6883–6886
134. Mishra A, Dwivedi AD, Shee S, Kundu S (2020) Cobalt-catalyzed alkylation of methyl-substituted N-heteroarenes with primary alcohols: direct access to functionalized N-heteroaromatics. Chem Commun 56:249–252
135. Midya SP, Pitchaimani J, Landge VG, Madhu V, Balaraman E (2018) Direct access to N-alkylated amines and imines via acceptorless dehydrogenative coupling catalyzed by a CobaltIJII-NNN pincer complex. Cat Sci Technol 8:3469–3473
136. Zhang SQ, Guo B, Xu Z, Li HX, Li HY, Lang JP (2019) Ligand-controlled phosphine-free Co (II)-catalysed cross-coupling of secondary and primary alcohols. Tetrahedron 75:130640
137. Prabha D, Pachisia S, Gupta R (2021) Cobalt mediated N-alkylation of amines by alcohols: role of hydrogen bonding pocket. Inorg Chem Front 8:1599–1609
138. Pradhan DR, Pattanaik S, Kishore J, Gunanathan C (2020) Cobalt-catalyzed acceptorless dehydrogenation of alcohols to carboxylate salts and hydrogen. Org Lett 22:1852–1857
139. Abubakar S, Bala MD (2020) Transfer hydrogenation of ketones catalyzed by symmetric Imino-N-heterocyclic carbene Co(III) complexes. ACS Omega 5:2670–2679
140. Ibrahim JJ, Reddy CB, Fang X, Yang Y (2020) Efficient transfer hydrogenation of ketones catalyzed by a phosphine-free cobalt-NHC complex. Eur J Org Chem:4429–4432
141. Gangwar MK, Dahiya P, Emayavaramban B, Sundararaju B (2018) Cp*Co(III)-catalyzed efficient dehydrogenation of secondary alcohols. Chem Asian J 13:2445–2448
142. Emayavaramban B, Chakraborty P, Manoury E, Poli R, Sundararaju B (2019) Cp*Co(III)-catalyzed N-alkylation of amines with secondary alcohols. Org Chem Front 6:852–857
143. Chakraborty P, Garg N, Manoury E, Poli R, Sundararaju B (2020) C-alkylation of various carbonucleophiles with secondary alcohols under Co(III)-catalysis. ACS Catal 10:8023–8031

144. Chakraborty P, Gangwar MK, Emayavaramban B, Manoury E, Poli R, Sundararaju B (2019) α-Alkylation of ketones with secondary alcohols catalyzed by well-defined Cp*CoIII-complexes. ChemSusChem 12:3463–3467
145. Chakraborty P, Manoury E, Poli R, Sundararaju B (2021) New borrowing hydrogen mechanism for redox-active metals. ACS Catal 11:11906–11920
146. Tian H, Xue W, Wu J, Yang Z, Lu H, Tang C (2020) A general and practical bifunctional cobalt catalytic system for N-heterocycle assembly via acceptorless dehydrogenation. Org Chem Front 9:4554–4560
147. Subaramanian M, Sivakumar G, Balaraman E (2021) Recent advances in nickel-catalyzed C–C and C–N bond formation via HA and ADC reactions. Org Biomol Chem 19:4213–4227
148. Vellakkaran M, Singh K, Banerjee D (2017) An efficient and selective nickel-catalyzed direct N-alkylation of anilines with alcohols. ACS Catal 7:8152–8158
149. Das J, Vellakkaran M, Banerjee D (2019) Nickel-catalysed direct α-olefination of alkyl substituted N-heteroarenes with alcohols. Chem Commun 55:7530–7533
150. Alanthadka A, Bera S, Vellakkaran M, Banerjee D (2019) Nickel-catalyzed double dehydrogenative coupling of secondary alcohols and β-amino alcohols to access substituted pyrroles. J Org Chem 84:13557–13564
151. Peng Y, Zhang C, Gao WC, Ma Y, Wang X, Zhang L, Yue J, Tang B (2019) Nickel-catalyzed borrowing hydrogen annulations: access to diversified N-heterocycles. Chem Commun 55: 7844–7847
152. Waiba S, Das A, Barman MK, Maji B (2019) Base metal-catalyzed direct olefinations of alcohols with sulfones. ACS Omega 4:7082–7087
153. Ramalingam BM, Ramakrishna I, Baidya M (2019) Nickel-catalyzed direct alkenylation of methyl heteroarenes with primary alcohols. J Org Chem 84:9819–9825
154. Donthireddy SNR, Pandey VK, Rit A (2021) [$(PPh_3)_2NiCl_2$]-catalyzed C−N bond formation reaction via borrowing hydrogen strategy: access to diverse secondary amines and quinolines. J Org Chem 86:6994–7001
155. Yu H, Fu K, Yang G, Liu M, Yang P, Tao Liu T (2023) Divergent upgrading pathways of sulfones with primary alcohols: nickel-catalyzed α-alkylation under N_2 and metal-free promoted β- olefination in open air. Chem Commun 59:615–618
156. Bains AK, Kundu A, Yadav S, Adhikari D (2019) Borrowing hydrogen-mediated N-alkylation reactions by a well-defined homogeneous nickel catalyst. ACS Catal 9:9051–9059
157. Bains AK, Singh V, Adhikari D (2020) Homogeneous nickel-catalyzed sustainable synthesis of quinoline and quinoxaline under aerobic conditions. J Org Chem 85:14971–14979
158. Bains AK, Biswas A, Adhikari D (2020) Nickel-catalysed chemoselective C-3 alkylation of indoles with alcohols through a borrowing hydrogen method. Chem Commun 56:15442–15445
159. Bains AK, Kundu A, Maiti D, Adhikari D (2021) Ligand-redox assisted nickel catalysis toward stereoselective synthesis of (n+1)-membered cycloalkanes from 1,n-diols with methyl ketones. Chem Sci 12:14217–14223
160. Bains AK, Adhikari D (2020) Mechanistic insight into the azo radical-promoted dehydrogenation of heteroarene towards N-heterocycles. Cat Sci Technol 10:6309–6318
161. Sharma V, Chavan KA, Mali G, Sarkar D, Lama P, Majumder M, Erande RD, Metre RK (2023) A catecholaldimine-based Ni(II)-complex as an effective catalyst for the direct conversion of alcohols to trans-cinnamonitriles and aldehydes. J Org Chem 88:7448–7453
162. Balamurugan G, Ramesh R, Malecki G (2020) Nickel(II)−N^N^O pincer type complex-catalyzed N-alkylation of amines with alcohols via the hydrogen autotransfer reaction. J Org Chem 85:7125–7135
163. Zhang X, Zhang J, Hao Z, Han Z, Lin J, Lu GL (2021) Nickel complexes bearing N,N,O-tridentate salicylaldiminato ligand: efficient catalysts for imines formation via dehydrogenative coupling of primary alcohols with amines. Organometallics 40:3843–3853

164. Panigrahi D, Mondal M, Gupta R, Ganesan Mani G (2022) Four- and five-coordinate nickel (II) complexes bearing new diphosphine–phosphonite and triphosphine–phosphite ligands: catalysts for N-alkylation. RSC Adv 12:4510–4520
165. Arora V, Dutta M, Das K, Das B, Srivastava HK, Kumar A (2020) Solvent-free N-alkylation and dehydrogenative coupling catalyzed by a highly active pincer-nickel complex. Organometallics 39:2162–2176
166. Arora V, Narjinari H, Kumar A (2021) Pincer-nickel catalyzed selective Guerbet-type reactions. Organometallics 40:2870–2880
167. Midya SP, Subaramanian M, Babu R, Yadav V, Balaraman E (2021) Tandem acceptorless dehydrogenative coupling−decyanation under nickel catalysis. J Org Chem 86:7552–7562
168. Benitez-Medina GE, García JJ (2019) Hydrogenation and N-alkylation of anilines and imines via transfer hydrogenation with homogeneous nickel compounds. Dalton Trans 48:17579–17587
169. Zhang C, Liang Q, Yang W, Zhang G, Hu M, Zhang G (2022) Nickel(I)-catalyzed (de)-hydrogenative coupling of amines and alkyl heteroarenes with alcohols. Green Chem 24: 7368–7375
170. Kallmeier F, Fertig R, Irrgang T, Kempe R (2020) Chromium-catalyzed alkylation of amines by alcohols. Angew Chem Int Ed 59:11789–11793
171. Narjinari H, Tanwar N, Kathuria L, Jasrabc RV, Kumar A (2020) Guerbet-type β-alkylation of secondary alcohols alcohols catalyzed by chromium chloride and its corresponding NNN pincer complex. Cat Sci Technol 12:4753–4762
172. Tan DW, Li HX, Zhang MJ, Yao JL, Lang JP (2017) Acceptorless dehydrogenation of alcohols catalyzed by CuI N-heterocycle thiolate complexes. ChemCatChem 9:1113–1118
173. Sankar V, Kathiresan M, Sivakumar B, Mannathan S (2020) Zinc-catalyzed N-alkylation of aromatic amines with alcohols: a ligand-free approach. Adv Synth Catal 362:4409–4414
174. Pal S, Das S, Chakraborty S, Khanra S, Paul ND (2023) Zn(II)-catalyzed multicomponent sustainable synthesis of pyridines in air. J Org Chem 88:3650–3665